DEFECTS AND DIFFUSION IN SOLIDS

AN INTRODUCTION

MATERIALS SCIENCE MONOGRAPHS

Advisory Editor: C. LAIRD

MATERIALS SCIENCE MONOGRAPHS, 5

DEFECTS AND DIFFUSION IN SOLIDS
AN INTRODUCTION

STANISŁAW MROWEC

Stanisław Staszic Academy of Mining and Metallurgy
Institute of Materials Engineering
Cracow, Poland

ELSEVIER SCIENTIFIC PUBLISHING COMPANY
Amsterdam — Oxford — New York

PWN — POLISH SCIENTIFIC PUBLISHERS
Warszawa
1980

Revised and enlarged translation from the Polish original
Defekty struktury i dyfuzja atomów w kryształach jonowych
published in 1974 by Państwowe Wydawnictwo Naukowe, Warszawa

Translated by *Stefan Marcinkiewicz*

Distribution of this book is being handled by the following publishers:

For the U.S.A. and Canada
ELSEVIER/NORTH-HOLLAND, INC.
52 Vanderbilt Avenue
New York, New York 10017

For Albania, Bulgaria, Chinese People's Republic, Cuba, Czechoslovakia, German Democratic Republic, Hungary, Korean People's Democratic Republic, Mongolia, Poland, Romania, the U.S.S.R., Vietnam and Yugoslavia

ARS POLONA
Krakowskie Przedmieście 7
00-068 Warszawa, Poland

For all remaining areas
ELSEVIER SCIENTIFIC PUBLISHING COMPANY
355 Jan van Galenstraat
P.O. Box 211, 1000 AE Amsterdam, The Netherlands

Library of Congress Cataloging in Publication Data

Mrowec, Stanisław.
Defects and diffusion in solids.
(Materials science monographs; 5)
Revised and enlarged translation from the Polish original Defekty struktury i dyfuzja atomów w kryształach jonowych.
Bibliography: p.
Includes index.
1. Diffusion. 2. Crystals—Defects. 3. Solid state physics. I. Title. II. Series.
QC176.8.D5M7613 530.4′1 79-16334
ISBN 0-444-99776-8 (vol. 5)
ISBN 0-444-41685-4 (series)

With 200 illustrations

Printed in Poland by D.R.P.

Preface to the English Edition

The theory of defects and diffusion in solids, which has been developed for over 40 years, is at the present time the cornerstone of physics and chemistry of solids and of materials science, since the reactivity of solids and their principal physical and chemical properties sometimes depend more on the kind and concentration as well as the mobility of defects than on the nature of the solids themselves. For this reason the reactivity of solids and their physical and chemical properties can be modified in a rational way using general rules governing defect structure and transport phenomena in the solid state.

The English edition of this book differs considerably from the first Polish edition. The introduction of the changes was mainly due to the fact that over 7 years have elapsed from the time when the Polish version of the book was printed. During this time considerable progress in experimental and theoretical studies in solid state science has been made. In particular, the Polish edition did not contain the chapter dealing with extended defects and the discussion of empirical relations between thermodynamic properties of defects, diffusion coefficients and other physical properties of solids. The tabulated data regarding self-diffusion and heterodiffusion in metals, alloys and oxides, sulphides and halides of metals have been brought up to date. The author hopes that the book will be well received by the readers.

Finally, the author wishes to thank his collaborators who have helped him in the collection of the numerical data presented in chapter four, in the correction of the typescript and in the preparation of the drawings.

Cracow, 1979 *Stanisław Mrowec*

Contents

Introduction

The great majority of technological processes of production and transformation of various materials are either confined to solid phases or are carried out with their participation. It is sufficient to mention such great branches of modern industry as metallurgy, heterogeneous catalysis, production of refractory materials and cement, and also modern technologies such as the electronics and microelectronics, production of semiconductors, lasers, thermistors, ferrites, etc. During these technological processes mass transfer takes place in the solid phases participating in the whole reaction. This often leads to the formation of new phases or products and is always a necessary step in processes constituting the total reaction. As transport of matter in solid phases is, as a rule, the slowest step in heterogeneous reactions constituting the overall technological process, the solid state diffusion is of great importance. For example, the rate of reduction of metal ores is determined by diffusion in the solid state. The same applies to the preparation of ferrites, refractory materials, semiconducting oxides, etc. Diffusion plays the principal part not only in the reactions leading to the formation of new solid phases but also in their transformations, in particular in the processes leading to modifications of physico-chemical properties of solids, such as the heat treatment of metals, the preparation of semiconducting sulphides and oxides having definite concentrations of point defects, etc.

It follows from the above considerations that diffusion in solids plays a significant role in many processes which are the basis of modern industries. For this reason, extensive studies leading to the elucidation of the diffusion mechanism in solids and to the elaboration of general theoretical rules making it possible to predict the most suitable conditions for various reactions of solids have been carried out over many years.

Although the basis of the classical theory of diffusion was elaborated almost a hundred years ago, the development of the theory of diffusion in solids has only taken place during the last twenty years. Diffusion in solids is namely a phenomenon incomparably more complex than diffusion in gases, and for this reason its theoretical treatment only became possible after the more exact elucidation of the defect structure of solids. It should be emphasised that at the present time the theory of diffusion in polycrystalline materials is in the initial stage of its development. Nevertheless, a brief discussion of the current state of our knowledge of this subject seems to be indicated, since it will enable the reader to see the practical importance of this branch of science and at the same time the incompleteness and imperfection of the existing theoretical solutions of various problems.

Since polycrystalline materials are usually employed in modern technological processes, the theory of diffusion should describe the transport of matter both in the bulk of crystal and along their grain boundaries. Unfortunately, in view of insufficient knowledge of the structure of grain boundaries, a semiquantitative model is still used in the theory of intergranular diffusion, whereas the lattice diffusion has a sound theoretical basis which is the direct consequence of the considerable progress in the studies on the thermodynamics of point defects in crystals.

The rate of diffusion in solids increases exponentially with temperature, since this process is activated both in the region of grain boundaries and in the bulk of crystal. However, the activation energy of lattice diffusion is much higher than that of intergranular diffusion, and as a result the participation of the two forms of diffusion in the total diffusion in polycrystalline materials changes rapidly with temperature. At moderately high temperatures the diffusion process in polycrystalline substances takes place almost exclusively along grain boundaries, since the transport of matter through the crystal lattice requires much more energy. However, the rate of intergranular diffusion increases with increasing temperature much more slowly than that of lattice diffusion, since in both processes the diffusion is an exponential function of temperature. This function can be expressed by the equation:

$$D = \text{const}\exp\left(-\frac{E}{RT}\right)$$

where D is the diffusion coefficient — a measure of the rate of this process, E is the activation energy, R is the gas constant and T is the absolute temperature. For this reason, at sufficiently high temperatures, the transport of matter in polycrystalline materials should take place mainly through the crystal lattice. This conclusion has been fully confirmed by the results of experimental studies. Figure 1 shows the results of studies on self-diffusion in silver monocrystals and in polycrystalline silver at 350–800°C. It can be seen from the graph that at temperatures below

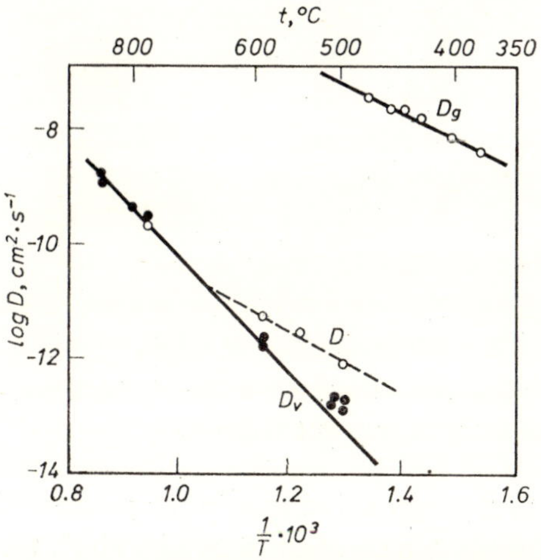

Fig. 1. The temperature dependence of diffusion coefficients of silver in silver monocrystals (white circles) and in polycrystalline material (black circles) (after [279]).

700°C the rate of diffusion in the polycrystalline material is faster than that observed in the monocrystals and the difference increases rapidly with decreasing temperature. At temperatures above 700°C, the rate of diffusion in silver monocrystals equals that observed in the polycrystalline material. This indicates that at high temperatures the diffusion of silver in the polycrystalline material takes place mainly through the crystal lattice, whereas at low temperatures it takes place practically exclusively along the grain boundaries.

In general, every polycrystalline substance has a definite limiting temperature above which the transport of matter takes place practically exclusively through the crystal lattice. This is illustrated by the scheme shown in Fig. 2. In this graph the continuous curve characterises the temperature dependence of the total (intergranular and lattice) diffusion rate in the polycrystalline material. The broken lines show the temperature dependence of the intergranular diffusion coefficient D_g and the lattice diffusion coefficient D_v. Obviously, the limiting temperature mentioned above depends on both the nature of the polycrystalline substance and the grain size. However, the results of many experiments indicate that the limiting temperatures are not very high and those of many metals, oxides and sulphides do not exceed 1000°C. This problem will be discussed further in Chapter 2.

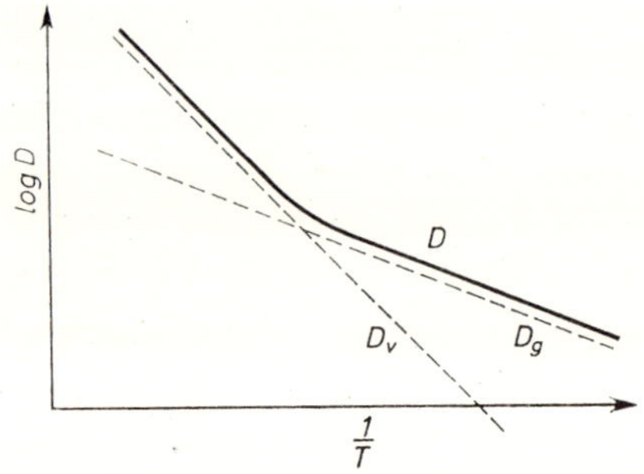

Fig. 2. Schematic plot of temperature dependence of diffusion coefficient for a polycrystalline material (continuous line) and for monocrystals (broken line).

The above considerations lead to the conclusion that in the process of matter transport in solids there are two temperature ranges in which the mechanisms of this phenomenon are completely different. This fact has the predominant effect on the reactivity of solids. In the nineteen twenties, Tammann put forward a hypothesis that a definite limiting temperature (later called the Tammann temperature) exists for every crystalline substance.

Above this temperature the sintering processes begin and other characteristic features of reactivity of solids appear. Tammann has shown that to a first approximation this temperature is equal to 2/3 of the absolute melting point. This was a purely empirical rule, but the Tammann temperature now has a definite physical sense. It is the limit above which the lattice diffusion rate becomes greater than the intergranular diffusion rate. Thus at high temperatures only is the lattice diffusion important in both monocrystals and polycrystalline substances. For this reason the theory of lattice diffusion plays an important part in the physics and chemistry of solids. This book is limited to considerations of the lattice diffusion, and only in the last chapter a few remarks regarding the intergranular and surface diffusion are made. The choice of subject matter is also due to the fact that the theory of lattice diffusion has been elaborated in detail and in some cases it has been checked experimentally, whereas the theory of intergranular and surface diffusion is still at the stage of qualitative and semiquantitative statements.

The principles of the theory of lattice defects are discussed in the first part of the book since the diffusive motion of atoms and ions in solids is possible as a result of structural disorder. The second part of the book deals with the mechanisms of diffusion, the correlation effect, the chemical diffusion and the temperature and pressure dependence of coefficients of self-diffusion, defects diffusion and tracer diffusion. Then the relation between the electrical conductivity of ionic crystals and the self-diffusion coefficient, as well as the Kirkendall and Frenkel effects, are discussed. The so-called reaction diffusion and finally the intergranular and surface diffusion are also briefly discussed. In the third part of the book, some of the methods of determination of defect equilibria and diffusion coefficients in ionic crystals are described, and in the fourth part the numerical values of selected coefficients of diffusion of metals, alloys, oxides and sulphides are given in tabular form. The book is limited to the description of phenomena taking place in the simplest inorganic ionic compounds, such as oxides, sulphides and halides of metals. The principal theoretical conclusions and the experimental methods can also be applied to more complex inorganic substances such as solid solutions, spinels, etc.

1. Defects in Crystalline Structure

The crystalline state is characterised by the orderly distribution of molecules, atoms or ions in space, the principal characteristic feature of crystal lattices being their periodic structure. However, the undisturbed infinite repetition of structural elements of crystal lattices exists only in hypothetical ideal crystals. The lattices of real crystals always contain imperfections which are called structural defects or lattice defects.

In the case of ionic crystals, two principal groups of defects can be distinguished:

(i) thermodynamically irreversible defects and

(ii) thermodynamically reversible defects.

The first group consists of defects occurring in crystals in amounts independent of the temperature and pressure of the surroundings. For the second group, the concentration of defects in crystals depends on both the temperature and the pressure of the surroundings and can be varied by suitable variation of the external conditions.

1.1 Typical crystal structures of compounds of the MeX type

Before considering the structure of defects and their mobility in crystal lattices of ionic compounds of the MeX type, we shall consider the types of crystal lattices frequently encountered in this group of substances [1–3].

Many oxides, sulphides and halides of various metals form crystals in which the anions constitute a closely packed hexagonal or cubic lattice and the cations fill the vacant spaces in this structure. In the closely packed hexagonal structure there are two kinds of such spaces, namely tetrahedral and octahedral "holes" formed by four and six anions respectively. In this structure, one oxygen ion corresponds to two tetrahedral and one octahedral holes. In many oxide and sulphide crystals, the cations are situated in either tetrahedral or octahedral positions or in both simultaneously.

1.1.1. The structure of NaCl

In lattices of the sodium chloride type, the anions form a closely packed cubic lattice and the smaller cations are situated in the octahedral positions. Each cation is surrounded by six anions. The structure of this type is shown schematically in Fig. 3. Oxides MgO, CaO, SrO, BaO, CdO, CoO, NiO,

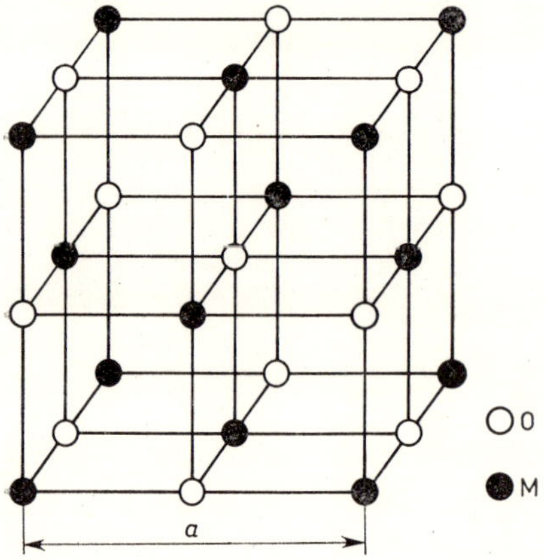

Fig. 3. The structure of the sodium chloride type.

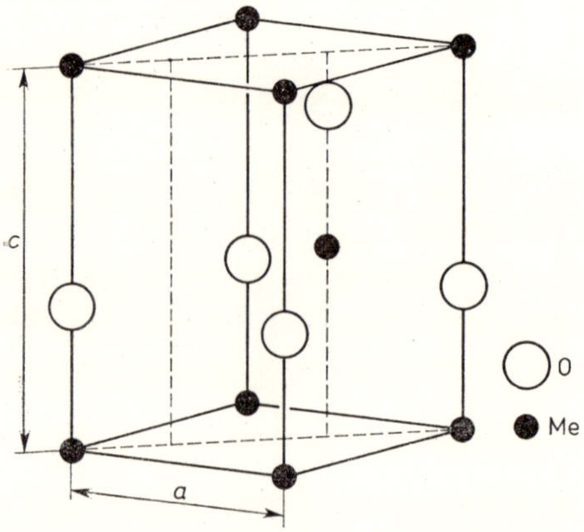

Fig. 4. The structure of the wurtzite type.

FeO, MnO, TiO, NbO and VO have this structure, but in some other oxides of the MeO type the cations are situated in the tetrahedral positions. For example, BeO and ZnO have the wurtzite structure (Fig. 4) in which both the anions and the cations have the tetrahedral configuration.

1.1.2. The structure of fluorite (CaF_2)

In the structure of fluorite the cations form a regular face centred lattice and the fluorine ions are situated in the middle of four imaginary cubes into which the elementary cell may be divided. Thus the anions occupy all the tetrahedral holes in the closely packed cubic structure of the cations.

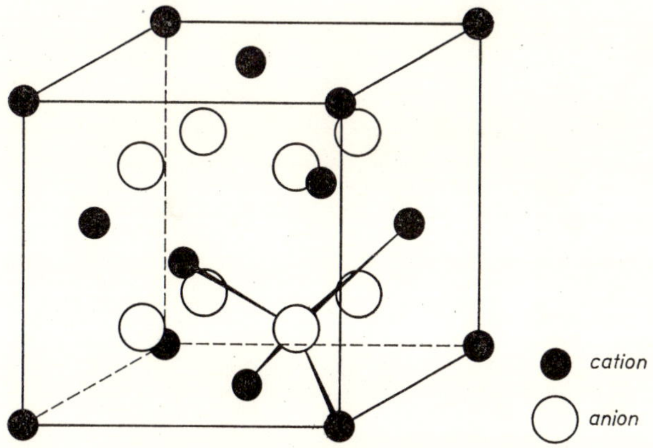

Fig. 5. The structure of the fluorite type (CaF_2).

In this situation every cation is surrounded by eight anions and there are four cations in the neighbourhood of each anion. This structure is shown schematically in Fig. 5.

The structure of fluorite is similar to that of zinc blende which could be obtained by removing one half of the anions from the fluorite lattice.

Many important high melting metal oxides, such as ZrO_2, HfO_2, UO_2, CeO_2, ThO_2, PnO_2, etc., have the fluorite structure.

It should be mentioned that alkali metal oxides K_2O, Na_2O and Li_2O have the anti-fluorite structure in which the oxygen anions form a closely packed cubic lattice and the cations fill all the tetrahedral holes.

1.1.3. The structure of rutile (TiO_2)

In the crystal lattice of rutile the cations have the octahedral configuration (Fig. 6). In this structure, oxygen octahedra TiO_6, which are the characteristic elements of crystal lattices of oxides of MeO_2, Me_2O_5 and MeO_3 types, can be distinguished. In the structure of rutile the octahedra are joined together by the apexes and edges so that every oxygen ion belongs to three neighbouring octahedra. Thus the coordination number of oxygen is 3. In the structure of rutile the octahedra are slightly deformed. Their shape and size are shown in Fig. 7 and their positions with respect to each other in Fig. 8.

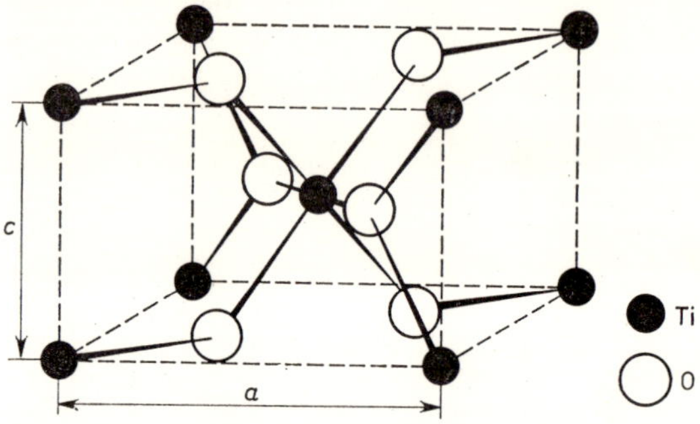

Fig. 6. The structure of the rutile type (TiO_2).

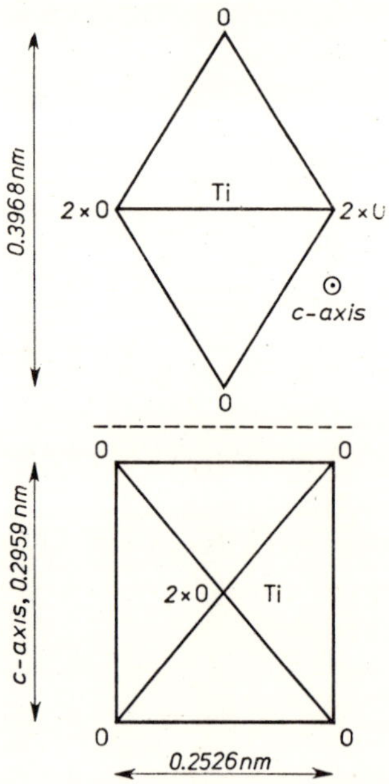

Fig. 7. The shape and dimensions of an octahedron in the crystal lattice of rutile.

Figure 8 shows that the structure of rutile contains channels parallel to the *c*-axis, which can constitute the paths of facilitated diffusion of atoms and internodal ions. We shall return to this problem in later chapters.

In addition to titanium dioxide, SnO_2, MnO_2, VO_2, MoO_2, WO_2, RnO_2, GeO_2 and certain other oxides also have the structure of rutile.

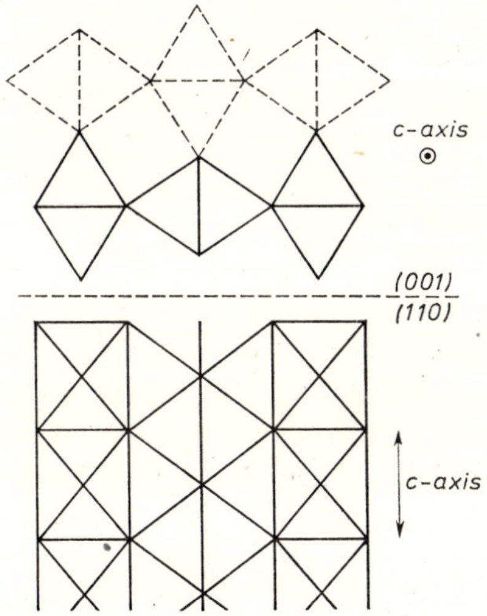

Fig. 8. Cross-section of the octahedral structure of rutile along the *c*-axis.

1.1.4 The structure of rhenium trioxide (ReO_3)

It has been mentioned that the oxygen octahedron MeO_6 is the fundamental structural unit of many oxides. In the structure of rutile these octahedra

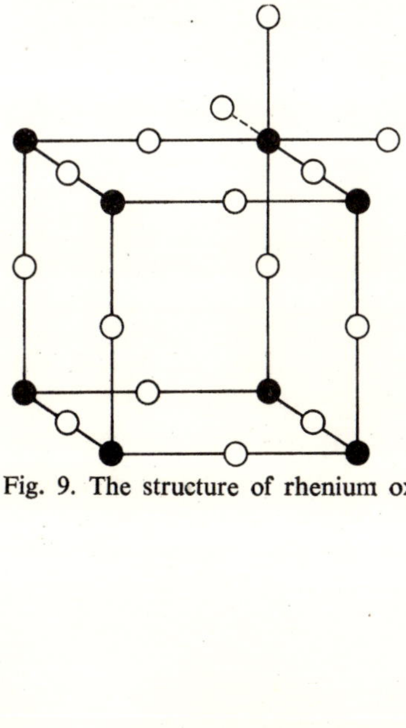

Fig. 9. The structure of rhenium oxide (RhO_3).

are joined together by their apexes and edges. In the structures of rhenium trioxide and certain other oxides, including important oxides of transition metals such as WO_3 and MoO_3, the oxygen octahedra are joined together by their apexes only and as a result a three-dimensional lattice is formed. This structure is shown schematically in Fig. 9. It is one of the "loosest" crystal structures. It will be shown in Chapter 2 that as a result of this property the lattice of the ReO_3 type has the tendency to collapse and consequently specific planar defects, called crystallographic shear planes, are formed. These defects occur not only in the structures of the ReO_3 type but also in almost all the crystals in which the MeO_6 octahedron is the fundamental structural unit, i.e. in TiO_2 and similar oxides.

1.1.5 The structure of corundum Al_2O_3

In the crystal structure of α-Al_2O_3 the anions form a closely packed hexagonal lattice in which the aluminium cations occupy 2/3 of all the octahedral positions. In this situation, every cation is surrounded by six oxygen anions and in the neighbourhood of every anion there are four cations. Hence

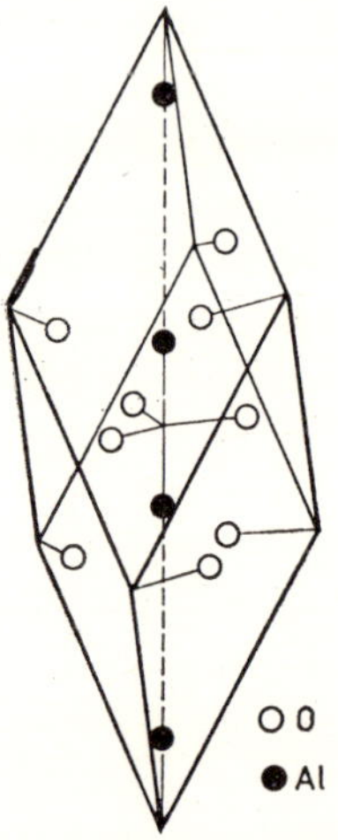

Fig. 10. The structure of corundum (α-Al_2O_3).

the coordination number of the cations is 6 and that of the anions is 4. The corundum lattice is shown in Fig. 10. Other oxides and sulphides of trivalent metals including α-Fe_2O_3, Cr_2O_3, Ti_2O_3 and V_2O_3 also have this structure. Many oxides of the $Me'Me''O_3$ type can have the corundum structure when the mean valency of metals Me′ and Me″ is 3 and when their ionic radii are comparable. For example, these conditions are fulfilled in the case of ferrous titanate (ilmenite — $FeTiO_3$) which has the corundum structure.

1.1.6 The structure of spinels (AB_2O_4)

In the structure of spinels, the oxygen anions form a closely packed cubic lattice in which the cations of both metals Me′ and Me″ occupy the octahedral and tetrahedral positions. The elementary cell of spinels contains 32 oxygen anions and hence it contains 32 octahedral and 64 tetrahedral positions. These positions can be filled by divalent and trivalent cations

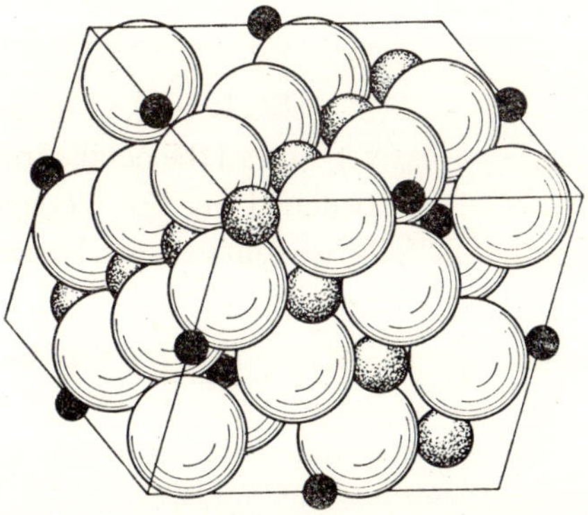

Fig. 11. The structure of the simple spinel.

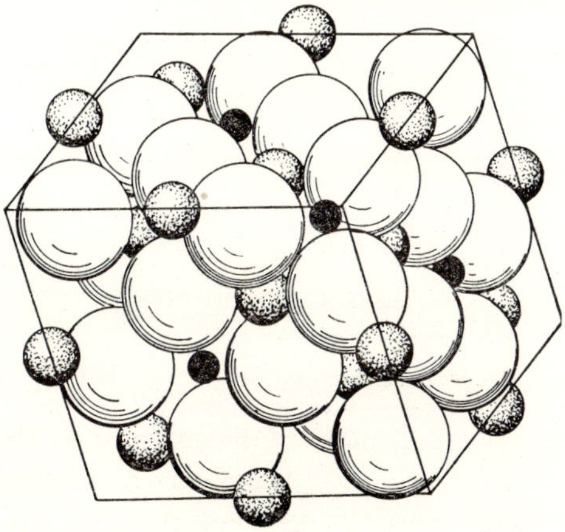

Fig. 12. The structure of the reverse spinel.

in different ways and as a result two spinel structures, the normal and the inverse, can be formed. In the normal spinel, e.g. $MgAl_2O_4$, one half of the octahedral positions is filled by trivalent aluminium ions (ions B^{3+}) and 1/8th of the tetrahedral positions is filled by divalent magnesium ions (ions A^{2+}). In the inverse spinel, 1/8th of the tetrahedral positions is filled by the trivalent cations and the rest of these cations as well as the divalent cations are situated in the octahedral positions. Magnetite, Fe_3O_4, is an example of this structure and for this reason its correct stoichiometric formula is Fe^{3+} $(Fe^{2+}, Fe^{3+})O_4$.

Both spinel structures are shown schematically in Figs. 11 and 12.

It should be mentioned that in compounds having the spinel structures and the general formula AB_2O_4, the valency of the cations of the two metals need not necessarily be 2 and 3 but the sum of their valencies must equal 8. Thus not only do the spinels of the $A^{2+}B_2^{3+}O_4$ type, but also those having the formulae $A^{4+}B_2^{2+}O_4$ and $A^{6+}B_2^{1+}O_4$, exist.

1.2 Thermodynamically irreversible defects

The thermodynamically irreversible defects can be divided into the following groups [1]:

1. Macroscopic defects having dimensions exceeding 10^{-3} cm. Various cracks, fissures and empty spaces in crystals belong to this group.

2. Microscopic defects of dimensions from 10^{-5} to 10^{-3} cm and physical character similar to that for the macroscopic defects.

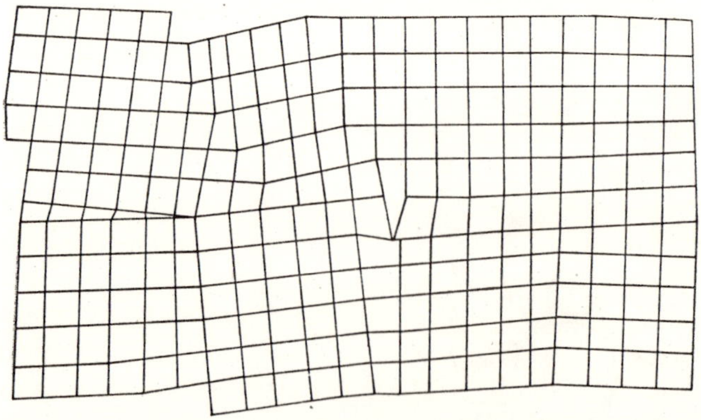

Fig. 13. The scheme of the mosaic structure (the deformations are exaggerated in order to give a better picture of the mosaic structure).

3. Submicroscopic defects smaller than 10^{-5} cm. They occur mainly as the so-called mosaic structure and are formed as a result of accidental deviations of straight lines of crystals from the average statistical direction by 1–2°. In such cases the so-called blocks, i.e. regions of ideal structure with size of the order of 10^{-5} cm and slightly tilted with respect to each other, can be observed in the crystal lattice. This is shown schematically in Fig. 13.

1.2.1 Line defects

Dislocations are one of the types of thermodynamically irreversible defects. In this type of defect, the disturbance of the periodic structure of crystal lattice occurs along some line of atoms or ions and for this reason dislocations are also called line defects [4–10]. The term dislocation had been introduced in order to explain the mechanism of plastic deformations of solids and in particular metals, but further studies have shown that the presence of dislocations also influences the kinetics of crystal growth and many of their properties, such as electrical conductivity and mobility of lattice components.

The plasticity of many solid substances is their specific mechanical property. For example, pure silver chloride melts at 455°C, but at room temperature it has a consistency similar to that of a paste and thin layers of this substance can be obtained by rolling it. Similarly, aluminium crystals are elastic only at negligible deformations less than thousandths parts of a per cent. Theoretical calculations of the limits of elasticity of crystals, based on the assumption of an ideal crystal lattice structure, gave values 10^3–10^4 times higher than those observed in experiments. This led to the interpretation of dislocations as structural defects causing a considerable decrease of elasticity of crystals. The presence of such defects in crystalline solid substances was later confirmed experimentally. The theory of dislocations was elaborated in detail. It will not be considered here except for the principal concepts which are essential for further discussion of the mechanism of diffusion in solid substances.

Two limiting types of dislocations, namely the edge and the screw, are distinguished. The formation of edge dislocations can be imagined as follows: when a crystal is pressed from the top, at a certain limiting stress (called Peierls stress) the upper part of the crystal becomes shifted with respect to the lower part and as a result a permanent deformation of the lattice takes place. Figure 14 shows schematically such deformation of the lattice of a simple regular crystal. The shift of the two fragments of the crystal with respect to each another takes place as a result of sliding along a definite lattice plane called the slip plane. It follows from Fig. 14 that the shift took place only in the right half of the crystal its left half was not

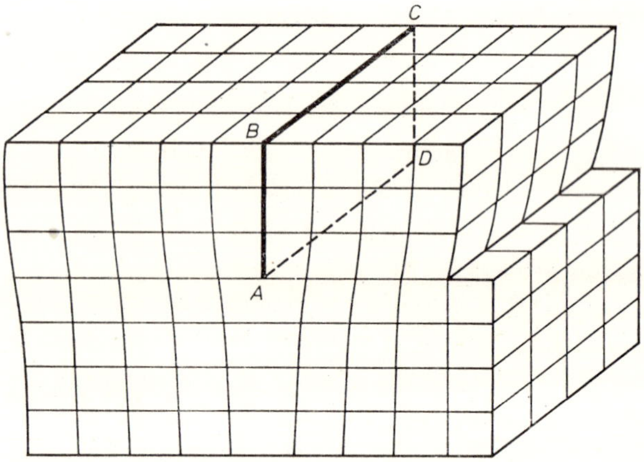

Fig. 14. The scheme of a simple regular lattice with edge dislocation.

deformed. Consequently, in the upper part of the crystal the atoms became crowded in the space reduced by one lattice parameter from the undeformed crystal state. Thus in the upper part of the crystal an additional lattice half-plane appeared. The limit of this plane inside the crystal is the dislocation line which is perpendicular to the slip plane. The additional half-plane is clearly visible in the perspective scheme shown in Fig. 15.

In the case of ionic crystals of the MeX type, the introduction of the additional half-plane of one kind of ion is not possible, since it would contradict the condition that the lattice should be electrically neutral. For this reason, in such cases the formation of edge dislocations results from the appearance of two neighbouring lattice half-planes of ions of the two signs as is schematically shown in Fig. 16. Obviously this does not change the picture of edge dislocation described above for the simplest crystals consisting of one kind of atom (e.g. metals).

Near the dislocation line *AD* (Fig. 14), the crystal is strongly deformed and as a result its elasticity is different.

The region in the direct neighbourhood of the dislocation line, where the deformation of the lattice is greatest, is called the nucleus of the dislo-

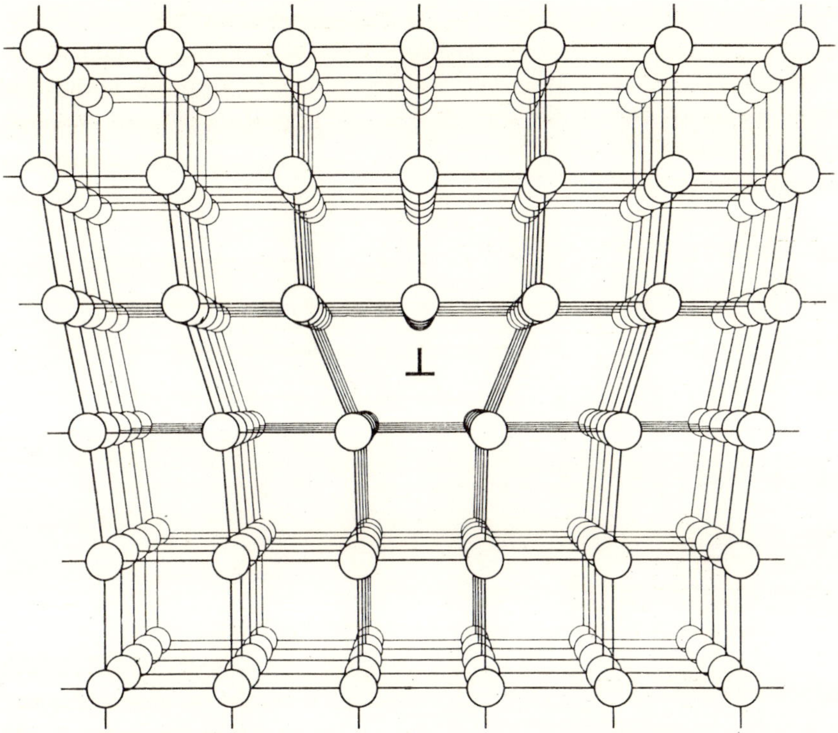

Fig. 15. Three-dimensional scheme of a lattice with an edge dislocation (⊥ shows the dislocation line) (after [6]).

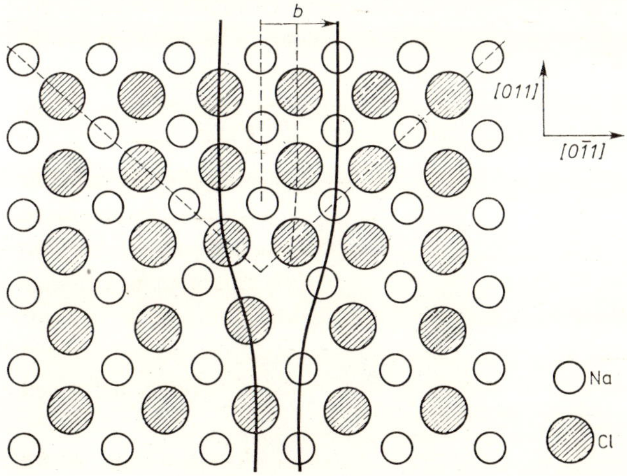

Fig. 16. The edge dislocation in the crystal lattice of sodium chloride.

cation and the relative crystal volume change caused by the edge dislocation is called the dilation of the lattice.

Internal stresses in crystals due to edge dislocations can cause shifts of these defects under the influence of external shear strain. Figure 17 shows schematically the movement of an edge dislocation in a simple regular crystal. Negative ⊤ and positive ⊥ dislocations corresponding to the pos-

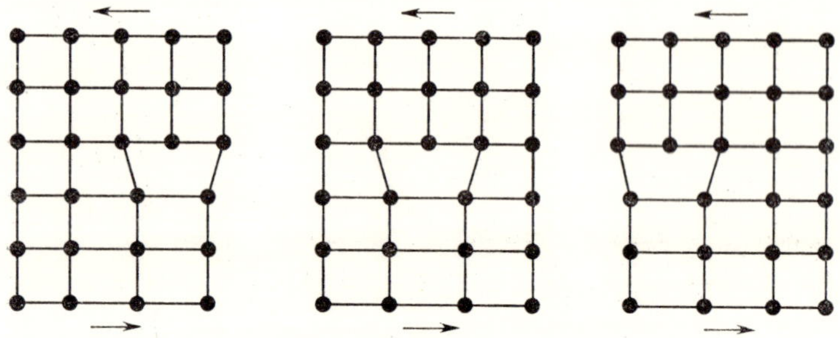

Fig. 17. The scheme of the movement of edge dislocation.

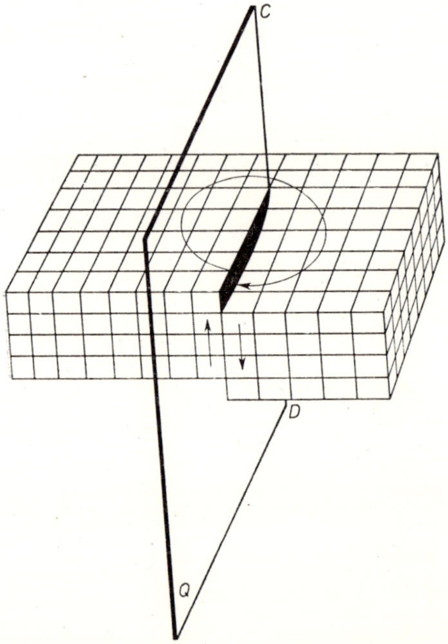

Fig. 18. The scheme of the lattice of a simple regular crystal containing a screw dislocation (Q is the slip plane; the arrow shows the direction of the movement of the dislocation).

ition of the additional half-plane of atoms below or above the slip plane are distinguished. A dislocation in a crystal is usually equivalent to a shift by one lattice constant, i.e. one so-called translational vector of the lattice.

The movement of an edge dislocation in a crystal can be compared to the movement of a fold in a carpet. The fold can be moved much more readily than the whole carpet but as a result of its movement the whole carpet is shifted.

The length of a dislocation can be equal to many tens of thousands of lattice parameters and its shape can be anything from a straight line to a spiral. A characteristic feature of edge dislocations is that the additional half-plane of atoms is perpendicular to the slip plane.

The second limiting type of line defect is the screw dislocation. Figure 18 shows schematically a simple regular crystal with a screw dislocation. The formation of this type of defect can be imagined as follows: if we cut a crystal along a plane *Q* and if we shift one of its parts parallel to edge *CD* in this plane by one lattice parameter, the horizontal planes in the crystal will become slightly bent and one edge of each of them will coincide with that of the nearest plane of atoms. As a result of the shift in plane *Q*

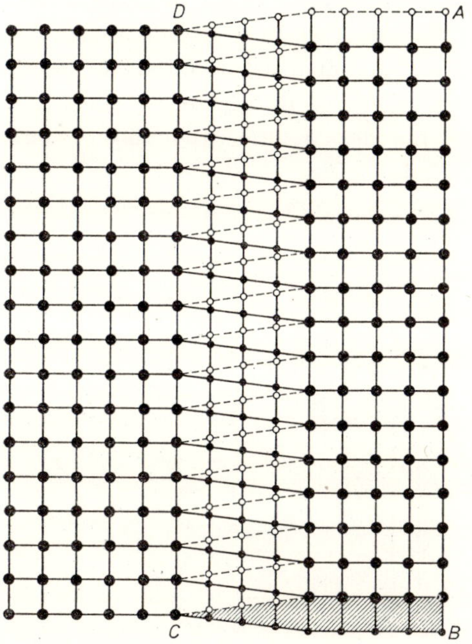

Fig. 19. The scheme of cross-section of a lattice containing a screw dislocation.

the atoms in the crystal become twisted in the direction of a screw line. The axis of this twist is dislocation line *CD*. Figure 19 shows schematically the positions of atoms above and below the slip plane *Q* which coincides with the plane of the drawing. The circles correspond to atoms situated above the slip plane and the discs to those situated below this plane. The screw line *CD* is called the dislocation line. It should be noted that the screw dislocation line is parallel to the slip vector in contrast to the edge dislocation in which the dislocation line is perpendicular to this vector.

As in the case of the edge dislocation, the degree of lattice deformation in the direct neighbourhood of the screw dislocation line is highest and rapidly decreases with increasing distance from this line, but in contrast to the edge dislocation the lattice is not dilated but is twisted around the nucleus of the screw dislocation.

Line defects can be also characterised by means of a closed loop surrounding the dislocation line. This loop, called the Burgers circuit, consists of sections which are integral multiples of lattice translations passing through the undisturbed region of the crystal surrounding the dislocation. The loop contains equal numbers of unit translations having the positive and negative directions and situated in the plane perpendicular to the dislocation line. The Burgers circuit is closed when the region which it surrounds does not contain dislocations. When this region does contain a dislocation the Burgers circuit is not closed, and a quantity called the Burgers vector is missing. Figure 20 shows schematically the Burgers circuit surrounding a screw

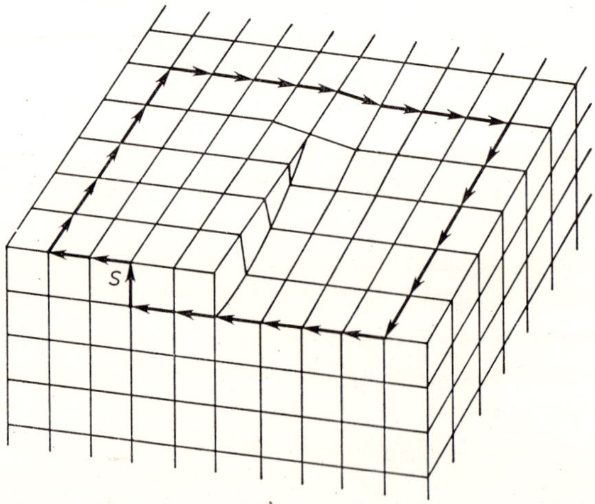

Fig. 20. The scheme of Burgers circuit for screw dislocation.

dislocation. It can be seen that in this case a single translation parallel to the dislocation line is still required to close the loop. This is the Burgers vector which always has a direction parallel to the screw dislocation line. It should be noted that when the loop is extended it becomes a spiral similar

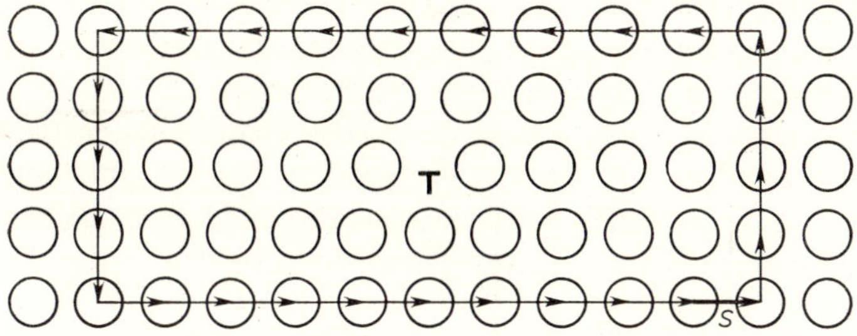

Fig. 21. The scheme of Burgers circuit for edge dislocation.

to the thread of a screw. Figure 21 shows schematically the Burgers circuit surrounding an edge dislocation. In this case the Burgers vector is perpendicular to the dislocation line.

The movement of an edge dislocation in a crystal is not connected with the shift of the additional lattice half-plane but is connected with its elimination and the appearance of a new half-plane which is formed from the neighbouring plane of atoms. When the external pressure is continued the dislocation moves in the direction of the slip vector to the border of the crystal where it forms a kink with a size corresponding to one lattice constant. This fact clearly indicates that the sliding motion of an edge dislocation is not connected with the transport of matter through the crystal.

The possibility of easy migration of dislocations, and consequently the ease with which solid substances are plastically deformed, is due to the following factors: as a result of deformation of crystal lattice caused by the presence of an edge dislocation the neighbouring atoms are slightly shifted from their normal positions corresponding to the ideal lattice. For this reason internal stresses leading to the elimination of the defect appear in the crystal in the immediate neighbourhood of the dislocation. The atoms situated in the crystal in front of the dislocation have the tendency to return to their original positions and consequently they interfere with the movement of the dislocation, whereas the atoms situated in the shifted part of the crystal, i.e. behind the dislocation, tend to occupy the positions corresponding to the shift by a complete lattice parameter and as a result they facilitate the movement of the dislocation. When the intensity of the influence of atoms situated in front of the dislocation is approximately equal to that of the atoms situated behind it, the defect can move in the crystal under the influence of a slight external pressure.

In the case of sliding movement of a screw dislocation the situation is analogous. Figure 22 shows schematically the displacement of atoms during the movement of a screw dislocation. In this case the slip plane

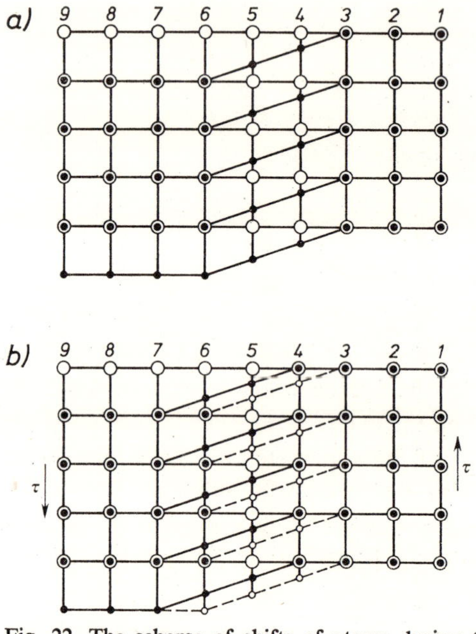

Fig. 22. The scheme of shifts of atoms during the shift of screw dislocation by one lattice parameter.

coincides with the plane of the drawing. The circles show the positions of the atoms situated behind the slip plane and the black discs those of the atoms situated before it. In Fig. 22*a*) the continuous line shows the original position of the dislocation and in Fig. 22*b*) its position after the shift by one lattice parameter. In order to make it possible for the dislocation to travel the distance equal to the distance between the atoms, the atoms situated in the range of the dislocation must move from their original positions in the direction perpendicular to that of the sliding movement of the dislocation. As in the case of edge dislocations, the forces of interaction of the atoms are almost balanced and the necessary shifts are slight. For this reason, the tangent stress necesssary for the movement of the dislocation in the crystal is relatively small.

The density of dislocations in crystals is determined by the number of dislocation lines appearing on unit surface area. It varies from 10^2 dislocations per cm^2 in the most perfect germanium and silicon crystal to 10^{12} dislocations per cm^2 in strongly defective crystals of metals. The most accurate measurements of dislocation density consists in the determination of the number of grooves etched in a polished surface of silicon crystals [8]. A microphotograph of a surface of silver bromide monocrystal with a visible network of dislocations is shown in Fig. 23. The etching method is

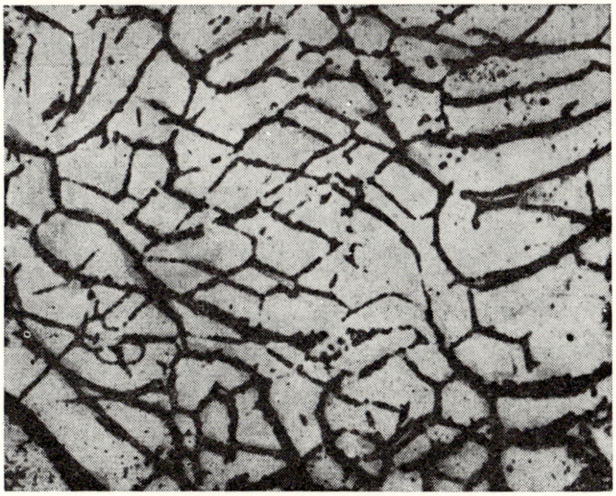

Fig. 23. Three-dimensional net of dislocations under the surface of silver bromide monocrystal (after [760]).

not suitable for the determination of dislocation density in strongly defective crystals in which the separation of the neighbouring lines is less than 10^{-4} cm. In this case, the density of dislocations is determined by calculations taking into account the increase of the internal energy of crystal caused by plastic deformation.

The presence of dislocations in crystals has a significant effect on the course of crystallisation from gas, liquid and solid phases. According to the theory of perfect growth of crystal, about 10-fold supersaturation is necessary for the formation of crystal nuclei from a gas phase. However, experimental studies have shown that the growth of iodine crystals takes place at a measurable rate when the supersaturation of iodine vapour is lower than 1%. In these experiments the crystallisation rate was faster by a factor e^{3000} than that calculated assuming an ideal crystal structure. Kittel [5] remarked that this result can be regarded as a world record of disagreement between theory and experiment. An explanation of this unusual disagreement was given by Frank [11] who pointed out that the high degree of supersaturation necessary for the growth of perfect crystals is connected with the difficulty (a high activation energy) of formation of a new lattice plane on a smooth and regular crystal surface. This difficulty is absent when the crystal contains screw dislocations, since in this case there is no need to form a new lattice plane and only the spiral extension of the step of the screw dislocation takes place. The individual stages of growth of a crystal with a screw dislocation are shown schematically in Fig. 24.

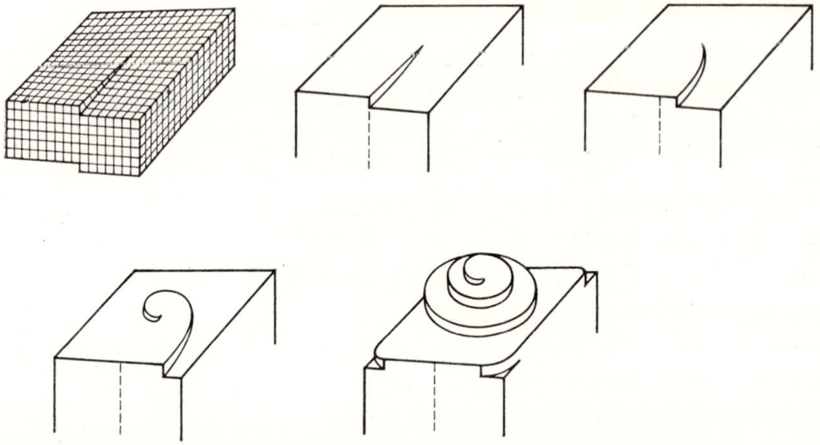

Fig. 24. The scheme of growth of a crystal containing screw dislocation.

It has been found that the results of calculations of rates of growth of crystals based on this mechanism of the crystallisation process are in complete agreement with the results of experimental studies, which can be regarded as an indirect proof of the existence of defects in crystal lattices. Direct proofs of validity of Frank's theory were provided by Griffin and collaborators who obtained optical and electron images of surfaces of many crystals [8]. A photograph of the surface of cadium iodide monocrystal is shown in Fig. 25. The spiral growth of the crystal is clearly visible.

So far we have considered two limiting types of linear defects, namely edge and screw dislocations. However, these defects seldom occur in real

Fig. 25. Microphotograph of the surface of growing cadmium iodide crystal (spira growth around the screw dislocation) (after [6]).

crystals in their pure forms. Usually the dislocations have partly screw and partly edge character [12]. In the limiting cases, the dislocation lines are straight lines shown in Figs. 14 and 18, but in reality the borders between the shifted and unshifted parts of the crystal are curves. This is shown schematically in Fig. 26 where the dislocation line has the shape of an arc of a circle. In the bulk of the crystal it is neither screw nor edge dislocation, but at point *a*, where it cuts the surface of the crystal, the dislocation line has pure edge dislocation character since it is perpendicular to the direction of displacement of atoms. On the other hand, the dislocation line cutting the surface of the crystal at point *b* has pure screw dislocation character since it is parallel to the direction of displacement of atoms at this point. Between these points the Burgers vector is neither parallel nor perpendicular to the direction of lattice deformation. Thus in this case the dislocation has a mixed character and the vector resulting from the Burgers circuit can be resolved into its components—perpendicular and parallel to the dislocation line. The relative values of these components are variable. The Burgers vector, obtained by tracing the Burgers circuit clockwise around point *a* (in which the dislocation has pure edge character) and shifting this circuit along the dislocation line to the point of intersection of this line with the second wall of the crystal (point *b*), has the same numerical value and direction as the vector obtained for point *a*.

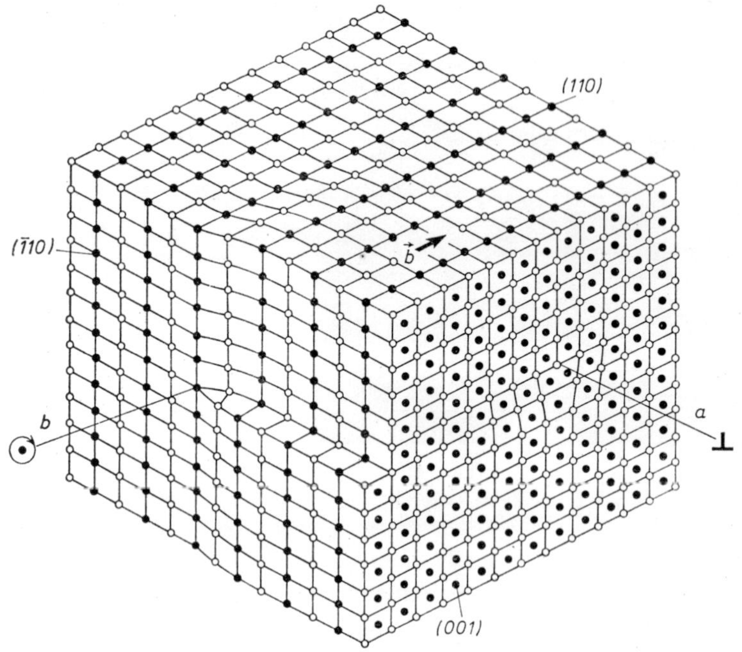

Fig. 26. Mixed dislocation in a hypothetical regular crystal; *a* and *b* are the points of intersection of the dislocation line with the surface of the crystal.

Since the same Burgers vector corresponds to any point of the dislocation, it cannot end at any point in the bulk of the crystal. For this reason, dislocations not appearing on the surface of the crystal and not forming complete loops in its bulk must pass into other dislocations and as a result a network of line defects is formed in the crystal. This is shown schematically in Fig. 27.

As a result of the possibility of occurrence of a dense network of dislocations in crystals and the ease with which they move under the influence of weak forces, the defects can react with each other and with other crystal elements. These reactions lead to the formation of breaks in the dislocation lines and as a result the dislocations are situated not in one but in several lattice planes. The breaks in the direction perpendicular to the slip plane are called jogs and those situated in the slip plane are called kinks of the dislocation line. They are shown schematically in Fig. 28. In ionic crystals,

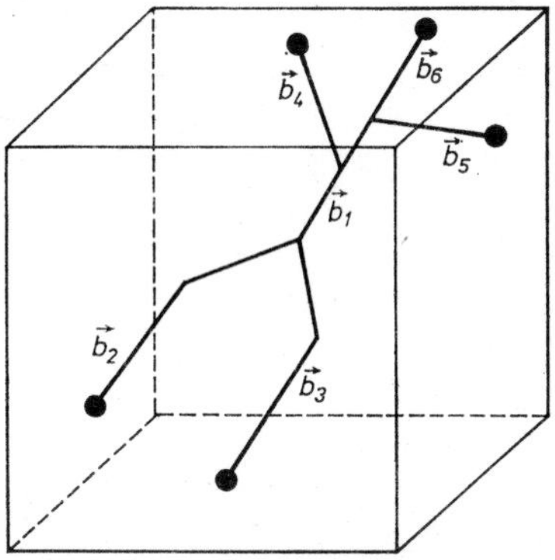

Fig. 27. A net of dislocations in a crystal. Black circles correspond to points of intersection of dislocation lines with the surface of the crystal.

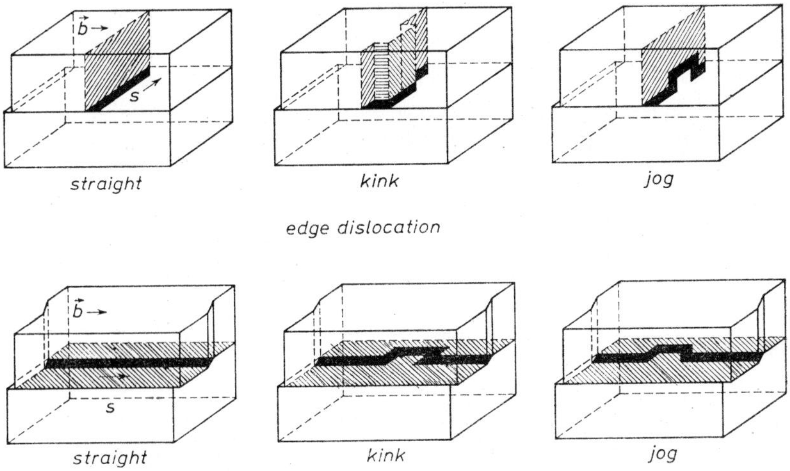

Fig. 28. Steps and kinks in edge and screw dislocation lines.

the breaks in the dislocation lines often lead to the formation of jogs which play an important part in the generation of point defects. A jog in the line of edge dislocation of NaCl crystal is shown in Fig. 29. When this jog is connected with the break of the dislocation line extending to only two neighbouring lattice planes (as shown in Fig. 29), it corresponds to only one uncompensated electrical charge which is localised on the lowest row of atoms of the half-plane. In this case, there are favourable conditions for extension of the jog in both directions by addition of ions of the opposite

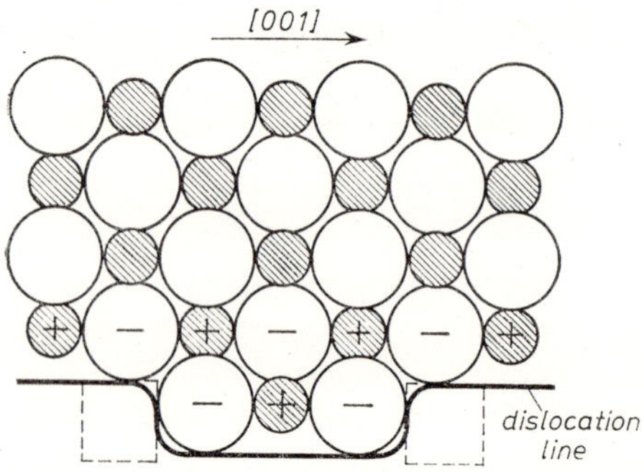

Fig. 29. A step in an edge dislocation line in (100) plane in sodium chloride structure

sign (in this case cations). This is equivalent to the generation of cation vacancies in the crystal or to the elimination of interstitial cations.

The formation and elimination of point defects need not occur as only jogs or kinks of dislocation lines, since a deformation of the structure of crystal around the line of edge dislocation is always a potential source of generation or absorption of point defects. This is shown schematically in Fig. 30. The scheme indicates that the additional lattice half-plane can be expanded or reduced as a result of addition or removal of atoms or ions. This

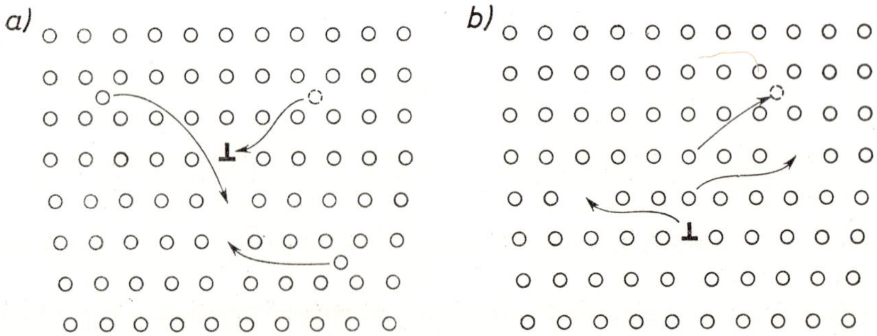

Fig. 30. The scheme of climbing of edge dislocations caused by: *a*—addition of atoms (annihilation of interstitial defects); *b*—release of atoms (generation of interstitial defects or annihilation of vacancies).

process is called climbing of the dislocation having the edge component. It consists of a shift of the dislocation line to the bulk crystal or in the opposite direction corresponding to the addition or removal of atoms, respectively. In particular, the reduction of the additional half-plane of atoms, i.e. the migration of the dislocation line towards the surface of crystal, can take place as a result of either the generation of interstitial atoms or the elimination of vacancies. On the other hand the expansion of the additional plane, i.e. the shift of the dislocation line to the bulk of the crystal, is accompanied by the process of elimination of interstitial atoms or ions or by the formation of vacancies in the crystal structure. Thus the mechanism of climbing of edge dislocations is fairly obvious. Every dislocation containing the edge component can move in this way. On the other hand, in the case of screw dislocation, the slip plane cannot be unambiguously defined since the Burgers vector is parallel to the dislocation line and as a result the process of climbing of the whole dislocation is not possible. This conclusion is in agreement with the fact that the screw dislocation does not contain an additional half-plane of atoms. However, in the case of screw dislocation, a process corresponding to the phenomenon of climbing takes place, but it is a fairly complex movement which is possible only in the case of dislocations having a particular shape. Spiral-shaped screw dislocations have edge components and for this reason they are capable

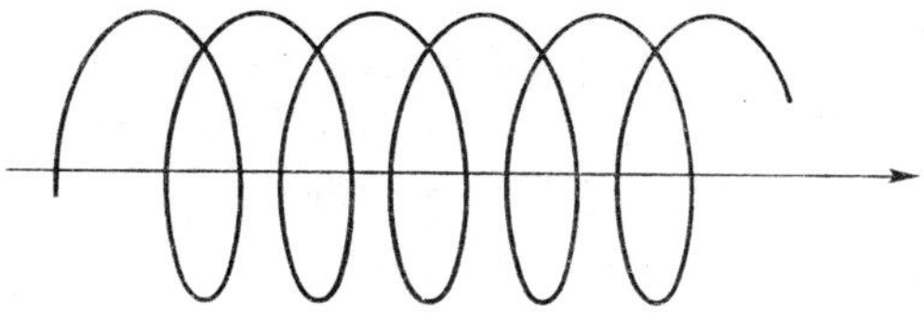

Fig. 31. The scheme of climbing of screw dislocation. (The screw dislocation forming the spiral has an edge component.)

of the climbing motion which is accompanied by radial expansion of the spiral. The configuration making this kind of movement possible is shown schematically in Fig. 31.

The above considerations lead to the conclusion that linear defects not only influence mechanical properties of solid substances and the mechanisms of their formation, but also have a considerable effect on the concentration of point defects.

1.2.2 Planar defects

So far our discussion was limited to linear defects occurring in separate single crystals. However, in practice we are usually dealing with polycrystalline materials, i.e. systems consisting of interconnected crystals (grains) with random spatial orientation. The structures of borders between neighbouring grains must be in some way adjusted to their orientation and structures. Obviously these borders constitute breaks in continuity of the periodic structure of lattices of the neighbouring crystals and for this

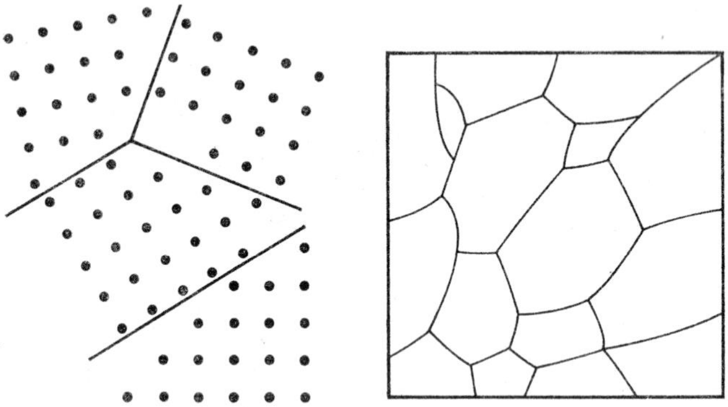

Fig. 32. The scheme of grain boundaries.

reason they can be regarded as lattice defects. Disturbances of this kind are called planar defects (Fig. 32).

The character of distribution of atoms on the boundaries between the grains can be described as follows: the range of interactions between the atoms is limited and as a rule does not exceed the sizes of the atoms. When two grains are separated by a relatively thick non-crystalline layer, each of them increases its size by forcing the atoms situated on its surface to occupy the positions dictated by their lattices, since this leads to a decrease of energy of the system. This process takes place until the time when the thickness of the layer between the grains becomes reduced to several atomic radii. Then each grain begins to act on the atoms of the border layer with a similar force and as a result these atoms occupy the positions which are a compromise between the structural requirements of the two grains.

The above conclusions of theoretical considerations have been fully confirmed by results of experimental studies. The energies of grain boundaries have been determined for many metals and they have been calculated on the assumption that the structure of grain boundaries corresponds to a series of correctly distributed dislocations. The agreement between experimental and theoretical data is very good when assuming that analogous planes in two adjacent grains are twisted with respect to one another by an angle up to 30°. At high values of this angle the structure of grain boundaries resembles that of a liquid. It has been found that the "viscosity" of grain boundaries in solid aluminium is similar to that observed in liquid aluminium, provided that the thickness of the border layer is 0.4 nm. This can be regarded as an additional confirmation of the validity of the assumption that the thickness of grain boundaries is of the order of several atomic diameters.

At high temperatures, the mobility of structural elements of polycrystalline substances is increased, particularly near grain boundaries where surface tension is an additional factor influencing this phenomenon. It has been found that under the influence of surface tension, the high angle grain boundaries have the tendency to form common apexes for three grains, so that the angles between them are 120°. Although such an ideal configuration can be formed in two dimensions, the complete filling of space with grains in such a way that all the angles between their planes are equal to 120° is not possible. For this reason the individual grains are slightly deformed.

The structure of grain boundaries has a significant effect on many properties of polycrystalline materials. During the growth of individual grains in these materials there is a tendency to eliminate various irregularities of the periodic lattice structure from the bulk of the grains, particularly foreign atoms, interstices, etc., which are shifted to the layer between the grains. As a result the grain boundaries contain a particularly large number of defects and it is difficult to evaluate their effect on mechanical and other physical properties of polycrystalline solids.

An exact model of grain boundaries has been elaborated only for the case when the difference between the mutual orientation of adjacent grains is equal to a few degrees (low angle grains). This model is based on the concept of two-dimensional distribution of edge dislocations. Let us assume that the structures of two adjacent grains are twisted with respect to one another by a few degrees, as shown schematically in Fig. 33*a*). When the twist angle is not very large, the grains can be connected in the way shown schematically in Fig. 33*b*). In this case a series of parallel edge dislocations is formed along the grain boundaries. The distances X between the neighbouring dislocations can be determined on the basis of the scheme shown in Fig. 33.

Let us consider the arc corresponding to twist angle ψ of a circle, radius X. When translation b perpendicular to the grain boundary is much smaller than radius X, the length of this translation is very similar to that of the arc. Hence:

$$b \cong X\psi \tag{1.2.1}$$

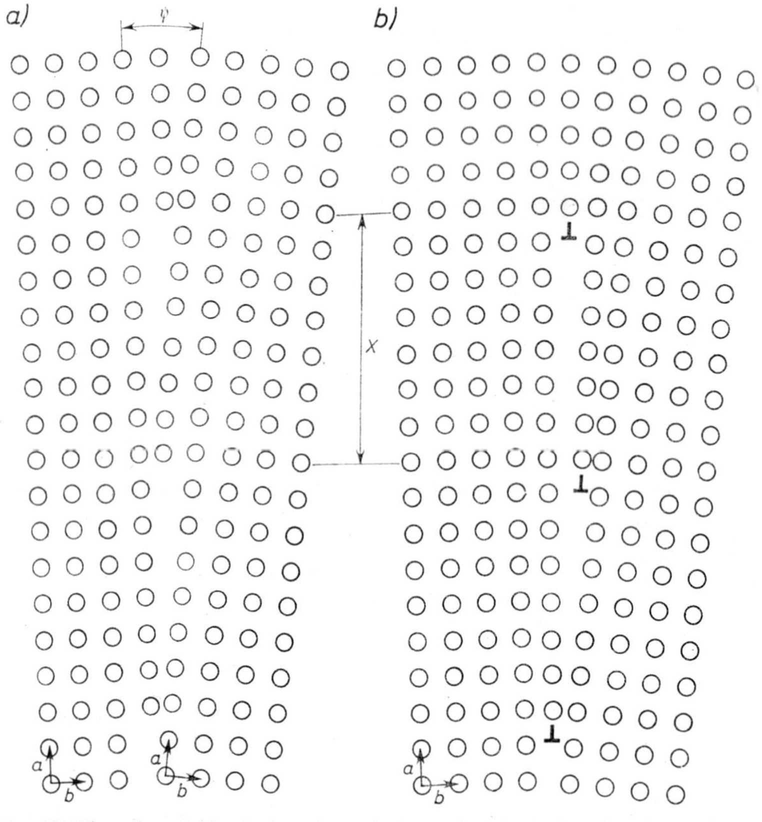

Fig. 33. The scheme of a low angle grain boundary (X is the distance between the two nearest dislocations).

where b and X are expressed in the same units of length and angle ψ is expressed in radians.

When a low angle grain boundary appears on the surface of the solid, the edge dislocations can be observed on the boundary after etching the surface so that grooves are formed along the dislocation lines. Figure 34 shows a micrograph of an etched surface of germanium crystal on which the grooves can be seen. Thus it is possible to check experimentally the relation (1.2.1) by measuring the distances X between the grooves. The relative values of angles of X-ray diffraction by two grains on both sides of the boundary corresponding to the same diffraction order can also be deter-

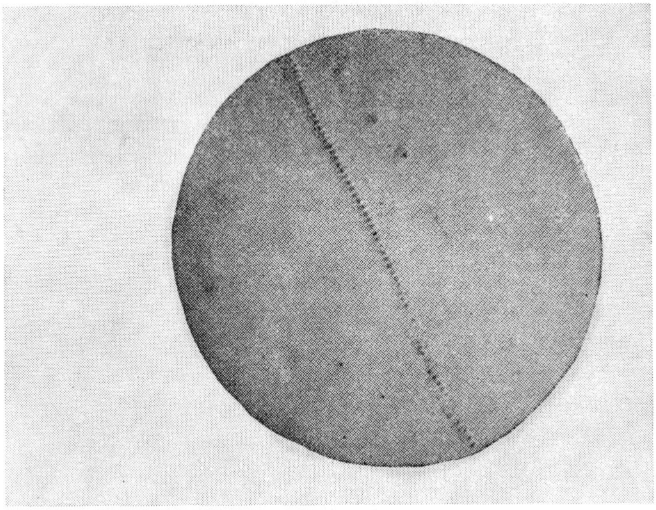

Fig. 34. Microphotograph of the surface of germanium monocrystal (after [761]).

mined. The difference between the two diffraction angles is the direct measure of angle ψ, i.e. the angle by which the two grains are twisted with respect to one another. Similarly, by means of X-ray diffraction, it is possible to determine the translation perpendicular to the grain boundary and to show that the distance X, calculated as the ratio b/ψ, is equal to that determined from the distances between the neighbouring etched grooves.

The mutual twist of two grains is partly compensated by elastic deformation which takes place at the grain boundary in the region several atomic diameters thick. On the basis of knowledge of the strain field surrounding a single dislocation, it is possible to calculate the elastic energy of a small angle boundary and hence that of the two-dimensional series of dislocations. It has been shown that the relation between the energy per unit surface area of grain boundaries and the degree of twist of these boundaries can be expressed by the following equation:

$$E = E_0 \psi (A - \ln \psi) \qquad (1.2.2)$$

where E_0 is the energy of elastic deformation caused by the dislocation and A is a constant determining the energy of twist of the grains. Using the slope of the straight line obtained by plotting E/ψ against $\ln\psi$, it is possible to determine E_0 and using the intercept of this line with the axis of ordinates it is possible to obtain the numerical value of A.

The two-dimensional grain boundaries can also be formed in monocrystals since the strain field surrounding every edge dislocation has a tendency to bend the crystal along the dislocation line in order to make the distances between the atoms as close as possible to those corresponding to the state of equilibrium in the crystal. Such a grain boundary is shown schematically in Fig. 35. The simplest method of joining together structural elements

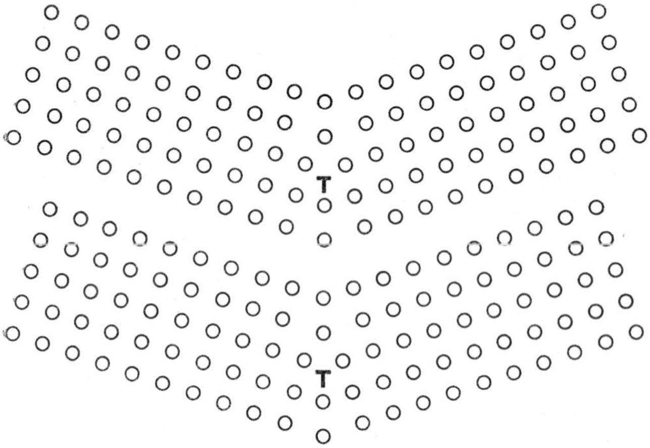

Fig. 35. The scheme of two-dimensional boundary in a monocrystal containing edge dislocations.

in order to obtain large crystals is fitting them in such a way that the breaking lines of atoms become overlapped. In this case, it can be presumed that the energy of a crystal containing many edge dislocations decreases when they become situated one over another in the same plane. Such a plane of dislocations is called a wall of dislocations. It shows a considerable similarity to a low angle grain boundary. Such walls are formed in elastically deformed crystals when they are heated. In these conditions quanta of heat (phonons) act on edge dislocations and supply them with energy necessary for the migration of dislocation lines through the crystal and for their accumulation on the walls where they form the so-called polygonal boundaries. The formation of this type of defect can usually be detected by X-ray methods.

It follows clearly from the above brief description of thermodynamically irreversible defects that their role is not limited to the effect on plastic properties of polycrystalline solids since they can also act as sources of point defects, i.e. thermodynamically reversible defects having the predominant effect on the principal physico-chemical properties of these substances.

1.3 Thermodynamically reversible defects

Irregularities of periodic structure of crystal lattices, caused by the lack of an atom or an ion in a definite site of the lattice or the presence of an additional atom in an interstitial position, are called point defects [6, 13–19]. The concentration of these defects in crystals is a function of temperature and pressure of the surroundings of the crystals and for this reason the defects of this type are called thermodynamically reversible defects. A foreign atom, and in the case of crystals of compounds consisting of two components an atom replacing the second component in its site, also constitute point defects. In contrast to the previously discussed types of crystal lattice defects, point defects have dimensions of the same order as those of the elementary cells. They can move relatively easily through crystals as a result of jumps of atoms or ions in interstitial positions or as a result of migration of vacant lattice sites, which is equivalent to shifts of structural elements in the direction opposite to that of diffusion of the defect.

Thus the phenomenon of matter transport in crystalline solids is closely connected with the existence of point defects. The theory of these defects is a basic part of modern physical chemistry of solids. The existence of point defects of crystals is also connected with many other important physico-chemical properties of solids, such as electrical conductivity of ionic crystals, their catalytic activity, semiconduting properties, optical properties, etc. The theory of defects provides the quantitative interpretation of all the physico-chemical phenomena connected with the occurrence of point defects in crystalline solids having the ionic structure. In this chapter, only the principal concepts and conclusions of the theory of defects, which are directly connected with the processes of transport of matter in solids, are discussed.

The principles of this theory were created in the nineteen thirties by Frenkel, Jost, Wagner and Schottky. In later years this theory was greatly expanded by Hauffe, Wagner, Verwey and their collaborators, which made it possible to explain the effect of impurities on the character and the degree of disorder of ionic crystal lattices. In the nineteen sixties, considerable progress in the theory of point defects was made as a result of the fundamental research carried out by Kröger and Vink [20].

However, during the last ten years the range of thermodynamical theory of point defects has been considerably limited, since it became clear that the structure of defects in crystals showing large deviations from stoichiometry is as a rule more complicated than that suggested by the classical point-defect model. When the concentration of defects exceeds a certain critical value, equal in the first approximation to about 0.1 at. %, interactions of defects lead to the formation of complexes and defect clusters and consequently to the formation of superstructure and separate intermediate phases. These problems will be discussed in Chapter 2. However, it should be stressed that in the majority of cases the concentration of point defects even at very high temperatures does not exceed the limiting value mentioned above. This remark is particularly valid in the case of oxides, sulphides and halides of metals belonging to the main groups of the periodic system but it also applies to many oxides of transition metals. For this reason, the theory of point defects still remains the main and constantly developing basis of interpretation of results of studies on many physico-chemical properties of solid substances.

It is generally known that many inorganic compounds and in particular oxides and sulphides of transition metals show deviations from stoichiometric composition, which are in some cases so large that they can be determined by the methods of classical chemical analysis. For example, ferrous oxide FeO shows deviations from the stoichiometric composition as a deficit of metal up to 0.12 moles of Fe per mole of FeO at 1000°C. Hence in these conditions the correct stoichiometric formula of this compound is $Fe_{0.88}O$. Cuprous sulphide ($Cu_{1.75}S$) shows even greater deviation from the stoichiometric composition. However, there is a group of inorganic compounds which have strict stoichiometric compositions within wide temperature ranges.

Deviations from the stoichiometric composition are caused by various defects occurring in the crystal lattices of these compounds. This does not mean that the crystal lattices of compounds having the correct stoichiometric composition are free from this type of defect. There are considerable differences between the character of defects occurring in the crystal lattices of these two groups of inorganic compounds and for this reason they will be considered separately.

One of the principal problems which arise in the consideration of the formation of point defects is the method of representing these processes by suitable quasi-chemical equations. Several kinds of such equations have been proposed. Until 1960 the system elaborated by Schottky and Wagner was commonly used, but currently the more convenient and easier to follow system proposed by Kröger and Vink is being accepted by an increasing number of authors. Since both methods of expressing defect equilibria are used in the literature, their comparison seems to be justified (Table 1). In this book, only the method proposed by Kröger and Vink is used, but the data shown in Table 1 will enable the reader to write any defect equation according to the Schottky and Wagner method.

In expressions describing defect equilibria in the form of equations of quasi-chemical reversible reactions, the following conditions must be fulfilled:

1. The ratio of cationic to anionic sites in the crystal lattice of compound MeX should remain constant. This applies to the strictly stoichiometric compounds and to those showing deviations from the stoichiometric composition. Thus in the crystal lattice of an oxide of a divalent metal (MeO) the ratio of cationic to anionic sites is 1:1 and this applies to the stoichiometric oxide as well as the oxide containing an excess of the metal ($Me_{1+y}O$) and that containing an excess of oxygen (MeO_{1+y}). Similarly in oxides or sulphides of pentavalent metals (Me_2O_5) this ratio must always be 2:5, etc.

2. As a result of formation or elimination of point defects, the total number of correct nodal positions (lattice sites) in the structure of compound MeX is variable. The corresponding defect equation must take into account this variation.

3. In all defect equations, the condition of conservation of mass must be observed. This means that the number of atoms or ions participating in a defect reaction must be the same before and after the reaction. Cation and anion vacancies representing unoccupied lattice sites have mass equal to zero. The same applies to electronic defects.

4. The reactions of formation and elimination of defects have no effect on the total electrical charge of the crystal, which is always equal to zero. Therefore the defect equations must always fulfil the condition of electrical

TABLE 1

The symbols employed by various authors for point defects in crystals of $Me^{2+}X^{2-}$ type

Type of defect	Wagner [21]	Baum-bach and Wagner [22]	Schottky [23]	Rees [24]	Kröger *et al.* [25]
Interstitial cation	$M^{\bullet\bullet}(z)$	—	—	$\|M^{2+}\|\triangle$	—
Interstitial anion	$X''(z)$	—	—	$\|X^{2-}\|\triangle$	—
Cation vacancy	$\square M^{\bullet\bullet}$	$\square(M^{2+})$	$\overline{\|M^{2+}\|}$	$\square+$	V_c
Anion vacancy	$\square(X'')$	—	$\overline{\|X^{2-}\|}$	$\square-$	V_A
Interstitial metal atom	M(z)	M(x)	—	$e_2M^{2+}\|\triangle$	—
Interstitial oxidant atom	X(z)	X(z)	—	$p_2\|X^{2-}\|\triangle$	—
Vacancy left by metal atom	$\square$(M)	—	—	$p_2\square^+$	V_c^{++}
Vacancy left by oxidant atom	$\square$(X)	—	—	$e_2\square^-$	V_A^{--}
Me^{2+} cation in a normal lattice site	—	$M^{2+}(g)$	$\overline{\|M^{2+}\|}$	—	—
X^{2-} anion in a normal lattice site	—	$X^{2-}(g)$	$\overline{\|X^{2-}\|}$	—	—
Electrons in the conduction band	$e^-(z)$	—	—	e	M^+
Electron holes in the conduction band	—	e^-	—	p	X^-
Trivalent cation of the dopant in the cationic sublattice site	—	—	—	$F^{3+}\square^+$	—
Monovalent cation of the dopant in the cationic sublattice site	—	—	—	$F^+\square^+$	—
Divalent cation of the dopant in the cationic sublattice site	—	—	—	$F^{2+}\square$	—

compounds

Besson [26]	Kröger and Vink [20]	Schottky and Stockmann [27]	Lander [28]	Thomas [30]	Schottky [30]	Albers [31]	Ong Ir. [32]
$(M^{2+})^{2+}$	$M_i^{\bullet\bullet}$	$M_o^{\bullet\bullet}$	$M_{\vert}^{++}$	$M_{\vert}^{++}$	$M^{\bullet\bullet}$	$M_i^{\bullet\bullet}$	$M_i^{\bullet\bullet}$
$(X^{2-})^{2-}$	X_i''	$X_\circ''$	$X_{\vert}^{=}$	$X_{\vert}^{=}$	X''	X_i''	$X_i^{\vert}$
$\diamond_{2+}^{2-}$	V_M''	$M_\square''$	$V_M^{=}$	$V_M^{=}$	$\vert M\vert''$	V_M''	$X^{=}$ or $(V_M X_X)^{=}$
$\diamond_{2-}^{2+}$	$V_X^{\bullet\bullet}$	$X_\square^{\bullet\bullet}$	V_X^{++}	V_X^{++}	$\vert X\vert^{\bullet\bullet}$	$V_X^{\bullet\bullet}$	M^{++} or $(M_M V_X)^{++}$
$(M)^0$	M_i^x	$M_\circ^x$	$M_{\vert}$	$M_{\vert}$	M^x	$M_{i=}^x$	M_i
$(X)^0$	X^x	$X_\circ^x$	$X_{\vert}$	$X_{\vert}$	X^x	X_i^x	X_i
$\diamond_{2+}^{0}$	V_M^x	$M_\square^x$	V_M	V_M	$\vert M\vert^x$	V_M^x	X or $(V_M^{2+}X_X^{2-})$
$\diamond_{2+}^{0}$	V_x^x	$X_\square^x$	V_X	V_X	$\vert X\vert^x$	V_X^x	M or $(M_M^{2+}V_X^{2-})$
$\langle M^{2+}\rangle_{2+}^{0}$	M_M^x	—	N.O.	M_M	—	—	—
$\langle X^{2-}\rangle_{2-}^{0}$	X_X^x	—	N.O.	X_X	—	—	—
$(e^-)^-$	e'	$\ominus$	e^-	e^-	e'	e	e
	$h^\bullet$	$\oplus$	e^+	e^+	$\vert e\vert^\bullet$	h	M^+
$\langle F^{3+}\rangle_{2+}^{+}$	$F_M^\bullet$	F●· (M)	F_s^+	F_c^+	$F\vert M\vert^\bullet$	$F_M^\bullet$	—
$\langle F^-\rangle_{2+}^{-}$	F_M'	F●′(M)	F_s^-	F_c^-	$F\vert M\vert'$	F_M'	—
$\langle F^{2+}\rangle_{2+}$	F_M	F●(Me)	F_s	F_c	$F\vert Me\vert$	F_{Me}	—

neutrality of the lattice, i.e. the algebraic sum of positive and negative effective charges involved in the defect reaction must be equal to zero.

According to the Kröger and Vink method (Table 1), an ion situated in the correct site bears zero charge with respect to the lattice when its degree of oxidation corresponds to the stoichiometric formula of the compound forming the crystal. For example, divalent positive nickel ions in the crystal lattice of NiO and the divalent negative oxygen ions in this lattice have effective zero charges with respect to the lattice (Ni_{Ni}, O_O, compare Table 1). On the other hand, trivalent nickel cations in this compound have effective +1 charges ($Ni^{\bullet}_{Ni}$), as those of trivalent cations of an impurity, e.g. aluminium or chromium, incorporated in the sites of the cationic sublattice ($Al^{\bullet}_{Ni}$, $Cr^{\bullet}_{Ni}$). Univalent cations of an impurity, e.g. lithium, have effective −1 charges (Li'_{Ni}). The absence of a cation or an anion in a lattice site is equivalent to the absence of the corresponding electrical charge in this position. Thus a cation vacancy has a negative effective charge and an anion vacancy a positive one. For example, in nickel oxide a cation vacancy has effective charge $-2(V''_{Ni})$ and an anionic vacancy $+2(V^{\bullet\bullet}_O)$. However, these vacancies can have other effective charges or they can be electrically neutral with respect to the lattice as a result of capture of positive or negative electronic defects. For example, the localisation of two positive electron holes in a cation vacancy of nickel oxide causes complete elimination of its effective charge (V^x_{Ni}). The same effect is obtained by the localisation of two quasi-free electrons in an anion vacancy (V^x_O).

The effective charges of atoms or ions situated in interstitial positions are equal to the real electrical charges of these species. Interstitial zinc atoms in the crystal lattice of ZnO have effective charge equal to zero (Zn^x_i), whereas mono- and divalent cations of this metal have charges +1 and +2 respectively ($Zn^{\bullet}_i$ and $Zn^{\bullet\bullet}_i$) with respect to the lattice. Similarly an interstitial oxygen atom or ion in the anionic sublattice has no effective electrical charge (O^x_i), or else it has the effective charge −1 or −2 (O'_i or O''_i).

The above considerations lead to the conclusion that the effective electrical charge of an ionic defect can be equal to or different from its real electrical charge. On the other hand, the effective charges of parent ions situated in the correct sites and the effective charges of ions of impurities situated in these positions are always different from their real electrical charges. For this reason in the Kröger and Vink method the effective charge is shown by dots (+) or dashes (−) and not by signs plus or minus, in order to stress the difference discussed above.

1.3.1 Types of point defects

Let us begin our consideration of types of point defects by discussing the limiting types of defects occurring in ionic crystals having correct stoichiometric composition, since these defects occur over and above those defects found in compounds, which deviate from stoichiometric composition. In the first group of compounds, the following five limiting types of crystal structure defect can be distinguished:

1. The cations situated in interstitial positions and the equivalent number of vacancies in the cationic sublattice (the Frenkel type). These defects are formed as a result of the passage of a certain number of cations from crystal lattice sites to interstitial positions.

2. The anions situated in interstitial positions and the equivalent number of vacancies in the anionic sublattice (the anti-Frenkel type). In this case, a certain number of anions leave the crystal lattice sites and pass to the interstitial positions.

3. The equivalent numbers of vacancies in the cationic and anionic sublattice (the Schottky type). Defects of this type are formed as a result of the diffusion of a certain number of cations and the equivalent number of anions from their sites in the lattice nodes to the surface of the crystal, where the formation of new lattice elements takes place.

4. The equivalent numbers of cations and anions situated in the interstitial positions (the anti-Schottky type). In this case a certain number of cations and anions situated in the surface layer of the crystal leave their positions in the lattice sites and become incorporated in the interstitial positions.

5. The cations and anions which exchange their positions in lattice sites.

The last type of defect is improbable in ionic crystals of the MeX type, since the forces of electrostatic repulsion of ions of the same sign prevent a considerable mutual approach of cations or anions in the crystal lattice. However, this type of defect often occurs in metal alloys. Figures 36 *a–c* show schematically models of defects of crystal lattices of compounds having the stoichiometric composition proposed by Frenkel and Schottky.

The formation of Schottky-type defects is accompanied by a decrease of density of the crystal, since the direct result of the formation of these defects is an increase of the volume of crystal. In the case of anti-Schottky defects, the reverse situation is observed. On the other hand, the formation of defects of Frenkel and anti-Frenkel type does not change, to the first approximation, the volume of crystal, and as a result its density remains unchanged.

The concentration of defects discussed above is a specific function of temperature. As a result of thermal vibrations of ions in lattice sites at temperatures above absolute zero, there is a particular probability that

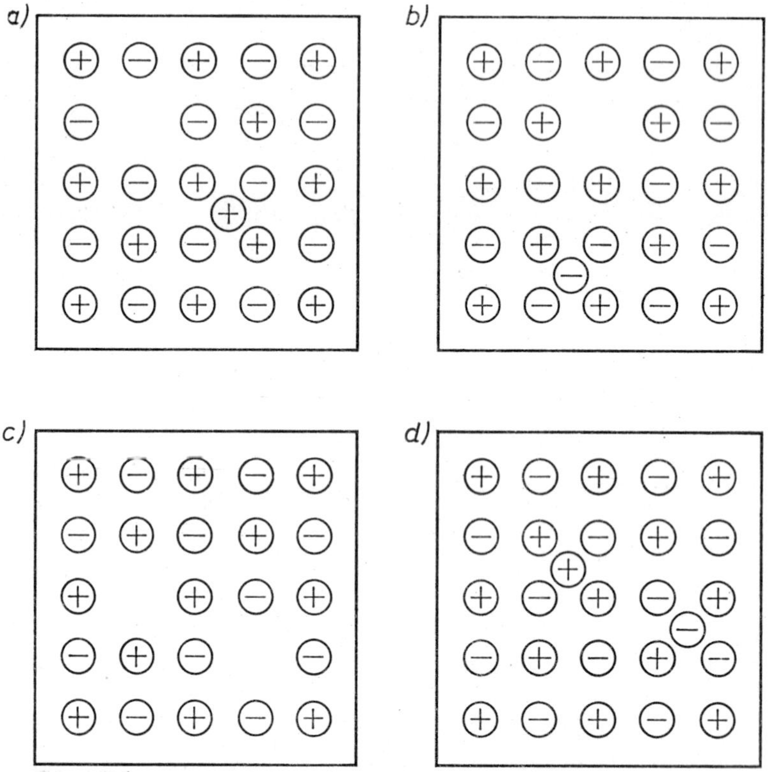

Fig. 36. Models of lattice disorder in MeX type compounds: *a*) the Frenkel type; *b*) the anti-Frenkel type; *c*) the Schottky type; *d*) the anti-Schottky type.

a given ion will jump to the interstitial position. It can be shown that this probability increases exponentially with temperature. The kind of defect which appears in the crystal lattice depends on the amount of energy required for removal of an ion from the lattice site and its incorporation in the interstitial position. When this energy is considerably lower for a cation than for an anion, the defects in the cationic sublattice (the Frenkel type) become predominant. In the opposite case, the defects appear mainly in the anionic sublattice (the anti-Frenkel type). When the energies required for the formation of defects in the cationic and anionic sublattices are comparable, the defects of the Schottky type appear. However, it should be stressed that the same compound showing predominance of defects of one type in a given temperature range may show predominance of another type in suitably changed thermodynamic conditions. This is due to differences in the activation energy of the formation of defects of different types and hence to differences in the temperature dependence of concentration of these defects. This phenomenon is analogous to diffusion along grain boundaries and through crystal lattices, as was discussed earlier.

In general, the formation of defects in the anionic sublattices of compounds of the MeX type occurs less frequently than in the cationic sublattice. This is due to the fact that the formation of a defect in the anionic sublattice, which consists in the formation of an interstitial anion, is much more difficult energetically than the formation of any other defect. The ionic diameters of anions are considerably larger that those of cations and as a result the incorporation of anions in the interstitial positions requires more energy than for the smaller cations. For this reason, anti-Frenkel and anti-Schottky defects have only been observed in isolated cases. This difficulty does not exist in the case of formation of anion vacancies, but in spite of this the large majority of inorganic compounds and in particular oxides, sulphides and halides of metals contain defects mainly in the cationic sublattice. Oxides of high-melting metals are an exception. The above considerations lead to the conclusion that the total energy balance of formation of defects in crystal lattices of this type of compound is more favourable for the cationic sublattice.

1.3.2 The concentration of point defects

The thermodynamic theory of point defects is based on the assumption that a real crystal can be regarded as a solution in which the lattice is the solvent and the defects are the solute. When the number of defects is very small in comparison with that of filled lattice sites, the solutions are, from the thermodynamic point of view, infinitely dilute. In this case, the law of mass action can be applied with a good approximation using the concentrations of suitable defects. Until recently it was believed that crystals containing up to 1 at. % of defects or even more can be regarded as very dilute solutions, but now it is known that when the concentration of defects exceeds 0.1 at. %, the interactions between them and the interactions between the lattice ions and the defects must be taken into account. These interactions lead to the formation of less numerous, but much larger defect clusters, called extended defects. Obviously in these cases the thermodynamic theory of point defects is not valid. For this reason, the quantitative statements and conclusions derived from this theory are valid only in the cases when the concentration of defects in crystal does not exceed several tenths of per cent. It has already been shown in Chapter 2 that in certain cases (e.g. oxides of high-melting metals) the concentration of point defects cannot exceed 10^{-2} at %.

Before considering defect equilibria, the causes of formation of defects in crystals and the factors determining their concentration must be discussed.

According to the rules of chemical thermodynamics, every spontaneous process taking place at constant pressure and temperature is accompanied by a decrease of thermodynamic potential of the reacting system:

$$\Delta G = \Delta H - T\Delta S < 0 \qquad (1.3.1)$$

where ΔG, ΔH and ΔS are the changes of thermodynamic potential, enthalpy and entropy of the system, respectively, caused by the process. In the case of reactions of solids, and in particular the reactions leading to the formation of defects, the crystal volume remains practically unchanged ($\Delta V \cong 0$). Therefore

$$\Delta H = \Delta U + p\Delta V \cong \Delta U \qquad (1.3.2)$$

Hence

$$\Delta G \cong \Delta F = \Delta U - T\Delta S \qquad (1.3.3)$$

where ΔF and ΔU are the changes of free energy and internal energy of the system respectively. When the system is in thermodynamic equilibrium its thermodynamic potential and free energy have their minimum values:

$$dG = 0; \quad dF = 0 \qquad (1.3.4)$$

The change of thermodynamic potential or free energy of a crystal caused by the formation of one mole of defects can be expressed by the following equation:

$$\Delta G_f = \Delta H_f - T\Delta S_V - T\Delta S_k \qquad (1.3.5)$$

where ΔH_f is the enthalpy (heat) of formation of one mole of defects, ΔS_V is the change of vibrational entropy of atoms situated in the neighbourhood of one mole of defects and ΔS_k is the change of configurational entropy (entropy of mixing) caused by the increase of the degree of disorder of the system.

Since the process of formation of defects in a crystal is spontaneous, $\Delta G_f < 0$. Therefore the question arises, which term of the right-hand side of equation (1.3.5) determines the sign of thermodynamic potential of the defect reaction. The process of formation of a defect in a crystal, i.e. the removal of an atom or an ion from the lattice site or their incorporation in the interstitial position, always requires a definite amount of external energy and for this reason $\Delta H_f > 0$.

The change of vibrational entropy ΔS_V is due to the difference between the frequency of vibrations of atoms or ions situated in the neighbourhood of defects and that of atoms or ions situated in the undisturbed parts of the crystal. The relation between this component of entropy of the system and the frequency of vibrations of atoms in the crystal can be expressed approximately by the following equation:

$$S_V = xk\ln\frac{\nu}{\nu'} \tag{1.3.6}$$

where ν and ν' are the frequencies of vibrations of atoms situated in the undisturbed part of crystal and in the neighbourhood of defects respectively and x is the number of atoms surrounding one of the defects. When this defect is a vacancy $\nu' < \nu$, and when it is an atom or an ion situated in an nterstitial position then $\nu' > \nu$. Therefore ΔS_V can have either negative or positive values depending on the type of defect. In the case when the disorder of the lattice consists of simultaneous occurrence of vacancies and interstitial atoms (e.g. Frenkel-type disorder) ΔS_V may be equal to zero. In general, the vibrational entropy changes do not exceed a few calories per degree per mole. Therefore the term determining the sign of the thermodynamic potential change ΔG_f is the configurational entropy ΔS_k. This is comprehensible in view of the fact that this component of entropy represents the tendency of the system to reach the state of maximum thermodynamic probability, i.e. the state with maximum disorder:

$$S_k = k\ln W \tag{1.3.7}$$

where W is the thermodynamic probability and k is the Boltzmann constant. It follows that the configurational entropy change accompanying the formation of defects is always positive ($\Delta S_k > 0$). Thus the equilibrium concentration of defects, corresponding to the minimum value of the thermodynamic potential of the system, is established as a result of two opposing tendencies existing in the crystal. This is illustrated by the graph shown in Fig. 37. It can be seen from the graph that the formation of defects is accompanied by an increase of energy of the crystal ($\Delta H > 0$), but at the same time this energy is evolved as a result of an increase of entropy ($-T\Delta S_k$). Therefore, at temperatures above absolute zero, an ideal crystal is thermodynamically unstable since the tendency to disorder is stronger

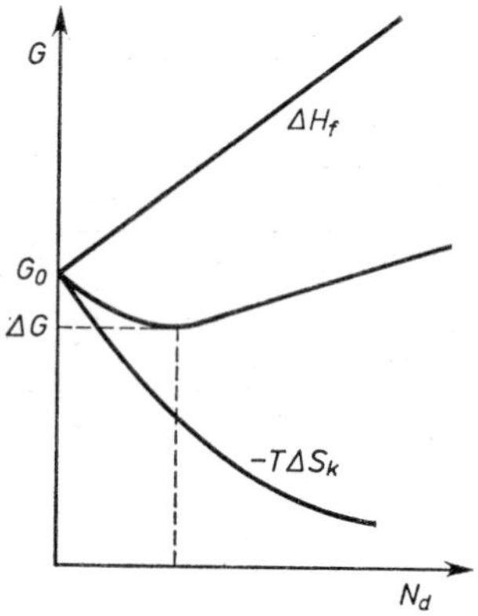

Fig. 37. The dependence of the thermodynamic potential of a crystal on the concentration of point defects.

than to perfect order. However, as the degree of disorder increases, the tendency to increase in entropy decreases, whereas the amount of energy required for the removal of an atom from the correct site remains the same. In other words, the increase of energy of the crystal is a linear function of concentration of defects, whereas the configurational entropy is an exponential function of this parameter (see Fig. 37). As a result, the thermodynamic potential of the crystal at first decreases:

$$\left|\frac{d(\Delta H_f)}{dn}\right| < \left|\frac{d(T\Delta S_f)}{dn}\right| \tag{1.3.8}$$

then reaches its minimum value:

$$\left|\frac{d(\Delta H_f)}{dn}\right| = \left|\frac{d(T\Delta S_f)}{dn}\right| \tag{1.3.9}$$

and finally increases:

$$\left|\frac{d(\Delta H_f)}{dn}\right| > \left|\frac{d(T\Delta S_f)}{dn}\right| \tag{1.3.10}$$

where n is the defect concentration. Obviously the minimum value corresponds to the equilibrium concentration of defects.

1.3.3 The chemical equilibrium and law of mass action

Before describing defect equilibria in ionic crystals, it is worthwhile expressing rigorously the concept of chemical equilibrium and the law of mass action, since they are constantly used in defect theory and in the theory of diffusion in solids.

In thermodynamic considerations of chemical reactions taking place in any system consisting of two or more components, it is convenient to use partial molar thermodynamic quantities. A partial molar thermodynamic function of a component of a system represents the increase of this function caused by the addition of one mole of the component to the system which has a volume so large that its composition remains practically unchanged. In considerations of chemical equilibria, the most important role is played by a function called the partial molar thermodynamic potential or the chemical potential. When the considered system contains $n_1+n_2+ \ldots +n_i$ moles of components 1, 2, ..., i respectively, the partial molar thermodynamic potential of component i (its chemical potential) is given by the expression:

$$\left(\frac{\partial G}{\partial n_i}\right)_{T,p,n_1\ldots} = G_i = \mu_i \qquad (1.3.11)$$

where G is the thermodynamic potential of the system, or free enthalpy. This function is defined by the equation:

$$G = H - TS \qquad (1.3.12)$$

where H and S are the enthalpy and entropy of the system, respectively, and T is the absolute temperature.

A characteristic feature of this function is the minimum value, which it reaches when the system is in the state of equilibrium under definite isobaric and isothermic conditions. Thus in the equilibrium state

$$dG_{T,p} = 0 \qquad (1.3.13)$$

The thermodynamic potential of any system can be expressed by means of chemical potentials in the form of the following equation:

$$G = n_1\mu_1 + n_2\mu_2 + \ldots + n_i\mu_i + \ldots \qquad (1.3.14)$$

In the case of an open system remaining under constant pressure and at constant temperature, a change of thermodynamic potential is given by the equation:

$$dG_{T,p} = \mu_1 dn_1 + \mu_2 dn_2 + \ldots + \mu_i dn_i + \ldots \qquad (1.3.15)$$

According to the condition expressed by equation (1.3.13) in the state of thermodynamic equilibrium, expression (1.3.15) becomes:

$$\sum \mu_i dn_i = 0 \qquad (1.3.16)$$

It can easily be shown that the following equation is also valid:

$$n_1 d\mu_1 + n_2 d\mu_2 + \ldots + n_i d\mu_i + \ldots = 0 \qquad (1.3.17)$$

This equation is one form of the Gibbs–Duhem equation.

It should be noted that when a heterogeneous system consisting of many phases is in a state of thermodynamic equilibrium, the chemical potentials of each component in all the phases are equal.

Let us now consider the change of thermodynamic potential of a reversible chemical reaction represented by the following general equation:

$$a\mathrm{A} + b\mathrm{B} \rightleftarrows c\mathrm{C} + d\mathrm{D} \qquad (1.3.18)$$

This change is the algebraic sum of chemical potentials of the reagents multiplied by their numbers of moles:

$$\Delta G = c\mu_{\mathrm{C}} + d\mu_{\mathrm{D}} - a\mu_{\mathrm{A}} - b\mu_{\mathrm{B}} \qquad (1.13.19)$$

This equation can be transformed further utilising the relation between the chemical potential of a component and its activity:

$$\mu_i = \mu_i^0 + RT \ln a_i \qquad (1.3.20)$$

where a_i is the activity of the component i in the system consisting of many components. The relation between the activity and the concentration of component i is given by the following relationship:

$$a_i = \gamma_i N_i \qquad (1.3.21)$$

where γ_i is the activity coefficient, equal to unity in ideal solutions and very dilute real solutions, and N_i is the concentration expressed in mole fractions; μ_i^0 in equation (1.3.20) is the chemical potential of component i in a standard state which is so chosen that the activity of this component is equal to unity. In heterogeneous systems, the activity of pure compounds is regarded as equal to unity when the pressure is equal to 1013.25 hPa (1 atm).

Combination of equations (1.3.20) and (1.3.19) gives:

$$\Delta G = c\mu_{\mathrm{C}}^0 + RT \ln a_{\mathrm{C}}^c + d\mu_{\mathrm{D}}^0 + RT \ln a_{\mathrm{D}}^d - a\mu_{\mathrm{A}}^0 - RT \ln a_{\mathrm{A}}^a - b\mu_{\mathrm{B}}^0 - RT \ln a_{\mathrm{B}}^b \qquad (1.3.22)$$

The algebraic sum of standard chemical potentials of the reactants multiplied by the numbers of their moles is equal to the standard change of thermodynamic potential of the reaction ΔG^0. Hence equation (1.3.22) can be expressed in the following form:

$$\Delta G = \Delta G^0 + RT \ln \frac{a_C^c \cdot a_D^d}{a_A^a \cdot a_B^b} \tag{1.3.23}$$

According to condition (1.3.4) in the equilibrium state $\Delta G = 0$ and since the standard change of thermodynamic potential ΔG^0 is a constant at constant temperature, equation (1.3.23) becomes:

$$\Delta G^0 = -RT \ln K \tag{1.3.24}$$

where K is the equilibrium constant. The relation between K and the activities of the reactants is given by the following equation:

$$K = \frac{a_C^c \cdot a_D^d}{a_A^a \cdot a_B^b} \tag{1.3.25}$$

Equation (1.3.25) expresses the law of mass action, according to which the ratio of the product of activities of the reaction products raised to suitable powers, to the product of activities of the substrates raised to suitable powers, is constant when the system is in the equilibrium state. This is one of the fundamental laws of chemical thermodynamics, which is constantly used in the theory of point defects. It applies to gases, liquid solutions and solid solutions. The activities of reactants can be replaced by their concentrations, when the solutions obey the Raoult law, i.e. when they behave like ideal solutions. In practice, this means that very dilute solutions can be regarded as ideal solutions. In general, defect ionic crystals can be regarded as very dilute solutions of defects in lattices, since the concentration of point defects does not usually exceed a fraction of an atom per cent. When it is higher, suitable corrections should be used in order to take into account differences between the concentrations and the activities. However, it has been found that at higher concentrations, point defects form complexes and clusters, excluding the possibility of the application of the thermodynamic theory of defects. These problems will be considered further in Chapter 2.

1.3.4 The band model of solid substances

In ionic crystals of the MeX type, point defects are always accompanied by electronic defects which occur in various concentrations. They are formed as a result of both spontaneous thermal excitation and the presence of point defects. Thus the presence of electronic defects in lattices of compounds of the MeX type is closely connected with ionic defects. The interdependence of the two kinds of defects obviously affects both the electronic and the ionic equilibrium. For this reason, before discussing ionic–electronic defect equilibria in crystals of the MeX type, a brief description of the electronic structure of these substances will be given.

In single, mutually isolated atoms of an element, electrons can occupy only certain definite energy levels, the so-called permitted levels, which are defined by four quantum numbers. In addition, it follows from the Pauli exclusion principle that each energy level can be filled by only two electrons having different fourth quantum numbers, i.e. having opposite spins. However, when the atoms approach each other so that the distances between them become comparable with their diameters, as in the case of solid substances, the electronic orbitals of the atoms overlap and as a result the splitting of energy levels and the formation of energy bands takes place.

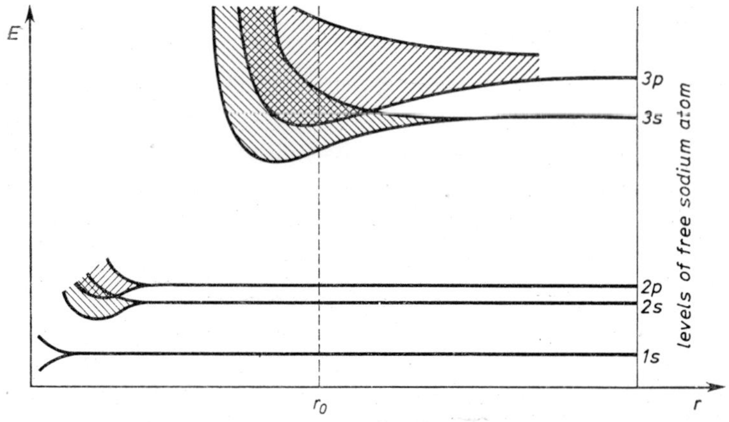

Fig. 38. The scheme of cleavage of energy levels of electrons caused by the mutual approach of six atoms (metal).

This is shown schematically in Figs. 38 and 39. The total number of energy levels in each band is exactly equal to the number of atoms, whereas the width of the band is independent of this number. Thus the distances between the energy levels in the band decrease with increasing numbers of neighbouring atoms. In solid substances the number of atoms in 1 cm^3 is of the order 10^{22}–10^{23}, and hence the number of energy levels in each band is of the same order. In view of the fact that the width of the bands is of the order 1 eV, the distances between the energy levels in the bands are of the order of 10^{-22} eV. These energy differences are so small that the electrons can move practically freely (without activation) within the bands.

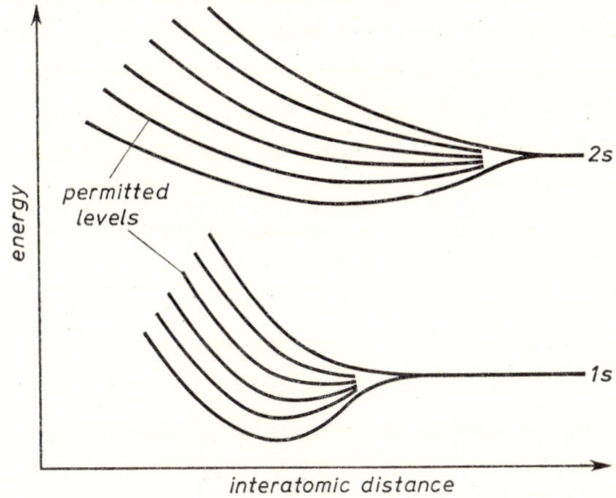

Fig. 39. The scheme of cleavage of energy levels of electrons caused by the mutual approach of six atoms (semiconductor).

The outer energy bands can overlap (e.g. in metals, Fig. 38) or can be separated by energy gaps (e.g. in insulating substances and semiconductors, Fig. 39). These gaps are called forbidded energy bands, since electrons cannot occupy any levels within such bands.

At absolute zero, the band containing the electrons of the outer electronic shell (the valence band) is completely filled, whereas the next higher energy

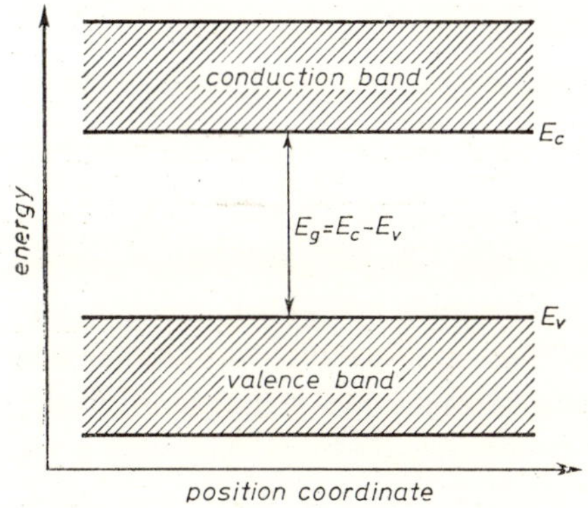

Fig. 40. The scheme of energy band in a pure semiconductor.

band (the conductance band) is completely empty. Under these conditions solid substances cannot conduct electric current. The positions of the bands with respect to each other are shown schematically in Fig. 40. The energy of electrons situated in the valence band increases with temperature and as a result at higher temperatures they can pass to a higher level, i.e. to the conductance band. This causes the appearance of vacant electronic levels in the valence band, which are called electron holes. They are shown schematically in Fig. 41. In this way, the spontaneous formation of elec-

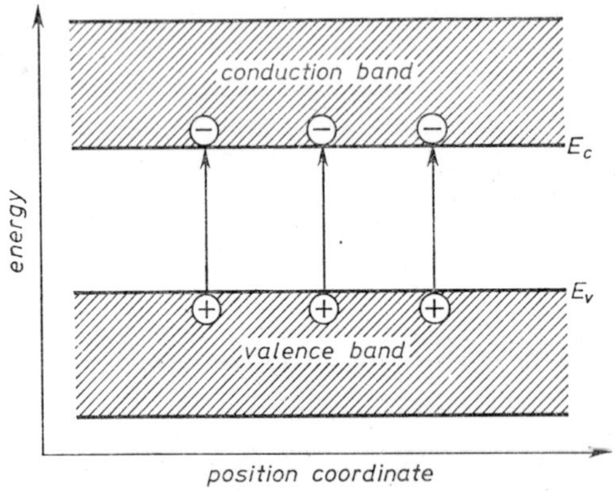

Fig. 41. The scheme illustrating the passage of electrons from the valence band to the conductivity band in pure semiconductor.

tronic defects, which is not due to the presence of point defects or impurities, takes place in pure semiconducting elements and in ionic crystals of the MeX type. The concentration of electrons in the conductance band (the so-called quasi-free electrons) is obviously equal in this case to that of electron holes in the valence band. The quasi-free electrons in the conductance band and the electron holes in the valence band can move under the influence of an applied electric field and as a result crystals conduct electric current. The movement of the two current carriers takes place simultaneously but in opposite directions, since electron holes have positive effective charges. The movement of the electron holes is obviously equivalent to the movement of electrons in the opposite direction, i.e. in the direction of movement of quasi-free electrons in the conductance band.

However, the transport of the electronic carriers of current within the energy bands does not take place without activation in any ionic crystal of the MeX type. In many compounds of this type, particularly at elevated temperatures, periodic fluctuations of electric potential of individual ions are so considerable in comparison with the width of energy bands, that the electronic defects become localised either on certain ions or on point defects. For this reason, their movement under the influence of applied electric field takes place in the form of jumps from one localised level to another, through energy barriers separating these levels. Thus in this type of crystal the transport of electricity has the activated character (compare Chapter 1.3).

Irrespectively of the mechanism of electrical conductivity, the equilibrium of spontaneous electronic defects depends on the distribution of electrons in the permitted energy levels of the crystal. It follows from Fermi statistics that the number of electrons in each level i having energy E_i is given by the following equation:

$$n_i = \frac{g_i}{1+\exp\dfrac{E_i-E_F}{kT}} \qquad (1.3.26)$$

where g_i is the degree of degeneracy of level i, i.e. the number of states having the same energy. E_F is called the Fermi level. It will be shown later that it is equal to the chemical potential of the electrons.

It is noteworthy that when $E_F = E_i$, $n_i = 1/2\, g_i$. This situation is observed when one half of the number of states with energy E_i is filled at temperature T. When the condition:

$$\frac{E_i-E_F}{kT} \gg 1 \qquad (1.3.27)$$

is fulfilled, only an insignificant number of states is occupied by electrons Then equation (1.3.26) becomes:

$$n_i = g_i \exp\left[-\left(\frac{E_i-E_F}{kT}\right)\right] \qquad (1.3.28)$$

It follows that condition (1.3.27) determines the limit, over which the statistical Fermi distribution (equation 1.3.26) becomes equivalent to the classical Boltzmann statistics (equation 1.3.28).

According to equation (1.3.28), the number of electrons in the conductance band of crystals of the MeX type is given by the following relationship:

$$n = N_C \exp\left[-\left(\frac{E_C-E_F}{kT}\right)\right] \qquad (1.3.29)$$

where N_C is the number of states (the effective density of states) in the conductance band, and E_C is the lowest permitted energy level in the conductance band (according to the scheme shown in Fig. 40), called the bottom of this band.

Assuming that the electrons occupy only a narrow portion of the band, near to its bottom, the density of states N_C is a parabolic function of energy, which leads to the following relationship:

$$N_C = \left(\frac{8m_e^x kT}{h^2}\right)^{3/2} \tag{1.3.30}$$

where m_e^x is the effective mass of electron and h is Planck's constant.

An equation analogous to equation (1.3.29) describes the number of electron holes in the valence band:

$$p = N_V \exp\left[-\left(\frac{E_F - E_V}{kT}\right)\right] \tag{1.3.31}$$

where N_V is the effective density of states in the valence band and E_V, according to Fig. 40, determines the highest energy level in this band, i.e. its top.

When the effective density of states is localised in a narrow region, near the peak of the valence band, N_V, like N_C, is a parabolic function of energy and can be expressed by the following relationship:

$$N_V = \left(\frac{8m_h^x kT}{h^2}\right)^{3/2} \tag{1.3.32}$$

where m_h^x is the effective mass of the electron hole.

The equilibrium constant of the spontaneous formation of electronic defects is equal to the product of the number of electrons in the conductance band and that of the electron holes in the valence band (compare Section 1.3). Therefore we can write:

$$K_i = n \cdot p \tag{1.3.33}$$

Substituting in this equation relationships (1.3.29) and (1.3.31) we obtain:

$$K_i = N_C N_V \exp\left(-\frac{E_g}{kT}\right) \tag{1.3.34}$$

where $E_g = E_C - E_V$ is the width of the forbidden energy band (the energy gap between the discussed bands, see Fig. 40). Thus from the thermodynamical point of view E_g is the enthalpy of spontaneous formation of electronic defects in the ionic crystals of the MeX type.

It should be stressed that expression (1.3.34) was derived on the assumption that condition (1.3.27) is valid, i.e. that the considered system (the crystal) obeys classical statistics (low densities of electrons and electron holes in both bands). It should also be noted that the Fermi level was eliminated during the derivation of the above expression of the law of mass action (equation (1.3.34)).

When quasi-free electrons in the conductance band and electron holes n the valence band are the predominant defects in the crystal (intrinsic semiconductors), the number of quasi-free electrons is equal to that of electron holes:

$$n = p \tag{1.3.35}$$

In this case, the solution of the set of equations (1.3.29)–(1.3.32) with respect to the Fermi level gives the following relationship:

$$E_F = \frac{E_V + E_C}{2} + \frac{3kT}{4}\ln\left(\frac{m_e^x}{m_h^x}\right) \tag{1.3.36}$$

When the effective mass of the electrons is equal to that of the holes, equation (1.3.36) is simplified to the following form:

$$E_F = \frac{E_V + E_C}{2} \tag{1.3.37}$$

It follows from this equation that when $m_e^x = m_h^x$, the Fermi level is exactly in the middle of the forbidden energy band.

When the compound in question is not a band semiconductor, but contains electronic defects localised on definite ions of the crystal lattice, equations (1.3.30) and (1.3.32) are not valid. On the other hand, equations (1.3.29) and (1.3.31) are also valid in this case, but here parameters N_V and N_C correspond to the total number of atomic (ions) on which the electrons can be localised, multiplied by the degeneracy of states of these atoms.

So far our considerations have been limited to the description of the spontaneous formation of electronic defects in pure crystals of the MeX type, not containing impurities or point defects. However, as already mentioned, the presence of these defects can have a considerable effect on the equilibrium of the spontaneous formation of electronic defects and in the case of non-stoichiometric compounds it can determine the concentration of quasi-free electrons in the conductance band or electron holes in the valence band. This is due to the fact that interstitial cations and anions as well as cation and anion vacancies constitute additional energy levels localised in the forbidden energy band. There are two kinds of such levels: the donor levels localised near the bottom of the conductance band and the acceptor levels situated near the top of the valence band (Fig. 42). The terms "donor" and "acceptor" reflect the fact that the donor levels readily release electrons to the conductance band and the acceptor levels readily accept electrons from the valence band (small energy differences between these levels and the edges of the corresponding bands; see Fig. 42).

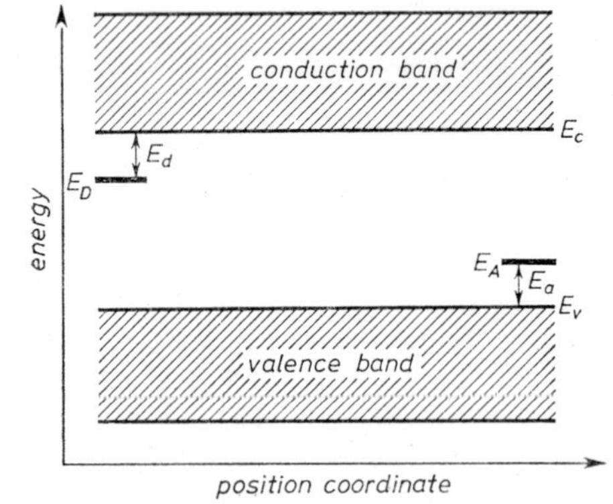

Fig. 42. The scheme of donor (E_D) and acceptor (E_A) levels in the forbidden energy band of a semiconductor.

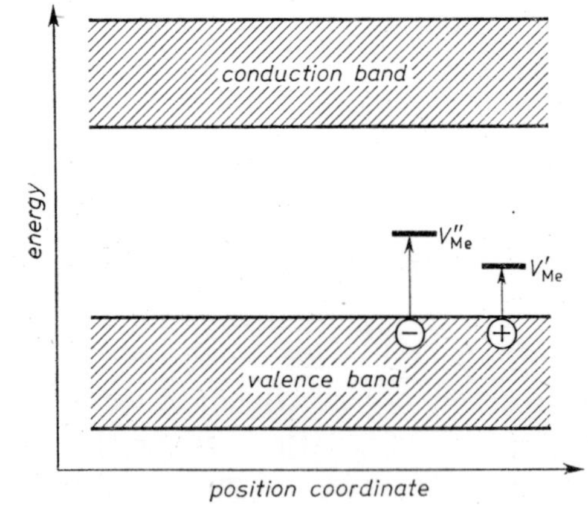

Fig. 43. The scheme of donor levels (interstitial cations) in the forbidden energy band of $Me_{1+y}X$ crystal.

The donor levels are interstitial atoms and ions of metals or anion vacancies, whereas cation vacancies and interstitial anions constitute electron acceptors. This is shown schematically in Figs. 43 and 44. When the interstitial defect is an atom of a divalent metal, as shown in Fig. 43, it in-

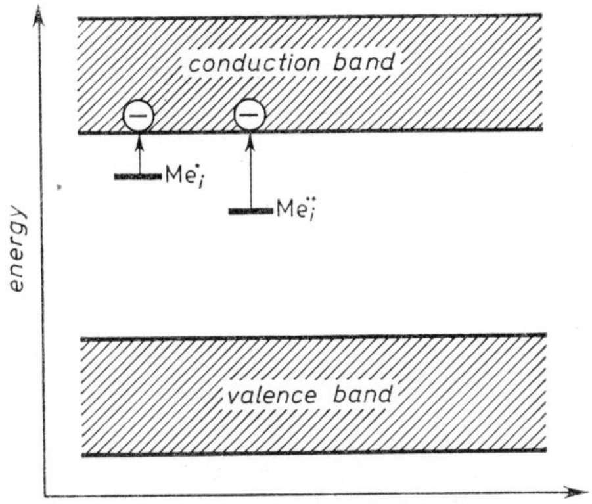

Fig. 44. The scheme of acceptor levels (cation vacancies) in the forbidden energy band of $Me_{1-y}X$ crystal.

troduces to the forbidden energy band two local donor levels, since it can release successively two electrons to the conductance band. However, release of the second valence electron requires a greater amount of energy than that of the first and as a result the second level is localised farther from the bottom of the conductance band than the first. In the case of acceptor levels, the situation is analogous (see Fig. 44).

From the point of view of defect equilibria (see Chapter 1.3.6), ionisation of donors can be represented by the following equation:

$$D = D^{\cdot} + e^- \qquad (1.3.38)$$

where D is a non-ionised donor centre (e.g. an interstitial atom of a metal or an anion vacancy electrically neutral with respect to the lattice) and $D^{\cdot}$ is a singly ionised donor level (e.g. $Me_i^{\cdot}$ or V_X').

Application of the law of mass action to this relationship gives:

$$K_D = \frac{[D^{\cdot}][e^-]}{[D]} \qquad (1.3.39)$$

where K_D is the equilibrium constant of the considered process. When the total number of donor levels is N_D, the following relationship is valid:

$$[D^{\cdot}] = N_D - [D] \qquad (1.3.40)$$

and equation (1.3.39) can be expressed in the following form:

$$K_D = \frac{\left(1-\frac{[D]}{N_D}\right)[e^-]}{\frac{[D]}{N_D}} \tag{1.3.41}$$

If the energy of the donor level is equal to E_D. Fermi statistics (equation (1.3.26)) leads to the following relationship:

$$\frac{[D]}{N_D} = \frac{1}{1+\exp\left(\frac{E_D-E_F}{kT}\right)} \tag{1.3.42}$$

By substituting this equation in equation (1.3.41) and utilising equation (1.3.29), the equilibrium constant of ionisation of donors can be expressed by the following equation:

$$K_D = N_C \exp\left[-\left(\frac{E_C-E_D}{kT}\right)\right] = N_C \exp\left(-\frac{E_d}{kT}\right) \tag{1.3.43}$$

where $E_d = E_C - E_D$ is the donor ionisation energy (see Figs. 41 and 43). From the thermodynamical point of view, this is the enthalpy of the defect reaction, which is given by equation (1.3.38).

Ionisation of acceptor centres can be considered analogously. From the point of view of the theory of defects, this process can be expressed by the following equation:

$$A = A' = h^{\bullet} \tag{1.3.44}$$

where A is a non-ionised acceptor centre (a cation vacancy electrically neutral with respect to the lattice or an atom of an oxidising agent in an interstitial position; see Figs. 42 and 44) and A′ is a singly ionised acceptor level (V'_{Me}, X'_X). Application of the law of mass action to this reaction gives the following equation:

$$K_A = \frac{[A'][h^{\bullet}]}{[A]} \tag{1.3.45}$$

where K_A is the equilibrium constant of ionisation of acceptors.

Transformations analogous to those used in the case of the donor processes lead to the following expression for the equilibrium constant of ionisation of acceptors:

$$K_A = N_V \exp\left[-\left(\frac{E_A-E_V}{kT}\right)\right] = N_V \exp\left(-\frac{E_a}{kT}\right) \tag{1.3.46}$$

where E_A is the energy of acceptor levels and $E_a = E_A - E_V$ is the energy of ionisation of these levels (see Fig. 42).

From the thermodynamical point of view, E_a is the enthalpy (heat) of the defect reaction expressed by equation (1.3.44).

The above considerations lead to the conclusion that in order to determine the numerical values of equilibrium constants of both the intrinsic and extrinsic formation of electronic defects it is necessary to first know the width of the forbidden energy band (E_g). The knowledge of this energy is also necessary for characterisation of several other parameters of semiconducting crystals of the MeX type. It is enough to mention that the width of the energy gap E_g determines whether the substance is an insulator or a semiconductor, since the difference between such substances has only this quantitative character. In the case of a wide energy gap (large E_g), the passage of electrons from the valence band to the conductance band requires a high energy and for this reason even at elevated temperatures the concentrations of quasi-free electrons and electron vacancies in both bands are negligibly small. If in a substance having a wide energy gap the energy levels due to impurities (point defects, foreign atoms, etc.) are practically absent, the substance has the properties of an insulator. On the other hand, when the forbidden energy band is relatively narrow, the concentrations of electrons and holes in both bands are correspondingly high and the substance has the properties of a semiconductor.

Experimental determination of the width of a forbidden energy band is a difficult problem. For this reason, attempts have been made for a long time to correlate this width with other properties of solids, which could be determined more easily experimentally or calculated theoretically. It has been found that there is such a correlation between E_g and several physico-chemical properties of crystals of the MeX type. In particular, E_g depends on bond energies, electronegativity and enthalpy of formation of these compounds. According to Manca [33], Vijh [34–36] and Kofstad [37] the correlation between bond energy and E_g in compounds of the MeX type is of particular importance. When the energy of bonds between atoms in a crystal of the MeX type is expressed as the so-called atomisation energy per one equivalent, i.e. the energy required for dissociation of the compound into atoms in the gas phase, this energy is a linear function of the width of the forbidden energy band E_g. (In this case the equivalent is the number of cations or anions in the binary compound MeX multiplied by their valence. For example, in niobium pentoxide, Nb_2O_5, the atomisation energy per equivalent is equal to 1/10-th of the atomisation energy of this compound. In nickel oxide (NiO) the atomisation energy per equivalent is equal to one half of the atomisation energy.)

Figure 45 shows this relationship graphically. It can be expressed analytically in the form of the following linear equation:

$$E_g = 2(E_s - b)[\text{eV}] \quad (1.3.47)$$

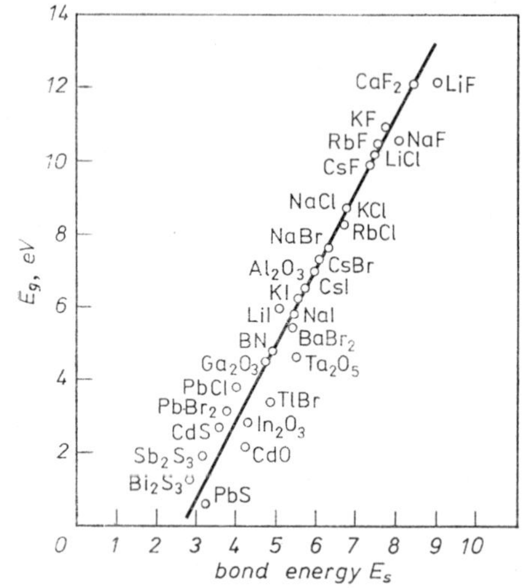

Fig. 45. The dependence of the width of forbidden energy band E_g on the bond energy in various MeX type compounds. (after [37]). The bond energy is expressed as the so-called atomisation energy gram-equivalent (see the text).

where E_s is the atomisation energy E_{at} divided by the equivalent ($E_s = E_{at}$/equivalent) and b is a constant approximately equal to 2.7.

At the end of this chapter it is worthwhile mentioning the direct relationship which exists between the band model of solid substances and the thermodynamic approach to this problem, particularly in view of the fact that the theory of defects is based on thermodynamical concepts.

It was mentioned in Chapter 1.3.3 that the state of chemical equilibrium is characterised by equality of chemical potentials of all the components of the system in all the phases. This is one of the most general laws of chemical thermodynamics.

With reference to electrons in the considered ionic crystals, this law stipulates that the chemical potential of electrons in a crystal remaining in the state of thermodynamic equilibrium with the surroundings μ_e is equal to that of quasi-free electrons in the conductance band μ_e^C and to that of valence electrons μ_e^V, although the energy of these electrons is different:

$$\mu_e = \mu_e^C = \mu_e^V \tag{1.3.48}$$

The relation between each of these chemical potentials, the temperature and the concentration (activity) of electrons is given by the general equation (1.3.20). For example, the chemical potential of quasi-free electrons in the conductance band can be expressed by the following equation:

$$\mu_e^C = \mu_e^{0C} + kT \ln \frac{n}{N_C} \tag{1.3.49}$$

where $\frac{n}{N_C}$ is the fraction of energy levels in the conduction band occupied by the electrons, i.e. the concentration of quasi-free electrons expressed in dimensionless terms analogous to mole fractions, and μ_e^{0C} is the chemical potential of quasi-free electrons in the conductance band of the standard state.

Equation (1.3.29) can be transformed to the following form:

$$E_F = E_C + kT \ln \frac{n}{N_C} \tag{1.3.50}$$

Comparison of equations (1.3.49) and (1.3.50) shows that the Fermi level, the principal concept in the band theory of solid substances, represents in fact the chemical potential of electrons in the crystal and in the concrete case the chemical potential of quasi-free electrons in the conductance band. Thus in this case the standard chemical potential is equal to the energy of the bottom of conductance band $\mu_e^{0C} = E_C$.

1.3.5 Defect equilibria in compounds having stoichiometric composition

As already mentioned, a crystal containing point defects can be regarded as a solution in which the lattice is the solvent and the solute are the defects. The kind and concentration of the latter have a much greater effect on many physico-chemical properties of the crystal than the kind of substance from which it is built. When the number of defects is very small in comparison with the number of occupied lattice sites, from the thermodynamic point of view the crystal is equivalent to a very dilute solution. For this reason the law of mass action can be applied with good approximation to reversible reactions of defect formation, using the concentrations of defects.

The formation of defects of the Frenkel type in a crystal of the general formula $Me^{2+}X^{2-}$ can be expressed by the following equation:

$$Me_{Me} \rightleftharpoons Me_i^{\bullet\bullet} + V_{Me}'' \qquad (1.3.51)$$

The formation of defects in the anionic sublattice can be expressed as follows:

$$X_X \rightleftharpoons X_i'' + V^{\bullet\bullet} \qquad (1.3.52)$$

The formation of defects of the Schottky and anti-Schottky types can be expressed by the following equations:

$$\text{zero} \rightleftharpoons V_{Me}'' + V_x^{\bullet\bullet} \qquad (1.3.53)$$

$$Me_{Me}X_x \rightleftharpoons Me_i^{\bullet\bullet} + X_i'' \qquad (1.3.54)$$

where zero stands for a crystal not containing defects.

Application of the law of mass action to these relationships gives the following expressions for the equilibrium constants of defect formation reactions:

$$K_F = [Me_i^{\bullet\bullet}][V_{Me}''] \qquad (1.3.55)$$

$$K_F' = [X_i''][V_x^{\bullet\bullet}] \qquad (1.3.56)$$

$$K_S = [V_{Me}''][V_x^{\bullet\bullet}] \qquad (1.3.57)$$

$$K_S' = [Me_i^{\bullet\bullet}][X_i''] \qquad (1.3.58)$$

where K_F, K_F', K_S and K_S' are equilibrium constants of formation of defects of the Frenkel, anti-Frenkel, Schottky and anti-Schottky types. The square brackets stand for the concentration of defects, which is usually expressed in mole fractions.

The relation between the equilibrium constant of a chemical reaction and its standard thermodynamic potential change is given by the following general equation (compare equation 1.3.24)):

$$K = \exp\left(-\frac{\Delta G}{RT}\right) \qquad (1.3.59)$$

Since

$$\Delta G = \Delta H - T\Delta S \qquad (1.3.60)$$

the expression for the equilibrium constant can be written in the following form:

$$K = \exp\left(\frac{\Delta S}{R}\right)\exp\left(-\frac{\Delta H}{RT}\right) \qquad (1.3.61)$$

Using this relationship, the expression for the equilibrium constant of the reaction of formation of Frenkel-type defects can be written as follows:

$$K_F = \exp\left(\frac{\Delta S_F}{R}\right)\exp\left(-\frac{\Delta H_F}{RT}\right) \tag{1.3.62}$$

where ΔS_F and ΔH_F stand for the changes of molar entropy and enthalpy of formation of pairs of Frenkel-type defects, respectively. Similarly, the equilibrium constant of formation of Schottky-type defects can be expressed by the following equation:

$$K_S = \exp\left(-\frac{\Delta G_S}{RT}\right) = \exp\left(\frac{\Delta S_S}{R}\right)\exp\left(-\frac{\Delta H_S}{RT}\right) \tag{1.3.63}$$

If the crystal has strictly stoichiometric composition and does not contain other defects, the concentration of cationic defects must be equal to that of anionic defects, since the condition of electrical neutrality of the lattice must be observed. In the case of the formation of Frenkel-type defects:

$$[\mathrm{Me}_i^{\bullet\bullet}] = [\mathrm{V}_{\mathrm{Me}}''] = K_F^{1/2} \tag{1.3.64}$$

and in the case of formation of Schottky-type defects:

$$[\mathrm{V}_{\mathrm{Me}}''] = [\mathrm{V}_{\mathrm{X}}^{\bullet\bullet}] = K_S^{1/2} \tag{1.3.65}$$

Analogous expressions can be obtained for the equilibria of formation of defects of the anti-Frenkel and anti-Schottky types.

Equations (1.3.62) and (1.3.63) are generally valid, since they represent the relation between the equilibrium constant of any reaction of formation of defects and the enthalpy and entropy of this process:

$$K_d = \exp\left(\frac{\Delta S_f}{R}\right)\exp\left(-\frac{\Delta H_f}{RT}\right) \tag{1.3.66}$$

where K_d is the equilibrium constant of the reaction of formation of defects and ΔS_f and ΔH_f are the entropy and the enthalpy of this process, respectively. The physical sense of ΔH_f is obvious, it is the heat of formation of one mole of defects, but ΔS_f is not the total change of entropy corresponding to the formation of one mole of defects. As already mentioned, this change is the sum of changes of vibrational and configurational entropies:

$$\Delta S_f = \Delta S_V + \Delta S_k \tag{1.3.67}$$

The change of configurational entropy, or the entropy of mixing, can be calculated by means of equation (1.3.37) resulting from statistical mechanics.

The thermodynamic probability appearing in this equation means in this case the number of possible distinct ways of distributing n defects in a crystal structure, containing $N+n$ possible sites:

$$W = \frac{(N+n)!}{N!n!} \tag{1.3.68}$$

Introduction of this equation into equation (1.3.7) gives:

$$\Delta S_k = k \ln \frac{(N+n)!}{N!n!} \tag{1.3.69}$$

Since $N \gg n$, the Stirling simplification can be used:

$$\ln x! = x \ln x - x \tag{1.3.70}$$

which leads to the following relationship:

$$\Delta S_k = -\left[kN \ln\left(\frac{N}{N+n}\right) + n \ln\left(\frac{n}{N+n}\right)\right] \tag{1.3.71}$$

Introduction of this expression into equation (1.3.60) gives:

$$\Delta G_f = \Delta H_f - T\Delta S_V + kT\left[N \ln\left(\frac{N}{N+n}\right) + n \ln\left(\frac{n}{N+n}\right)\right] \tag{1.3.72}$$

In the equilibrium state the thermodynamic potential of the system reaches its minimum value. Hence:

$$\frac{\partial \Delta G}{\partial n} = 0 \tag{1.3.73}$$

Therefore differentiation of equation (1.3.72) and further mathematical transformations lead to the following expression for the equilibrium concentration of defects:

$$\frac{n}{N+n} = \exp\left(\frac{\Delta S_V}{R}\right) \exp\left(-\frac{\Delta H_f}{RT}\right) \tag{1.3.74}$$

The left hand side of this equation represents the concentration of defects expressed in mole fractions N_d. Hence:

$$N_d = \exp\left(\frac{\Delta S_V}{R}\right) \exp\left(-\frac{\Delta H_f}{RT}\right) \tag{1.3.75}$$

It follows from this equation that when the concentration of defects is expressed in mole fractions the entropy term contains only the change of vibrational entropy. It should be remembered that when the concentrations of defects are expressed in mole fractions, the entropy terms in all the defect equations contain only the vibrational entropy.

1.3.6 Defect equilibria in non-stoichiometric compounds

Defects of the Frenkel and Schottky type occur in a relatively small group of inorganic compounds, particularly in metal halides. A much larger group consists of ionic crystals containing compounds showing deviations from stoichiometry. To this group belong metal oxides and sulphides.

Deviations from stoichiometry in compounds of the MeX type result from the presence of point defects in the crystal lattice of these compounds. The deviations are manifested as an excess or a deficit of a definite component of the lattice. Since these differences determine the kind of disorder, the compounds discussed can be divided into two main groups. To one of them belong crystals consisting of compounds containing an excess of metal or—what is equivalent—a deficit of the electronegative component. To the other group belong compounds containing an insufficient amount of metal or an excess of the electronegative component. In contrast to the group of crystals previously described having stoichiometric composition, the point defects in non-stoichiometric compounds are formed on the crystal surface as a result of chemical interactions between the solid phase and the atmosphere, and for this reason they are also called chemical defects. At sufficiently high temperatures the defects that form on the crystal surface begin to spread (as a result of diffusion) to the bulk of the lattice at such a rate that thermodynamic equilibrium between the defect crystal and its surroundings is established. At low temperatures the movement of defects in the crystal caused by diffusion is so slow that equilibrium between the bulk of crystal and its surface is in practice never established.

Let us first consider the defects in compounds showing deviations from the stoichiometric composition in the direction of an excess of metal or a deficit of the electronegative component. In this group two kinds of defects can occur. The deviation from stoichiometry can be caused by a real excess of metal resulting from the presence of interstitial cations when all the sites in both sublattices are filled. In this case the correct stoichiometric formula of a compound of the MeX type is $Me_{1+y}X$. However, the same type of deviation from stoichiometric composition can be caused by a real deficit of the electronegative component as a result of the occurrence of empty sites in the ionic sublattice when the cationic sublattice has a perfectly ordered structure. In this case the correct formula of the compound is MeX_{1-y}.

An example of a compound containing a real excess of metal is zinc oxide, ZnO. The formation of defects in its crystal lattice can be imagined as follows: when a strictly stoichiometric ideal crystal of zinc oxide has a temperature higher that the absolute zero it is thermodynamic unstable and suffers partial decomposition into zinc metal and oxygen gas. This process takes place on the surface of the crystal and can be expressed by the following equation:

$$ZnO \rightleftharpoons Zn(s)+\tfrac{1}{2}O_2(g) \qquad (1.3.76)$$

where Zn(s) stands for an atom of zinc on the crystal surface.

At sufficiently high temperatures the atoms of zinc metal, which appear on the crystal surface, "dissolve" in the lattice and occupy interstitial positions. At the same time ionisation of these atoms, leading to the formation of cations bearing one or even two positive charges, may take place. The free electrons which are formed during this process remain in the crystal as so-called quasi-free electrons.

From the point of view of the band theory of solids, an interstitial zinc atom represents a donor centre, since it occupies a position in the forbidden energy band close to the bottom of the conduction band. For this reason the ionisation of this centre takes place relatively readily and as a result the electron jumps to the conduction band. Monovalent zinc cation is also a donor level, but it occupies a position in the forbidden energy band further from the bottom of the conduction band than that occupied by an electrically neutral zinc atom. For this reason further ionisation of monovalent zinc cations requires a greater amount of energy and consequently takes place at correspondingly higher temperatures.

Thus at sufficiently high temperatures, interstitial mono- and divalent cations and the equivalent number of quasi-free electrons are formed in a ZnO crystal and, as a result, the lattice remains electrically neutral in spite of the deviation from stoichiometric composition. The whole process of formation of this type of defect in the crystal lattice of zinc oxide can be expressed by the following equations:

$$ZnO \rightleftharpoons Zn_i^{\cdot}+e^- +\tfrac{1}{2}O_2(g) \qquad (1.3.77)$$

$$Zn_i^{\cdot} \rightleftarrows Zn_i^{\cdot\cdot}+e^- \qquad (1.3.78)$$

As the temperature increases the equilibrium of reaction (1.3.78) shifts to the right.

It has already been mentioned that in the writing of the defect equilibria discussed and all other processes of this type, attention should be paid to the condition that the lattice must be electrically neutral and to the condition of conservation of mass. This means that the algebraic sum of positive and negative charges in the above equation must be always equal to zero and that the sums of masses of the individual lattice components on both sides of the equation must be equal. It is assumed that the masses of electron defects and those of cationic and anionic vacancies are equal to zero.

The general case of formation of defects in crystal lattices of compounds of the $ZnO(Me_{1+y}X)$ type can be expressed by the following equations:

$$MeX \rightleftharpoons Me_i + \tfrac{1}{2}X_2(g) \tag{1.3.79}$$

$$Me_i \rightleftharpoons Me_i^{\bullet} + e^- \tag{1.3.80}$$

$$Me_i^{\bullet} \rightleftharpoons Me_i^{\bullet\bullet} + e^- \tag{1.3.81}$$

As has already been mentioned, the deviation from stoichiometry in the direction of an excess of metal can also be caused by a real deficit of the component $X(MeX_{1-y})$. Examples of compounds of this type are zirconium dioxide ZrO_{2-y} and niobium pentoxide Nb_2O_{5-y}. In this case oxygen ions leave the sites situated at the crystal surface and, abandoning their electrons in the lattice, they form molecules which pass to the gas phase. The process of formation of this type of defect can be expressed as follows:

$$X_X \rightleftharpoons V_X + \tfrac{1}{2}X_2(g) \tag{1.3.82}$$

The empty site in the anionic sublattice, which later we will call an anion vacancy, has no effective charge with respect to the lattice, since the two electrons which had been previously associated with the anion have been left in the vacancy. Thus this defect is a donor centre like an interstitial metal atom in compounds of the ZnO type. At sufficiently high temperatures this centre can be ionised and as a result the vacancy acquires an effective positive charge with respect to the lattice. At the same time quasi-free electrons appear in the conductivity band:

$$V_X \rightleftharpoons V_X^{\bullet} + e^- \tag{1.3.83}$$

$$V_X^{\bullet} \rightleftharpoons V_X^{\bullet\bullet} + e^- \tag{1.3.84}$$

It follows from these considerations that deviations from stoichiometric composition in the direction of excess of metal can be caused by the presence of interstitial atoms or metal ions in the crystal or by the presence of a definite number of vacancies in the anionic sublattice. In both cases the condition of electrical neutrality of the lattice is preserved as a result of the appearance of an aquivalent number of quasi-free electrons. For this reason both kinds of compounds mentioned above have the properties of semiconductors of the *n* type (conduction of electric current due to

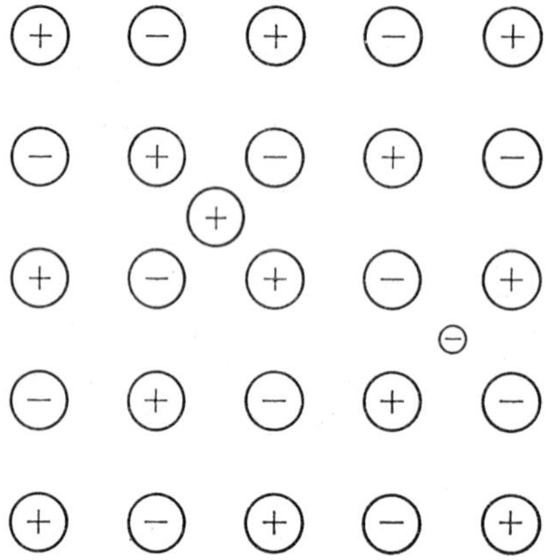

Fig. 46. A model of lattice disorder in the cation sublattice ($Me_{1+y}X$ type). The small circle with the minus sign stands for a quasi-free electron.

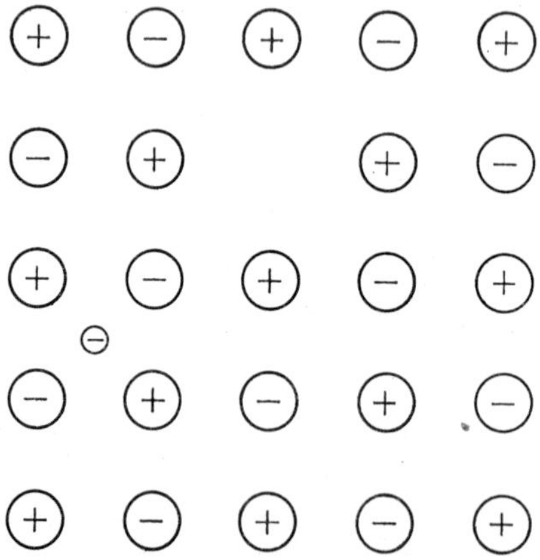

Fig. 47. A model of lattice disorder in the anion sublattice (MeX_{1-y}) type.

migration of quasi-free electrons in the conduction band). The defects of crystal lattices of the compounds discussed are shown schematically in Figs. 46 and 47.

The law of mass action can be applied to the defect equilibria represented by equations (1.3.77)–(1.3.84) in the same way as to the equilibria of defects of Frenkel and Schottky type. It should be mentioned that the defects of Frenkel and Schottky type are called intrinsic ionic defects, since unlike the chemical defects, now being considered, they are formed without the participation of the gas phase and their equilibria depend only on the temperature of the crystal. Another difference between the two types of defects is that the concentration of the chemical defects is a definite function of pressure of component X in the atmosphere surrounding the crystal [compare relationships (1.3.79) and (1.3.85)]. This problem will be further discussed in the next part of this book.

Another large group of inorganic substances consists of compounds showing deviations from stoichiometric composition in the direction of deficit of metal. In the compounds of this type a real deficit of metal can occur as a result of the presence of empty sites in the cationic sublattice. The deviation from stoichiometry can be also due to a real excess of the electronegative component which occupies the interstitial positions in the anionic sublattice, but this is rarely observed, since the size of the anions is relatively large. Thus the group of substances discussed consists mainly of compounds of the $Me_{1-y}X$ type. A typical compound of this group is nickel oxide, NiO.

The formation of chemical defects in crystals of this type also takes place at the gas–solid phase boundary. At a sufficiently high temperature the defects spread throughout the whole crystal and reach a state of thermodynamic equilibrium with the surroundings. In this case the process of defect formation is initiated by adsorption of the electronegative component from the gas phase on the surface of a crystal without defects. The adsorbate attracts a certain number of electrons from the bulk of the crystal and as a result is converted into ions that are chemisorbed on suitable centres in the lattice. At the same time electron defects, called electron holes, consisting of cations of higher valency or, less frequently, anions of lower valency, appear in the crystal. These defects are formed as a result of migration of a certain number of electrons to the chemisorbed layer of ions of the oxidant. This leads to the formation of a strong electrostatic field between the bulk of the crystal lattice and the surface layer of chemisorbed anions, which pulls the anions into the bulk of the crystal and pushes the cations to the surface. It has already been mentioned that the incorporation of anions into the interstitial sites is seldom observed (UO_2) and that therefore the step usually occurring in the process of chemical defect formation, in crystals with a real deficit of metal, is the diffusion of a certain number of cations from the lattice sites to the surface where they combine with the chemisorbed anions. As a result of this process new lattice elements are formed. The vacancies appearing in the cationic sublattice correspond to the number of electron holes and as a result the lattice remains electrically neutral.

A cationic vacancy has two effective negative charges with respect to the lattice in which a divalent cation is missing. However in the neighbourhood of the vacancy there are electron holes which are positive with respect to the lattice and which can interact with cationic vacancies. This interaction can lead to the formation of complex defects. Thus at not too high temperatures cationic vacancies which are electrically neutral with respect to the lattice can appear in the crystal and, as the temperature increases, they can become gradually ionised with the liberation of the associated electron holes.

In the case of compounds of the MeX_{1+y} type the anions formed on the crystal surface by the mechanism described above are drawn by the electrostatic field to interstitial positions and, at sufficiently high temperatures, they diffuse throughout the crystal. Since the interstitial anions, like the cationic vacancies, have negative charges, they become associated with the electron holes and form complexes which become gradually ionised when the temperature increases.

From the point of view of the band theory a cationic vacancy or an interstitial anion are acceptor levels, since they represent defects localised in forbidden energy bands near the top of the valence band. For this reason electrons can pass relatively readily from the valence band to the acceptor levels and as a result electron holes appear in the valence band. In the case of incomplete association with the acceptor centres the electron holes can move in the crystal like quasi-free electrons in the conduction band of n-type semiconductors. The hole semiconductors are also called p-type semiconductors.

The process of formation of defects in crystal lattices of compounds of the $Me_{1-y}X$ type can be represented by the following equations:

$$\tfrac{1}{2}X_2(g) \rightleftarrows V_{Me} + X_X \tag{1.3.85}$$

$$V_{Me} \rightleftarrows V'_{Me} + h^{\cdot} \tag{1.3.86}$$

$$V'_{Me} \rightleftarrows V''_{Me} + h^{\cdot} \tag{1.3.87}$$

The reactions for defect formation in the anionic sublattice of compounds of the MeX_{1+y} type can be expressed as follows:

$$\tfrac{1}{2}X_2(g) \rightleftarrows X_i \tag{1.3.88}$$

$$X_i \rightleftarrows X_i' + h^{\cdot} \tag{1.3.89}$$

$$X_i' \rightleftarrows X_i'' + h^{\cdot} \tag{1.3.90}$$

Models of defect formation in crystal lattices of compounds of this type ($Me_{1-y}X$ and MeX_{1+y}) are shown in Figs. 48 and 49.

It follows from comparison of equations (1.3.79) and (1.3.82) with equations (1.3.85) and (1.3.88) that the effect of partial pressure of the

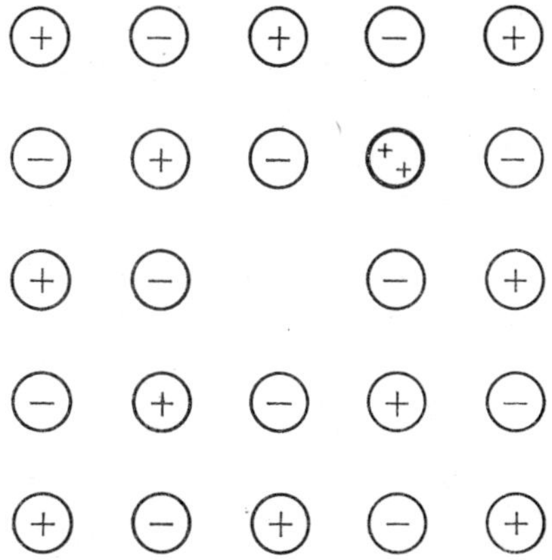

Fig. 48. A model of lattice disorder in the cation sublattice ($Me_{1-y}X$ type). The circle with two plus signs stands for an electron hole.

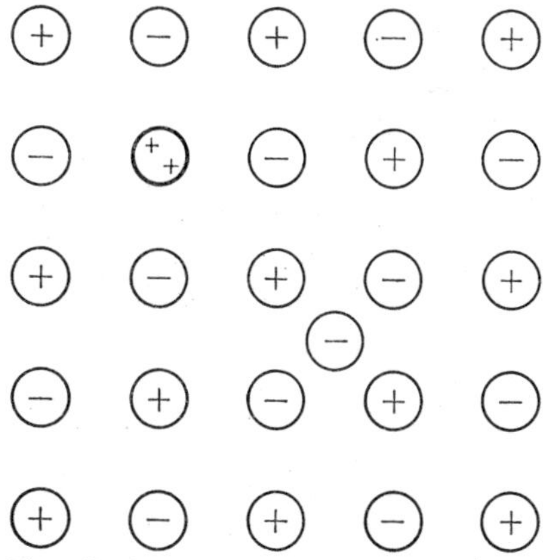

Fig. 49. A model of lattice disorder in the anion sublattice (MeX_{1+y} type).

oxidant on the concentration of defects in compounds containing an excess of metal or a deficit of the oxidant is the opposite of that observed in the case of compounds having a deficit of metal or an excess of component X. In the former case the concentration of defects decreases with increasing pressure of component X whereas in the case of compounds of the $Me_{1-y}X$ or MeX_{1+y} type the concentration of defects increases with increasing partial pressure of component X in the atmosphere.

So far we have been considering the limiting types of disorder in the crystal lattices of compounds having an ionic structure. However it should be pointed out that in practice the disorder has a much more complex character as the defects predominating under the conditions mentioned above are accompanied by defects of other kinds which are present in equilibrium with the predominant ones. The equilibrium constants of the individual types of defects differ considerably from each other and for this reason any change in the thermodynamic conditions causes a shift of the equilibria in a definite direction which may lead to a change of the predominant type of disorder. As a result *n* type semiconductor may become *p* type and vice versa. Since this problem is directly connected with the mechanisms of transport of matter in solid phases, in later chapter we will discuss the effect of temperature and pressure of the oxidant together on defect equilibria in non-stoichiometric compounds.

1.3.6.1 Defect equilibria in compounds of the $Me_{1+y}X$ *type*

Let us first consider the effect of the degree of ionisation of defects, i.e. the effect of temperature, on defect equilibria in a crystal with a large deviation from the stoichiometric composition. In such a crystal the principal disorder, as expressed by equations (1.3.79)–(1.3.81), is accompanied by intrinsic ionic defects of the Frenkel type expressed by equation (1.3.51) and intrinsic electronic defects, the formation of which can be represented by the following equation:

$$\text{zero} \rightleftarrows e^- + h^\bullet \qquad (1.3.91)$$

The latter process consists of the passage of electrons from the valence band to the conduction band leading to the formation of quasi-free electrons and electron holes. Since the forbidden energy band is considerably wider than the energy gap between the donor levels and the bottom of the conduction band, the process expressed by equation (1.3.91) requires much more energy than the process of ionisation of the excess metal atoms. At not too high temperatures and at high concentrations of the donors the contribution of reaction (1.3.91) to the process of formation of quasi-free electrons and electron holes is negligibly small and can be omitted in the consideration of defect equilibria. Also the formation of intrinsic ionic defects expressed by equation (1.3.51) can be neglected, since at large deviations from stoichiometry the concentration of interstitial atoms and cationic vacancies resulting from Frenkel type disorders is negligibly small in comparison with the concentration of the excess interstitial atoms. Also the disorder in the anionic sublattice need not be taken into account, since the compound considered contains a real excess of metal. Although defects of the anti-Frenkel type can occur in the anionic sublattice according to equation (1.3.52), their concentration, like that of the Frenkel type defects, is negligible in comparison with the concentration of the predominant type of disorder.

On taking into account the above limitation, it can be stated that the defect equilibria in the case discussed can be expressed to a good approximation by equations (1.3.79)–(1.3.81). For the complete description of these equilibria a simplified condition of electrical neutrality of the lattice is applies. In the case discussed it can be reduced to the following form:

$$[e^-] = [\mathrm{Me}_i^{\bullet}]+2[\mathrm{Me}_i^{\bullet\bullet}] \tag{1.3.92}$$

since, when these simplifying assumptions are made, the predominant electrically charged defects in a crystal are quasi-free electrons and interstitial metal ions.

By applying the law of mass action to reactions (1.3.79)–(1.3.81) the following expressions are obtained:

$$K_{79} = [\mathrm{Me}_i]p_{x_2}^{1/2} \tag{1.3.93}$$

$$K_{80} = \frac{[\mathrm{Me}_i^{\bullet}][e^-]}{[\mathrm{Me}_i]} \tag{1.3.94}$$

$$K_{81} = \frac{[\mathrm{Me}_i^{\bullet\bullet}][e^-]}{[\mathrm{Me}_i^{\bullet}]} \tag{1.3.95}$$

where the equilibrium constants subscripts stand for the corresponding defect equilibria. Combination of equations (1.3.92)–1.3.95) gives the following relationship between the electron defect concentration and oxygen pressure:

$$[e^-]^3 = K_{79}K_{80}p_{x_2}^{-1/2}(2K_{81}+[e^-]) \tag{1.3.96}$$

This expression has two limiting solutions. When the concentration of quasi-free electrons is considerably higher than $2K_{81}$, equation (1.3.96) becomes:

$$[e^-] = (K_{79}K_{80})^{1/2}p_{X_2}^{-1/4} \qquad (1.3.97)$$

On the other hand, when the concentration of electron defects is much lower than $2K_{81}$, equation (1.3.96) becomes:

$$[e^-] = (2K_{79}K_{80}K_{81})^{1/3}p_{X_2}^{-1/6} \qquad (1.3.98)$$

The first condition is fulfilled when the interstitial cations are mostly monovalent. On the other hand, the second condition requires that mainly divalent cations should be present in the interstitial positions. Since the degree of ionisation of defects increases with increasing temperature, relationship (1.3.97) is valid at low temperatures and relationship (1.3.98) at high. Obviously, in the intermediate temperature range, in which the concentrations of mono- and divalent interstitial cations are comparable, the concentration of quasi-free electrons in the crystal must depend on the pressure of component X_2 raised to a power between $-1/4$ and $-1/6$.

The solution of equations (1.3.92)–(1.3.95) with respect to the concentration of ionic defects leads to three limiting cases:

1. The concentration of non-ionised defects predominates $[Me_i] \gg [Me_i^{\bullet}]+[Me_i^{\bullet\bullet}]$. In this case:

$$[Me_i]_c = K_{79}\, p_{X_2}^{-1/2} \qquad (1.3.99)$$

2. The concentration of singly ionised defects predominates $[Me_i^{\bullet}] \gg [Me_i]+[Me_i^{\bullet\bullet}]$. In this case:

$$[Me_i]_c = (K_{79}K_{80})^{1/2}p_{X_2}^{-1/4} \qquad (1.3.100)$$

3. The concentration of doubly ionised defects predominates $[Me_i^{\bullet\bullet}] \gg [Me_i]+[Me_i^{\bullet}]$. In this case:

$$[Me_i]_c = (\tfrac{1}{4}K_{79}K_{80}K_{81})^{1/3}p_{X_2}^{-1/6} \qquad (1.3.101)$$

where $[Me_i]_c$ is the total concentration of interstitial ions.

It follows from comparison of relationships (1.3.97) and (1.3.98) with equations (1.3.99)–(1.3.101) that the pressure dependence of the concentration of electron defects in compounds of the $Me_{1+y}X$ type can vary with temperature between $p_{X_2}^{-1/4}$ and $p_{X_2}^{-1/6}$, whereas the analogous dependence of ionic defects can vary between $p_{X_2}^{-1/2}$ and $p_{X_2}^{-1/6}$.

Analogous relationships are entirely valid for compounds having defect anionic sublattices (anion vacancies and quasi-free electrons) and the general formula MeX_{1-y}. This leads to the conclusion that the concentration of ionic and electron defects in compounds having a real excess of metal or a real deficit of the oxidant decreases with increasing equilibrium pressure of component X in the gas phase.

1.3.6.2 Defect equilibria in compounds of the $Me_{1-y}X$ *type*

Defect equilibria in compounds having a real deficit of metal are expressed by equations (1.3.85)–(1.3.87). Since we are now considering large deviations from stoichiometric composition, we can, as before, neglect intrinsic ionic and electronic disorder. Hence, by applying the law of mass action we obtain:

$$K_{85} = \frac{[V_{Me}]}{p_{X_2}^{1/2}} \qquad (1.3.102)$$

$$K_{86} = \frac{[V'_{Me}][h^{\bullet}]}{V_{Me}} \qquad (1.3.103)$$

$$K_{87} = \frac{[V''_{Me}][h^{\bullet}]}{V'_{Me}} \qquad (1.3.104)$$

The simplified condition of electroneutrality for the lattice in such cases has the form:

$$[h^{\bullet}] = [V'_{Me}]+2[V''_{Me}] \qquad (1.3.105)$$

The solution of equations (1.3.102)–(1.3.105) leads to the following relationship for the concentration of electron holes:

$$[h^{\bullet}]^3 = K_{85}K_{86}p_{X_2}^{1/2}(2K_{87}+[h^{\bullet}]) \qquad (1.3.106)$$

This equation, like relationship (1.3.96), has two limiting solutions. When $[h^{\bullet}] \gg 2K_{87}$, equation (1.3.106) is reduced to the following form:

$$[h^{\bullet}] = (K_{85}K_{86})^{1/2}p_{X_2}^{1/4} \qquad (1.3.107)$$

On the other hand, when $[h^{\bullet}] \ll 2K_{87}$ we obtain:

$$[h^{\bullet}] = (2K_{85}K_{86}K_{87})^{1/3}p_{X_2}^{1/6} \qquad (1.3.108)$$

With respect to concentrations of ionic defects, equations (1.3.102)–(1.3.108) have the three limiting solutions, as follows:

$$[V_{Me}] \gg [V'_{Me}]+[V''_{Me}] \qquad [V_{Me}]_c = K_{85}p_{X_2}^{1/2} \qquad (1.3.109)$$

$$[V'_{Me}] \gg [V_{Me}]+[V''_{Me}] \qquad [V_{Me}]_c = (K_{85}K_{86})^{1/2}p_{X_2}^{1/4} \qquad (1.3.110)$$

$$[V''_{Me}] \gg [V_{Me}]+[V'_{Me}] \qquad [V_{Me}]_c = (\tfrac{1}{4}K_{85}K_{86}K_{87})^{1/3}p_{X_2}^{1/6} \qquad (1.3.111)$$

It follows from equations (1.3.107)–(1.3.111) that the concentration of electron defects can increase with increasing pressure of the oxidant according to the function $p_{X_2}^{1/4}$–$p_{X_2}^{1/6}$ and that of ionic defects according to the function $p_{X_2}^{1/2}$–$p_{X_2}^{1/6}$. This leads to the conclusion that the relation between the concentration of ionic and electron defects and the pressure of the oxidant in compounds of the ZnO type is the reverse of that of compounds of the NiO type.

The equilibrium constants in equations (1.3.93)–(1.3.95) and (1.3.102)–(1.3.104) are related to the free energy of formation of defects by the general expression (1.3.61). For this reason the dependence of defect concentrations on the equilibrium pressure of component X in the atmosphere can be related to the temperature dependence by means of more general equations taking into account the enthalpy and the entropy of formation of the given type of defect. For example in equation (1.3.109), constant K_{85} can be replaced with a suitable expression relating the equilibrium constant to the entropy and enthalpy changes accompanying to reaction (equation (1.3.61)). This gives following relationship:

$$[V_{Me}] \cong [V_{Me}]_c = p_{X_2}^{1/2} \exp\left(\frac{\Delta S_{V_{Me}}}{R}\right) \exp\left(-\frac{\Delta H_{V_{Me}}}{RT}\right) \qquad (1.3.112)$$

where $[V_{Me}]_c$ is the total concentration of the cation vacancies expressed in mole fractions and $\Delta S_{V_{Me}}$ and $\Delta H_{V_{Me}}$ are the molar vibrational entropy and enthalpy of formation of non-ionised vacancies.

Equations (1.3.110) and (1.3.111) can be transformed analogously but they contain products of two or three equilibrium constants raised to fractional powers. Hence the total entropy and enthalpy of formation of defects is the sum of the partial entropies and enthalpies of formation of non-ionised vacancies as well as the entropy and enthalpy of ionisation of these defects given by equations (1.3.86) and (1.3.87). These quantities must be divided by 2 or 3, i.e. the number appearing in the denominator of the exponent of the equilibrium constant product in equations (1.3.110) and (1.3.111). Thus by transforming equation (1.3.110) in this way we obtain the following expression:

$$[V'_{Me}] \cong [V_{Me}]_c = p_{X_2}^{1/4} \exp\left(\frac{\Delta S_{V'_{Me}}}{2R}\right) \exp\left(-\frac{\Delta H_{V'_{Me}}}{2RT}\right) \qquad (1.3.113)$$

where $\Delta S_{V'_{Me}} = \Delta S_{V_{Me}} + \Delta S_{h^{\bullet}}$ and $\Delta H_{V'_{Me}} = \Delta H_{V_{Me}} + \Delta H_{h^{\bullet}}$.

The terms containing subscript $h^{\bullet}$ represent the entropy and enthalpy of dissociation for an electron hole in a cation vacancy, which is electrically neutral with respect to the lattice. Thus equation (1.3.113) can finally be written in the following, more detailed form:

$$[V'_{Me}] \cong [V_{Me}]_c = p_{x_2}^{1/4} \exp\left(\frac{\Delta S_{V_{Me}} + \Delta S_{h^\cdot}}{2R}\right) \exp\left(-\frac{\Delta H_{V_{Me}} + \Delta H_{h^\cdot}}{2RT}\right) \tag{1.3.114}$$

Analogous transformation of equation (1.3.111) gives the following relationship:

$$[V''_{Me}] \cong [V_{Me}]_c = p_{x_2}^{1/6} \exp\left(\frac{\Delta S_{V_{Me}} + \Delta S_{h^\cdot} + \Delta S_{2h^\cdot}}{3R}\right) \times \exp\left(-\frac{\Delta H_{V_{Me}} + \Delta H_{h^\cdot} + \Delta H_{2h^\cdot}}{3RT}\right) \tag{1.3.115}$$

where the terms with the subscript $2h^\cdot$ stand for the entropy and enthalpy of dissociation for an electron hole from a singly ionised vacancy, and hence they are related to the second step in the ionisation of the vacancy. Analogous expressions can be derived for the other defect equilibria.

These expressions can be transformed to general equations. For this purpose the dependence of the concentration of ionic and electron defects on the pressure of the oxidant and the temperature for compounds of $Me_{1-y}X$ and MeX_{1+y} types can be expressed by the following equations:

$$[j_d] = p_{x_2}^{1/n} \exp\left(\frac{\frac{2}{n}\Delta S_f}{R}\right) \exp\left(-\frac{\frac{2}{n}\Delta H_f}{RT}\right) \tag{1.3.116}$$

and

$$[h^\cdot] = p_{x_2}^{1/n} \exp\left(\frac{\frac{2}{n}\Delta S_f}{R}\right) \exp\left(-\frac{\frac{2}{n}\Delta H_f}{RT}\right) \tag{1.3.117}$$

The numerical value of n depends on the degree of ionisation of the ionic defects, ΔS_f and ΔH_f stand for the entropy and the enthalpy of formation of one mole of defects and $[j_d]$ and $[h^\cdot]$ stand for the concentrations of ionic defects and electron holes expressed in mole fractions.

For compounds of $Me_{1+y}X$ and MeX_{1-y} types these relationships can be expressed by the following equations:

$$[j_d] = p_{x_2}^{-1/n} \exp\left(\frac{\frac{2}{n}\Delta S_f}{R}\right) \exp\left(-\frac{\frac{2}{n}\Delta H_f}{RT}\right) \tag{1.3.118}$$

and

$$[e^-] = p_{X_2}^{-1/n} \exp\left(\frac{\frac{2}{n}\Delta S_f}{R}\right) \exp\left(-\frac{\frac{2}{n}\Delta H_f}{TR}\right) \tag{1.3.119}$$

where $[e^-]$ stands for the concentration of quasi-free electrons expressed in mole fractions. In the above equations the value of n can vary from 2 to 6 depending on the degree of ionisation of the defects.

Figure 50 schematically shows the dependence of the concentration of singly and doubly ionised cation vacancies and electron holes on the pressure

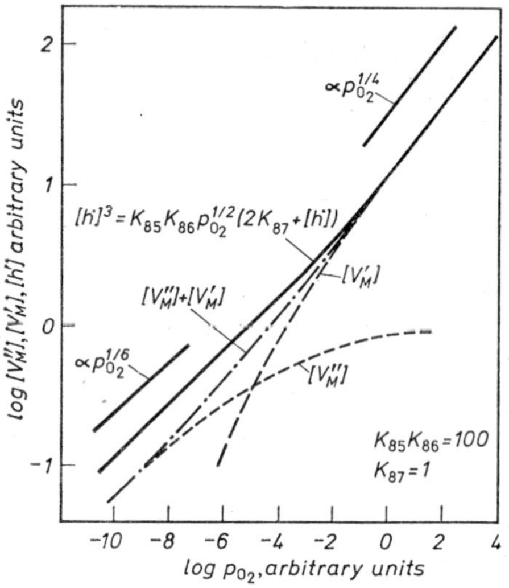

Fig. 50. Concentrations of singly and doubly ionised cation vacancies and electron holes in $Me_{1-y}X$ compound as a function of equilibrium oxidant pressure (after [37]).

of oxygen for a hypothetical crystal of an oxide $Me_{1-y}O$. It follows from this plot that the degree of association of cation vacancies and electron holes increases with increasing oxygen pressure and consequently with increasing equilibrium concentration of defects. For this reason the dependence of the total concentration of ionic defects and of electron defects on oxygen pressure change. When completely ionised vacancies predominate, the dependence of their concentration on oxygen pressure is given by the following equation:

$$[V''_{Me}] = \text{const}\ p_{O_2}^{1/6} \tag{1.3.120}$$

When singly ionised vacancies predominate, this dependence has the form:

$$[V'_{Me}] = \text{const}\, p_{O_2}^{1/4} \tag{1.3.121}$$

In the transition region, i.e. in the range of oxygen pressures close to the critical pressure, at which the concentrations of the two kinds of cation vacancies are comparable ($[V'_{Me}] \cong [V''_{Me}]$), the exponential coefficient of the oxygen pressure becomes approximately equal to 1/5.

In order to illustrate the usefulness of equations (1.3.112)–(1.3.119) in the interpretation of the results of studies on structures of defects in semiconducting crystals we shall consider cobaltous oxide. This oxide is a *p* type semiconductor containing a real deficit of the metal ($Co_{1-y}O$). Many authors have described thermogravimetric studies on cobaltous oxide of the dependence of the deviation from stoichiometry on temperature (920–1350°C) and oxygen pressure (from 1013.25 hPa (1 atm) up to the decomposition pressure of the oxide). These studies have shown that the dependence of the concentration of cation vacancies on oxygen pressure is different in different ranges of oxygen pressure. It has been found that at oxygen pressure of about 1013.25 hPa (1 atm) the concentration of cation vacancies is proportional to the pressure raised to a power greater than 1/4 (the values of the exponential coefficient vary from 1/4 to 1/2). This indicates that under these conditions both singly ionised and non-ionised cation vacancies occur (the intermediate state between the equilibria given by equations (1.3.112) and (1.3.113)).

At lower oxygen pressures [of the order of 101.325–10.1325 Pa (10^{-3}–10^{-4} atm)] the exponential coefficient is almost exactly equal to 1/4. This means that, in this range of pressures, practically all the cationic vacancies are singly ionised. At pressures lower than 10.1325 Pa (10^{-4} atm) the exponential coefficient is equal to 1/6. This indicates that, under these conditions, doubly ionised cations vacancies are practically the only ones to occur in the $Co_{1-y}O$ crystal.

The results of these studies are in agreement with theoretical predictions, since the concentration of cation vacancies decreases with decreasing oxygen pressure and as a result their degree of ionisation must increase (this is analogous to the degree of dissociation of aqueous solutions of acids and bases which increases with increasing dilution).

The temperature dependence of the concentration of cationic vacancies in cobaltous oxide, determined by Fisher and Tannhauser [38, 39], can be expressed for the above two ranges of oxygen pressure by the following empirical equations:

$$[V'_{Co}] = \text{const}\, p_{O_2}^{1/4} \exp\left(-\frac{7500}{RT}\right) \qquad 10^{-4} < p_{O_2} < 10^{-2} \tag{1.3.122}$$

and

$$[V''_{Co}] = \text{const}\, p_{O_2}^{1/6} \exp\left(-\frac{10\,000}{RT}\right) \qquad p_{O_2} < 10^{-4} \tag{1.3.123}$$

Comparison of these relationships with equations (1.3.114) and (1.3.115) indicates that $\Delta H_{V_{Me}} + \Delta H_{h^\cdot} = 2 \times 7500 = 15000$ cal/mole $= 62\ 802$ J/mole and that $\Delta H_{V_{Me}} + \Delta H_{h^\cdot} + \Delta H_{2h^\cdot} = 3 \times 10\ 000 = 30\ 000$ cal/mole $= 125{,}6$ kJ/mole. It follows that $\Delta H_{2h^\cdot} = 62{,}8$ kJ/mole, i.e. that the energy of dissociation of singly ionised cation vacancies is 62,8 kJ/mole or 0.65 eV.

The determination of the energy of ionisation of an electrically neutral vacancy, i.e. the energy required for cleavage of the first electron hole, requires knowledge of the temperature dependence of the concentration of cation vacancies in the range of pressures in which this concentration is a function of oxygen pressure raised to the power 1/2 (equation (1.3.112)).

Comparison of equations (1.3.122) and (1.3.123) with equations (1.3.114) and (1.3.115) indicates that the pre-exponential factor in equations (1.3.122) and (1.3.123) contains entropy terms which can be calculated using the known numerical values for these constants.

1.3.6.3 Defect equilibria in compounds having nearly stoichiometric composition

So far we have considered the extreme cases in which compounds of the $Me_{1+y}X$ and $Me_{1-y}X$ have large deviations from stoichiometry in the chosen range of temperature and pressure of the X component. This simplifying assumption has made it possible to neglect the effect of intrinsic electronic and ionic disorder on the defect equilibria considered. Experimental studies have shown that, in many cases, this assumption is fully justified as even considerable changes of temperature and pressure do not cause significant changes in the character of disorder of the crystal lattice of the compound. However, from a theoretical point of view for each compound of the types $Me_{1+y}X$ or MeX_{1-y} and $Me_{1-y}X$ or MeX_{1+y}, a definite set of parameters T and p_{X_2} exists, at which the composition of these substances is strictly stoichiometric and any changes in these parameters lead to the inversion of the type of predominant disorder. For example an increase of oxygen pressure over zinc oxide causes a decrease in the concentration of interstitial zinc ions and quasi-free electrons and as a result the chemical composition of this oxide approaches the stoichiometry. At a suitably high oxygen pressure the oxide reaches the stoichiometric composition, at which the concentration of interstitial metal ions is equal to that of cation vacancies according to equation (1.3.51) for intrinsic Frenkel type ionic disorder. Similarly the concentration of quasi-free electrons under these conditions is equal to that of electron holes according to equation (1.3.91) describing intrinsic electron disorder. A further increase of oxygen pressure causes an increase of the concentration of cation vacancies, which is accompanied by the formation of electron holes, i.e. it leads to deviation from stoichiometry in the direction of deficit of metal. Consequently in zinc oxide a new type of disorder appears and can become predominant. As a result the oxide changes its character from *n* type semiconductor to *p* type.

In the case of compounds of the $Me_{1-y}X$ type an analogous inversion of the type of predominant disorder can be obtained by suitably decreasing the pressure of the X component in the gas phase surrounding the crystal.

Obviously in many cases such changes cannot be carried out in practice. For example the decrease of the partial pressure of oxygen over ferrous oxide ($Fe_{1-y}O$) as far as the decomposition pressure of this compound does not cause the inversion of the type of disorder, since even at this very low oxygen pressure the oxide still has a significant metal deficit (at 1000°C and at $p_{O_2} = 1.01325 \cdot 10^{-10}$ Pa (10^{-15} atm) the composition corresponds to the formula $Fe_{0.995}O$). However, this is an exceptional case. There are experimental proofs of the possibility of inversion of the predominant type of disorder in crystal lattices of compounds of the $Me_{1+y}X$ and $Me_{1-y}X$ types. For this reason it is necessary to discuss defect equilibria near the inversion point, since it will be seen later that under these conditions the dependence of the concentration of defects on the pressure of component X changes radically.

In order to avoid complications we will assume that only completely ionised defects occur in the crystal. When this assumption is made, the following defect equilibria must be taken into account in the crystal having a defect cationic sublattice:

$$\tfrac{1}{2}X_2(g) \rightleftharpoons V''_{Me} + 2h^{\cdot} + X_X \tag{1.3.124}$$

$$MeX \rightleftharpoons Me_i^{\cdot\cdot} + 2e^- + \tfrac{1}{2}X_2(g) \tag{1.3.125}$$

$$Me_{Me} \rightleftharpoons Me_i^{\cdot\cdot} + V''_{Me} \tag{1.3.126}$$

$$\text{zero} \rightleftharpoons e^- + h^{\cdot} \tag{1.3.127}$$

Applying the law of mass action to these equilibria we obtain:

$$K_{124} = \frac{[V''_{Me}][h^{\cdot}]^2}{p_{X_2}^{1/2}} \tag{1.3.128}$$

$$K_{125} = [Me_i^{\cdot\cdot}][e^-]^2 p_{X_2}^{1/2} \tag{1.3.129}$$

$$K_{126} = [Me_i^{\cdot\cdot}][V''_{Me}] \tag{1.3.130}$$

$$K_{127} = [e^-][h^{\cdot}] \tag{1.3.131}$$

It follows from comparison of equations (1.3.102)–(1.3.104) with equation (1.3.128) and equations (1.3.93)–(1.3.95) with equation (1.3.129) that $K_{124} = K_{85}K_{86}K_{87}$ and $K_{125} = K_{79}K_{80}K_{81}$.

In the region of large deviations from stoichiometric composition in the direction of metal deficit ($Me_{1-y}X$) the condition of electroneutrality for the lattice is reduced to the following:

$$[V''_{Me}] = \tfrac{1}{2}[h^{\cdot}] \gg [Me_i^{\cdot\cdot}],\ [e^-] \tag{1.3.132}$$

Utilising equation (1.3.132), the dependence of the concentration of cationic vacancies on the oxidant pressure can be expressed by the following equation:

$$[V''_{Me}] = \left(\frac{K_{124}}{4}\right)^{1/3} p_{X_2}^{1/6} \tag{1.3.133}$$

Similarly the relation between the concentration of electron holes and the pressure p_{X_2} can be obtained:

$$[h^{\cdot}] = (2K_{124})^{1/3} p_{X_2}^{1/6} \tag{1.3.134}$$

All the defects in the crystal are in equilibrium with one another. For this reason, the presence of an excess of cation vacancies in the lattice has an effect on the equilibria for intrinsic ionic and electronic disorder. Hence, by introducing equation (1.3.133) in equation (1.3.130) we obtain an expression giving the effect of the oxidant pressure on the concentration of interstitial metal ions:

$$[Me_i^{\cdot\cdot}] = K_{126}\left(\frac{4}{K_{124}}\right)^{1/3} p_{X_2}^{-1/6} \tag{1.3.135}$$

By introducing equation (1.3.134) in equation (1.3.131) we obtain the expression for the dependence of the concentration of quasi-free electrons on the pressure of component X in the atmosphere:

$$[e^-] = \frac{K_{127}}{(2K_{124})^{1/3}} p_{X_2}^{-1/6} \tag{1.3.136}$$

It follows from equations (1.3.133) and (1.3.134) and from (1.3.135) and (1.3.136) that the concentration of cation vacancies and electron holes decreases with decreasing oxidant pressure and that this decrease is accompanied by an increase of the concentration of interstitial cations and quasi-free electrons. Hence, at a sufficiently low pressure, the compound approaches the stoichiometric composition and the simplified electroneutrality condition for the lattice given by equation (1.3.132) loses its validity.

However, the consideration of defect equilibria in the compound discussed with a composition close to stoichiometry first requires discussion of the situation in which the metal is present in excess ($Me_{1+y}X$). In this case the simplified electroneutrality condition of the lattice has the following form:

$$[Me_i^{\bullet\bullet}] = \tfrac{1}{2}[e^-] \gg [V_{Me}''], [h^\bullet] \qquad (1.3.137)$$

By introducing this equation in equation (1.3.129) we obtain the following relationship between the concentration of interstitial cations and the pressure of the oxidant:

$$[Me_i^{\bullet\bullet}] = \left(\frac{K_{125}}{4}\right)^{1/3} p_{X_2}^{-1/6} \qquad (1.3.138)$$

Analogous transformations of equation (1.3.129) give the following relationship for the concentration of quasi-free electrons:

$$[e^-] = (2K_{125})^{1/3} p_{X_2}^{-1/6} \qquad (1.3.139)$$

By introducing equations (1.3.138) and (1.3.139) in equations (1.3.130) and (1.3.131) respectively, we obtain the following expressions giving the dependence of the concentrations of cation vacancies and electron holes on the oxidant pressure:

$$[V_{Me}''] = K_{126}\left(\frac{4}{K_{125}}\right)^{1/3} p_{X_2}^{1/6} \qquad (1.3.140)$$

$$[h^\bullet] = \frac{K_{127}}{(2K_{125})^{1/3}} p_{X_2}^{1/6} \qquad (1.3.141)$$

These equations are valid at very low pressures of the oxidant where the excess of the metal is predominant. Let us now return to the equilibria at intermediate pressures of X_2 where the composition of the compound is nearly stoichiometric. In these conditions two extreme cases are possible.

1. Intrinsic electron disorder predominates. Then:

$$[e^-] = [h^\bullet] = K_{127}^{1/2} \gg [V_{Me}''] \text{ and } [Me_i^{\bullet\bullet}] \qquad (1.3.142)$$

Thus the concentration of electron defects does not depend on the equilibrium pressure of the oxidant. For this reason the concentrations $[h^{\cdot}]$ and $[e^-]$ in equations (1.3.128) and (1.3.129) can be replaced by $K_{127}^{1/2}$, leading to the following relationships:

$$[V''_{Me}] = \left(\frac{K_{124}}{K_{127}}\right) p_{X_2}^{1/2} \tag{1.3.143}$$

and

$$[Me_i^{\cdot\cdot}] = \left(\frac{K_{125}}{K_{127}}\right) p_{X_2}^{-1/2} \tag{1.3.144}$$

2. The intrinsic ionic disorder predominates. In this case the electroneutrality condition of the lattice is simplified to the following form:

$$[V''_{Me}] = [Me_i^{\cdot\cdot}] = K_{126}^{1/2} \gg [h^{\cdot}] \text{ and } [e^-] \tag{1.3.145}$$

In this case the concentration of ionic defects does not depend on the pressure of the oxidant and as a result the concentrations of the electronic defects and holes are the following functions of p_{X_2}:

$$[h^{\cdot}] = \left(\frac{K_{124}^{1/2}}{K_{126}^{1/4}}\right) p_{X_2}^{1/4} \tag{1.3.146}$$

and

$$[e^-] = \left(\frac{K_{125}^{1/2}}{K_{126}^{1/4}}\right) p_{X_2}^{-1/4} \tag{1.3.147}$$

It follows from equations (1.3.128)–(1.3.131) that compound MeX has a strictly stoichiometric composition at the oxidant pressure given by the following equation:

$$p_{X_2}(\text{stoich.}) = \frac{K_{125}}{K_{124}} \tag{1.3.148}$$

since under these conditions:

$$[V''_{Me}] = [Me_i^{\cdot\cdot}] \tag{1.3.149}$$

and

$$[h^{\cdot}] = [e^-] \tag{1.3.150}$$

It is noteworthy that all the above relationships contain products of constants and oxidant pressure rather than sums. This is a consequence of the use of the simplified electroneutrality conditions for the crystal, and it is one of the principal advantages of the method of Kröger and Vink which made it possible to solve many problems in the field of physical chemistry of defects. This method offers wide possibilities for the interpretation of results of studies on defect equilibria and for developing models of defect structures by means of suitable plots of these equilibria, since by taking logarithms of the above equations we obtain a set of linear equations which are very convenient for graphical interpretation:

$$\log[V''_{Me}] = \tfrac{1}{3}\log K_{124}/4 + \tfrac{1}{6}\log p_{X_2} \quad (1.3.151)$$

$$\log[h^\cdot] = \tfrac{1}{3}\log 2K_{124} = \tfrac{1}{6}\log p_{X_2} \quad (1.3.152)$$

$$\log[Me_i^{\cdot\cdot}] = \log K_{126} + \tfrac{1}{3}\log 4/K_{124} - \tfrac{1}{6}\log p_{X_2} \quad (1.3.153)$$

$$\log[e^-] = \log K_{127} - \tfrac{1}{3}\log 2K_{124} - \tfrac{1}{6}\log p_{X_2} \quad (1.3.154)$$

$$\log[Me_i^{\cdot\cdot}] = \tfrac{1}{3}\log K_{125} - \log 4 - \tfrac{1}{6}\log p_{X_2} \quad (1.3.155)$$

$$\log[e^-] = \tfrac{1}{3}\log 2K_{125} - \tfrac{1}{6}\log p_{X_2} \quad (1.3.156)$$

$$\log[V''_{Me}] = \log K_{126} - \tfrac{1}{3}\log 4/K_{125} + \tfrac{1}{6}\log p_{X_2} \quad (1.3.157)$$

$$\log[h^\cdot] = \log K_{127} - \tfrac{1}{3}\log 2K_{125} + \tfrac{1}{6}\log p_{X_2} \quad (1.3.158)$$

$$\log[V''_{Me}] = \log K_{124} - \log K_{127} + \tfrac{1}{2}\log p_{X_2} \quad (1.3.159)$$

$$\log[Me_i^{\cdot\cdot}] = \log K_{125} - \log K_{127} - \tfrac{1}{2}\log p_{X_2} \quad (1.3.160)$$

$$\log[h^\cdot] = \tfrac{1}{2}\log K_{124} - \tfrac{1}{4}\log K_{125} + \tfrac{1}{4}\log p_{X_2} \quad (1.3.161)$$

$$\log[e^-] = \tfrac{1}{2}\log K_{125} - \tfrac{1}{4}\log K_{126} - \tfrac{1}{4}\log p_{X_2} \quad (1.3.162)$$

The first four of these equations ((1.3.151)–(1.3.154)) express the dependence of concentrations of ionic and electronic defects on the oxidant pressure at relatively high pressures, i.e. when cation vacancies and electron holes are the predominant defects. The next four equations ((1.3.155)–(1.3.158)) describe the relationship discussed above at low oxidant pressures when interstitial cations and quasi-free electrons predominate in the crystal. Finally the last four equations ((1.3.159)–(1.3.162)) express the relationships at intermediate oxidant pressures when either electron holes and quasi-free electrons or cation vacancies and interstitial cations are the predominant defects.

All these relationships are linear functions with slopes varying from $-1/6$ to $+1/6$. Figures 51 and 52 show plots of these functions for the two extreme cases which were discussed above. It is clear from the plots that there are three ranges of the oxidant pressure in which the dependence of the concentrations of various kinds of defects on the equilibrium pressure of the oxidant is different. The limits of these regions can be readily read from the plots, since they correspond to the discontinuities in and the intersections of the straight lines for the dependence of log (concentrations of defects) on $\log p_{X_2}$. In addition, plots of this type make it possible to predict semiconducting properties of materials in given thermodynamic conditions.

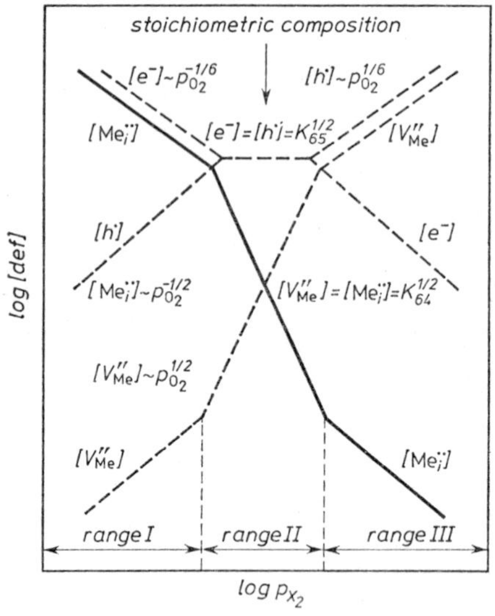

Fig. 51. Concentrations of ionic and electronic defects as a function of oxidant pressure. The first case: intrinsic electronic defects predominate (after [37]).

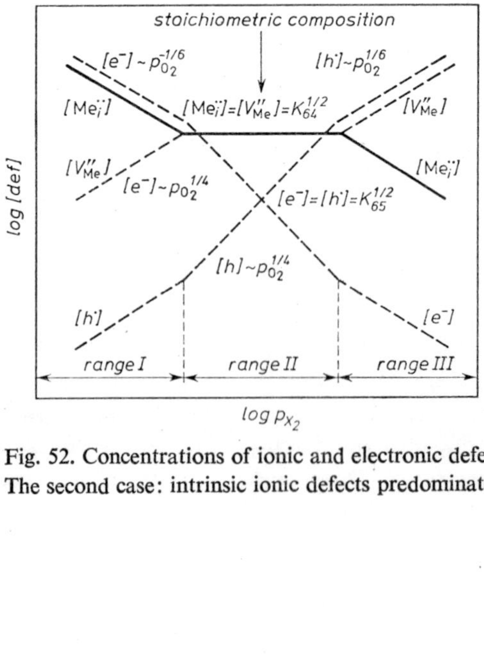

Fig. 52. Concentrations of ionic and electronic defects as a function of oxidant pressure. The second case: intrinsic ionic defects predominate (after [37]).

For example, it follows from Fig. 51 that, in region I, our hypothetical compound MeX is an *n*-type semiconductor and in region III it is a *p*-type semiconductor, whereas in region II it is an intrinsic electronic semiconductor. It follows from Fig. 52 that the material considered in this case is an intrinsic ionic semiconductor in region II, whereas in the other two regions its semiconducting properties do not differ from those of the material considered in Fig. 51.

In addition it is readily seen that the plots are symmetrical and for this reason they can be constructed when only a few values of the equilibrium constants are known. However, it should be stressed that the example discussed represents the simplest case, since the pure compound MeX considered contained defects in only one sublattice and, more importantly, it was assumed that the defects were completely ionised. The situation becomes much more complex when varying degrees of defect ionisation and the possibility of defects in both sublattices are taken into account. Obviously the problem becomes incomparably more complex in the case of complex systems such as compounds MeX containing various impurities or double compounds having the spinel structure. This is the reason that the method of Kröger and Vink plays such an important part in studies on the structure of point defects and semiconducting properties of inorganic compounds. The experimental basis for these studies are measurements of concentrations of various kinds of defects and their dependence on temperature and oxidant pressure. These measurements are very difficult but, in view of their fundamental importance, the methods for the determination of defect concentrations and their mobility in ionic crystals have been developed for many years. Some of these methods are described in Chapter 3.

1.3.7 The effect of impurities on defect equilibria in ionic crystals

So far our discussion has been limited to the description of defect equilibria in a system consisting of a pure compound of the MeX type and the oxidant. However, the concentrations of the individual kinds of defects depend not only on the thermodynamic parameters of the system but also, to a large degree, on the presence of small amounts of foreign ions in the crystal lattice having an effective charge different from that of the lattice ions. The effect of impurities is so large that it can lead to the inversion of the type of the predominant disorder. For this reason the physico-chemical properties of the parent compound can be modified within wide limits by introducing additives in its crystal lattice. The theory of defects makes it possible to predict the changes caused by impurities depending on their type and concentration. This is of considerable practical importance, in particular in technology of semiconductors and in other branches of modern industry.

We will begin consideration of these problems by discussing the effect of impurities on the disorder in crystal lattices of compounds with stoichiometric composition. When a small amount of a trivalent metal compound F_2X_3 (where F stands for an atom of a foreign metal) is incorporated in the crystal of compound MeX with Frenkel type disorder (equation (1.3.51)) and when it is assumed that the foreign cations are situated in the cationic sublattice sites, the process can be represented by the following equation:

$$F_2X_3 \rightleftarrows 2F^{\bullet}_{Me} + V''_{Me} + 3X_X \tag{1.3.163}$$

It follows from this equation that the process of incorporation of F^{3+} ions in the cationic sublattice is accompanied by an increase in the concentration of cationic vacancies, which is equivalent to a decrease in the concentration of interstitial cations. Hence the electroneutrality condition of the lattice in this case has the following form:

$$[F^{\bullet}_{Me}] + 2[Me^{\bullet\bullet}_i] = 2[V''_{Me}] \tag{1.3.164}$$

By utilising equation (1.3.55) we can represent relationship (1.3.164) in the following form:

$$[V''_{Me}] = \frac{K_F}{[V''_{Me}]} + \frac{1}{2}[F^{\bullet}_{Me}] \tag{1.3.165}$$

This last equation has two limiting solutions. When the concentration of the impurity considerably exceeds that of the intrinsic defects

$$\frac{K_F}{[V''_{Me}]} \ll [F^{\bullet}_{Me}] \tag{1.3.166}$$

which means that the relationship is simplified to the following form:

$$[V''_{Me}] = \tfrac{1}{2}[F^{\bullet}_{Me}] \tag{1.3.167}$$

Thus, in these conditions, the concentration of interstitial cations is negligibly small in comparison with that of cationic vacancies determined by the concentration of the impurity. When the concentration of the impurity is lower than that of the intrinsic defects:

$$\frac{K_F}{[V''_{Me}]} \gg \frac{1}{2}[F^{\bullet}_{Me}] \tag{1.3.168}$$

and, as a result, equation (1.3.165) takes the following limiting form:

$$[V''_{Me}] = [Me_i^{\cdot\cdot}] = K_F^{1/2} \qquad (1.3.169)$$

These considerations lead to the conclusion that there are two ranges of concentration of the impurity, in which its effect on the disorder in the crystal is quite different. This is shown schematically in Fig. 53. In the first

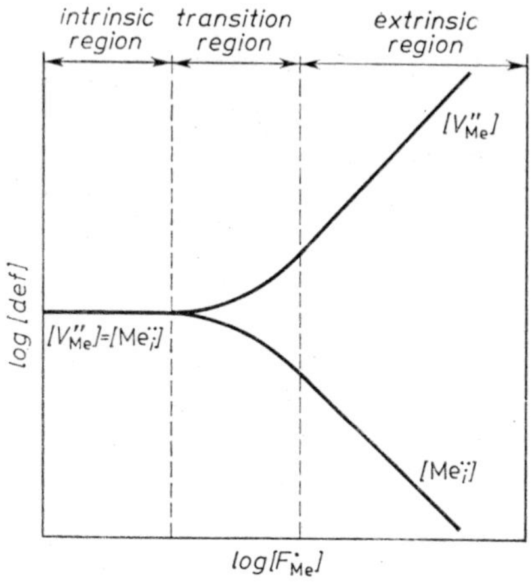

Fig. 53. The effect of trivalent dopant on the Frenkel type disorder in a stoichiometric MeX-type crystal.

region, in which the impurity has practically no effect on the disorder in the crystal, its concentration is lower than that of the intrinsic defects [the limiting case given by equation (1.3.168)]. This region is called the intrinsic region. In the second region, called the extrinsic region, the concentration of the impurity considerably exceeds that of the intrinsic defects (the limiting case given by equation (1.3.166)).

It should be pointed out that the boundary between the regions largely depends on the temperature, since the equilibrium constant K_F in equation (1.3.165), like any other constant for a defect equilibrium, increases exponentially with increasing temperature. This means that the concentration of intrinsic defects rapidly increases with increasing temperature and as a result the boundary between the intrinsic and the extrinsic regions is shifted to the right at higher impurity contents. However, it should be remembered that the concentration of intrinsic defects is as a rule so low, that the increase of the temperature up to the melting point of the crystal does not cause an increase of the first region for impurity contents exceeding a few atom per cent.

An additive of a compound F_2X, i.e. ions of a metal having a valency lower than that of the lattice cations, has the opposite effect on the disorder in the crystal. The process of incorporation of this type of impurity can be represented by the following equation:

$$F_2X \rightleftharpoons 2F'_{Me} + Me_i^{\bullet\bullet} + X_X \qquad (1.3.170)$$

From this equation it follows that the incorporation of ions F^{1+} in the cationic sublattice is accompanied by an increase of the concentration of interstitial cations, which is equivalent to a decrease in the concentration of cation vacancies. Hence, the condition of electrical neutrality of the lattice takes the following form:

$$[F'_{Me}] + 2[V''_{Me}] = 2[Me_i^{\bullet\bullet}] \qquad (1.3.171)$$

By introducing expression (1.3.55) in this equation we obtain:

$$[Me_i^{\bullet\bullet}] = \frac{1}{2}[F'_{Me}] + \frac{K_F}{[Me_i^{\bullet\bullet}]} \qquad (1.3.172)$$

Equation (1.3.172), like equation (1.3.165), has two limiting solutions:

1. When the concentration of the impurity considerably exceeds that of intrinsic defects, the following inequality is valid:

$$\frac{K_F}{[Me_i^{\bullet\bullet}]} \ll \frac{1}{2}[F'_{Me}] \qquad (1.3.173)$$

and hence

$$[Me_i^{\bullet\bullet}] = \tfrac{1}{2}[F'_{Me}] \qquad (1.3.174)$$

2. When the concentration of the impurity is lower than that of intrinsic defects, we have:

$$\frac{K_F}{[Me_i^{\bullet\bullet}]} \gg \frac{1}{2}[F'_{Me}] \qquad (1.3.175)$$

which means that:

$$[Me_i^{\bullet\bullet}] = [V''_{Me}] = K_F^{1/2} \qquad (1.3.176)$$

This leads to the conclusion that in the first region (Fig. 54) the presence of the impurity has no effect on the disorder in the crystal, as in the previous case, whereas in the second region (the extrinsic region) interstitial cations predominate and their concentration is determined by that of the impurity.

However, it should be stressed that these conclusions are of purely theoretical importance and the real systems not always obey these rules, since cations of the impurity can be incorporated in interstitial positions rather than in lattice sites. In this case the effect of impurity on the degree of

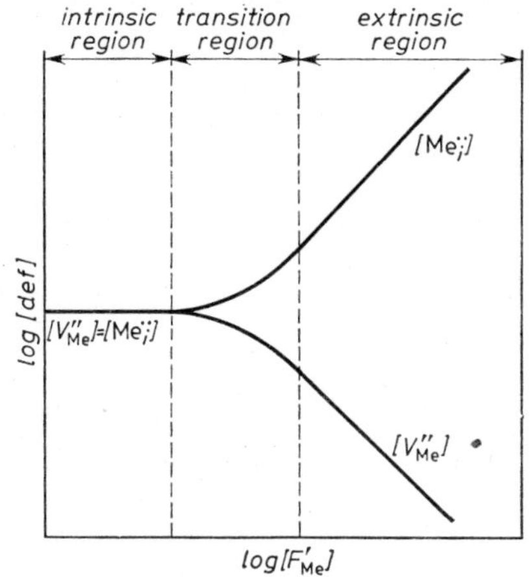

Fig. 54. The effect of monovalent dopant on the Frenkel type disorder in stoichiometric MeX-type crystal.

disorder of the lattice would be different. However, as a rule the mechanism of incorporation of atoms of the additive does not differ from that presented above.

When a compound with stoichiometric composition has a Schottky type defect [(equation (1.3.53)], the incorporation of mono- and trivalent cations of the dopant in the crystal lattice can be expressed by the following equations:

$$F_2X_3 \rightleftharpoons 2F^{\cdot}_{Me} + V''_{Me} + 3X_X \tag{1.3.177}$$

and

$$F_2X \rightleftharpoons 2F'_{Me} + V^{\cdot\cdot}_X + X_X \tag{1.3.178}$$

It follows from equations (1.3.177) and (1.3.178) that the introduction of ions of the impurity, F^{3+}, in the lattice sites of the crystal causes an increase in the concentration of cation vacancies, whereas the incorporation of F^{1+} ions causes an increase in the concentration of anion vacancies.

In the first case the condition of electrical neutrality of the lattice has the following form:

$$[F^{\cdot}_{Me}] + 2[V^{\cdot\cdot}_X] = 2[V''_{Me}] \tag{1.3.179}$$

Utilising equation (1.3.57) we obtain:

$$[V''_{Me}] = \frac{K_S}{[V''_{Me}]} + \frac{1}{2}[F^{\cdot}_{Me}] \tag{1.3.180}$$

In the second case the electroneutrality condition of the lattice is simplified to the following form:

$$[F'_{Me}]+2[V''_{Me}] = 2[V^{\cdot\cdot}_{X}] \tag{1.3.181}$$

By introducing equation (1.3.57) we obtain:

$$[V^{\cdot\cdot}_{X}] = \frac{K_S}{[V^{\cdot\cdot}_{X}]} + \frac{1}{2}[F'_{Me}] \tag{1.3.182}$$

Equations (1.3.180) and (1.3.182) have two limiting solutions, like equations (1.3.165) and (1.3.172).

1. When the concentration of tri- or monovalent impurity cations considerably exceeds that of intrinsic defects, the following inequalities are valid:

$$\frac{K_S}{[V''_{Me}]} \ll \frac{1}{2}[F^{\cdot}_{Me}] \tag{1.3.183}$$

and

$$\frac{K_S}{[V^{\cdot\cdot}_{X}]} \ll \frac{1}{2}[F'_{Me}] \tag{1.3.184}$$

Hence expressions (1.3.180) and (1.3.182) are simplified to the following forms:

$$[V''_{Me}] = \tfrac{1}{2}[F^{\cdot}_{Me}] \tag{1.3.185}$$

and

$$[V^{\cdot\cdot}_{X}] = \tfrac{1}{2}[F'_{Me}] \tag{1.3.186}$$

2. When the concentration of the impurity is lower than that of intrinsic defects we have:

$$\frac{K_S}{[V''_{Me}]} \gg \frac{1}{2}[F^{\cdot}_{Me}] \tag{1.3.187}$$

and

$$\frac{K_S}{[V^{\cdot\cdot}_{X}]} \gg \frac{1}{2}[F'_{Me}] \tag{1.3.188}$$

Hence:

$$[V''_{Me}] = [V^{\cdot\cdot}_{X}] = K^{1/2} \tag{1.3.189}$$

These considerations lead to the conclusion that, as in the case discussed previously of Frenkel type disorder, there are two concentration regions for the impurity. In the first region the impurity has no effect on the disorder of the crystal lattice, whereas in the second (the extrinsic region) the solid solution F_2X_3–MeX practically only gives rise to vacancies in the cationic sublattice (Fig. 55) and the solid solution F_2X–MeX only produces vacancies in the anionic sublattice (Fig. 56).

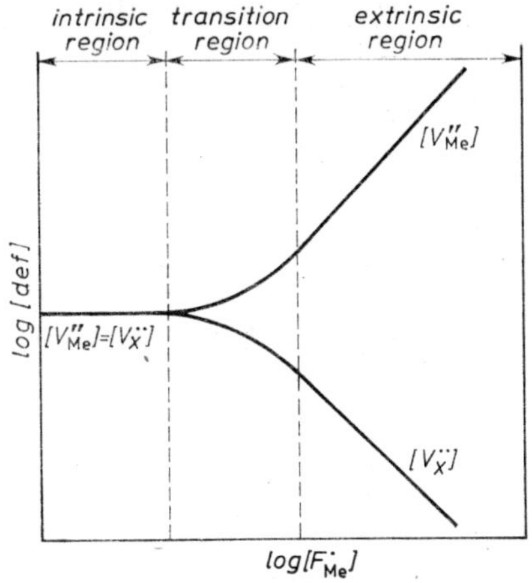

Fig. 55. The effect of trivalent dopant on the Schottky type disorder in a stoichiometric MeX-type crystal.

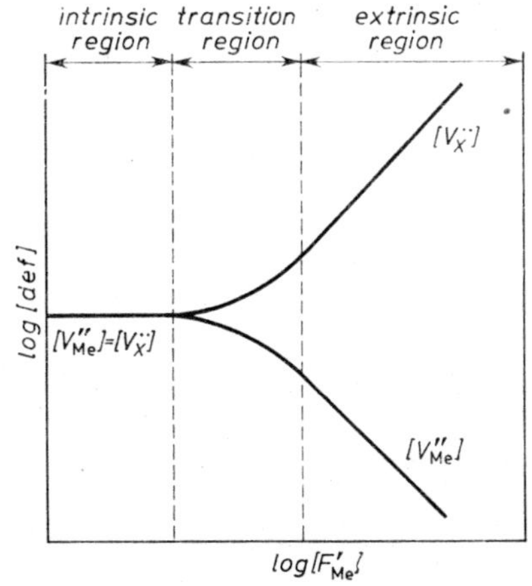

Fig. 56. The effect of monovalent dopant on the Schottky type disorder in a stoichiometric MeX-type crystal.

It should be pointed out that not only the temperature, but especially the concentration of the impurity determine the effect of the latter on the disorder of the crystal lattice of MeX. For this reason the effect desired is obtained only if the concentration of the dopant exceeds that of the defects present in a pure crystal of MeX. The limiting concentration of the dopant, above which it exerts the effect required on the disorder in the crystal at the given temperature, can be evaluated on the basis of the equilibrium constant for the predominant disorder. In the case of the Schottky type disorder the constant K_S, given by equation (1.3.57), can be expressed by means of the following relationship:

$$K_S = [V''_{Me}][V^{\cdot\cdot}_X] = [V''_{Me}]_0^2 \qquad (1.3.190)$$

where $[V''_{Me}]_0$ is the concentration of cation vacancies in pure compound MeX.

Analogous considerations make it possible to predict the effect of the kind of dopant on the character and concentration of defects in crystals with deviations from stoichiometric composition. When a small amount of trivalent cations is introduced in the crystal lattice of a compound of the $Me_{1+y}X$ type, the concentration of interstitial cations decreases, whereas that of quasi-free electrons increases. This process can be represented by the following quasi-chemical equations:

$$F_2X_3 + Me_i^{\cdot\cdot} \rightleftharpoons 2F^{\cdot}_{Me} + Me_{Me} + 3X_X \qquad (1.3.191)$$

or

$$F_2X_3 \rightleftharpoons 2F^{\cdot}_{Me} + 2e^- + 2X_X + \tfrac{1}{2}X_2 \qquad (1.3.192)$$

It follows from equation (1.3.191) that the introduction of the trivalent dopant in the lattice of compound $Me_{1+y}X$ causes a decrease of the concentration of interstitial cations. Since the concentration of ionic defects is related to that of electron defects by the law of mass action, this decrease must be accompanied by an increase of the concentration of quasi-free electrons. This process corresponds to equation (1.3.192). By adding the two equations together the following expression, showing the simultaneous effects of the dopant on the concentrations of ionic and electronic defects, is obtained:

$$2F_2X_3 + Me_i^{\cdot\cdot} \rightleftharpoons 4F^{\cdot}_{Me} + 2e^- + Me_{Me} + 5X_X + \tfrac{1}{2}X_2 \qquad (1.3.193)$$

In this case the electroneutrality condition of the lattice is as follows:

$$[F^{\cdot}_{Me}] + [2Me_i^{\cdot\cdot}] = [e^-] \qquad (1.3.194)$$

When $[F^{\cdot}_{Me}] \gg [Me_i^{\cdot\cdot}]$, we have $[e^-] = [F^{\cdot}_{Me}]$

Utilising equation (1.3.129) we can express the concentration of interstitial cations in the solid solution by the following relationship:

$$[Me_i^{\cdot\cdot}] = \frac{K_{124}}{[F_{Me}^{\cdot}]^2} p_{X_2}^{-1/2} \tag{1.3.195}$$

It follows from equation (1.3.195) that the concentration of ionic defects decreases, as expected (1.3.193), with both the oxidant pressure and the concentration of dopant. When this concentration and the oxidant pressure are high, the concentration of interstitial cations can be so low that the intrinsic ionic disorder occurring in crystal as a whole can no longer be neglected in the consideration of the defect equilibria. In the case discussed this disorder is of the Frenkel type [equation (1.3.51)]. Therefore, by using equation (1.3.130), we can express the concentration of cation vacancies with the following equation:

$$[V_{Me}''] = \frac{K_{126}[F_{Me}^{\cdot}]^2}{K_{125}} p_{X_2}^{1/2} \tag{1.3.196}$$

The above considerations lead to the conclusion that the concentration of cation vacancies increases with the oxidant pressure and the concentration of the dopant. For this reason the concentration of these defects, which is negligibly small in a pure crystal of $Me_{1+y}X$, under suitable conditions and at sufficiently high concentrations of dopant F_2X_3, can become higher than that of interstitial cations. Thus inversion of the predominant type of ionic disorder may occur.

It is noteworthy that, in contrast to stoichiometric crystals with disorder of the Schottky or Frenkel types, in the case of non-stoichiometric compounds the effect of dopant significantly depends not only on the temperature, but also on the equilibrium pressure of the oxidant. In the case of compounds of the $Me_{1+y}X$ type the concentration of intrinsic defects decreases with increasing pressure of the oxidant and, for this reason, small amounts of the dopant can have an effect only at high pressures of this gas. On the other hand, at low pressures of the oxidant the concentration of intrinsic defects can be much higher than that of the dopant and as a result the dopant has practically no effect on the disorder in the crystal. Thus there are two oxidant pressure ranges, where the disorder in a solid solution F_2X_3–$Me_{1+y}X$ is different (Fig. 57). In one region, corresponding to low oxidant pressures, the dopant has practically no effect on the concentration of defects in the crystal, given by equations (1.3.98) and (1.3.101) (the intrinsic region). On the other hand, in the region of high pressures the dopant determines the disorder in the crystal (the extrinsic region), since the concentration of intrinsic defects in this region is much lower. It follows from Fig. 57 that in this region the concentration of quasi-free electrons is constant and equal to that of the dopant, whereas

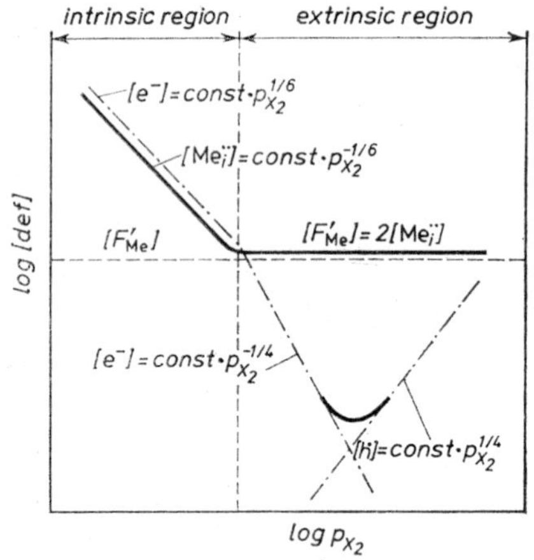

Fig. 57. The effect of a trivalent dopant on the ionic and electronic disorder in a $Me_{1+y}X$ type crystal as a function of equilibrium oxidant pressure.

the concentration of interstitial cations rapidly decreases and that of cation vacancies increases with increasing pressure of the oxidant. This means that at a high pressure of the oxidant the inversion of the predominant type of ionic disorder may occur and this may cause a change of the deviation from the stoichiometric composition of the solid solution from F_2X_3–$Me_{1+y}X$ to F_2X_3–$Me_{1-y}X$.

A dopant of cations having a valency lower than that of the lattice cations has the reverse effect. The process of incorporation of this dopant in the lattice of compound $Me_{1+y}X$ can be represented by the following equations:

$$F_2X+Me_{Me} \rightleftharpoons 2F'_{Me}+Me_i^{\cdot\cdot}+X_X \tag{1.3.197}$$

or

$$F_2X+2e^- +\tfrac{1}{2}X_2 \rightleftharpoons 2F'_{Me}+2X_X \tag{1.3.198}$$

By adding these equations together, we obtain the following general relationship:

$$2F_2X+Me_{Me}+2e^- +\tfrac{1}{2}X_2 \rightleftharpoons 4F'_{Me}+Me_i^{\cdot\cdot}+3X_X \tag{1.3.199}$$

In this case the condition of electrical neutrality of the lattice is as follows:

$$[F'_{Me}]+[e^-] = 2[Me_i^{\bullet\bullet}] \tag{1.3.200}$$

When $[F'_{Me}] \gg [e^-]$ we have:

$$[Me_i^{\bullet\bullet}] = \tfrac{1}{2}[F'_{Me}] \tag{1.3.201}$$

It follows from equations (1.3.197)–(1.3.199) that the introduction of dopant of the F_2X type to the crystal lattice of compound $Me_{1+y}X$ causes an increase of the concentration of interstitial cations which is accompanied by a decrease of the concentration of quasi-free electrons. When the concentration of the dopant is sufficiently large, the concentration of interstitial cations is strictly dependent on that of the foreign ions.

By introducing equation (1.3.201) in equation (1.3.129) the concentration of quasi-free electrons can be expressed as follows:

$$[e^-] = \left(\frac{2K_{125}}{[F'_{Me}]}\right)^{1/2} p_{X_2}^{-1/4} \tag{1.3.202}$$

It follows from this relationship that the concentration of quasi-free electrons decreases not only with increasing concentration of the dopant, but also with increasing pressure of the oxidant. For this reason, at a high pressure of oxidant, the concentration of quasi-free electrons may become so low that the intrinsic electronic defects [equation (1.3.127)] can no longer be neglected. In this case the concentration of electron holes must be taken into account in the electroneutrality equation for the lattice:

$$2[Me_i^{\bullet\bullet}]+[e^-]+[h^\bullet] = [F'_{Me}] \tag{1.3.203}$$

By using this relationship and equation (1.3.131) we can express the concentration of electron holes in solid solution F_2X–$Me_{1+y}X$ by the following equation:

$$[h^\bullet] = \frac{[F'_{Me}]^{1/2}K_{127}}{(2K_{125})^{1/2}} p_{X_2}^{1/4} \tag{1.3.204}$$

It follows from this equation that the concentration of electron holes increases with both the concentration of the dopant and the partial pressure of the oxidant. Thus at a high pressure of this gas the solid solution can change its semiconducting properties from *n* type to *p* type.

The above considerations lead to the conclusion that, in this case, as in the case of solid solutions F_2X_3–$Me_{1+y}X$, there are two ranges of oxidant pressure, in which the disorder in a solution F_2X–$Me_{1+y}X$ is different (Fig. 58). In the first region the dopant has practically no effect on the disorder in the crystal (the intrinsic region, in which the dependence of the concentrations of interstitial cations and quasi-free electrons on the oxidant pressure is given by equations (1.3.98) and (1.3.101)). On the other hand in the second region (the extrinsic region) the concentration of in-

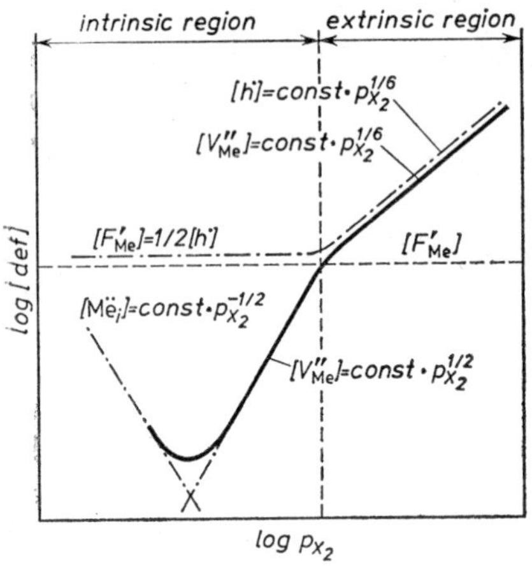

Fig. 58. The effect of a monovalent dopant on the ionic and electronic disorder in a $Me_{1+y}X$ type crystal as a function of equilibrium oxidant pressure.

terstitial cations is constant and equal to one half of the concentration of the dopant, whereas the concentration of quasi-free electrons rapidly decreases and that of electron holes increases with increasing pressure of the oxidant. This may cause inversion of the predominant type of electronic disorder and as a result a change in the semiconducting properties of the solid solution.

The effect of dopants on the type of disorder in crystal lattices of compounds with a real deficit of metal ($Me_{1-y}X$) can be similarly predicted. However, in this case, the effect is opposite. Ions of a metal having a valency higher than that of the lattice cations cause an increase of the concentration of cation vacancies and a simultaneous decrease of the concentration of electron holes. Addition of cations having a valency lower than that of the lattice cations causes a decrease of the concentration of ionic defects and an increase of electronic ones. The processes of incorporation of dopant in crystal lattices of compounds of the $Me_{1-y}X$ type can be represented by the following equations:

$$F_2X_3 \rightleftharpoons 2F^{\cdot}_{Me}+V''_{Me}+3X_X \quad (1.3.205)$$

or

$$F_2X_3+2h^{\cdot} \rightleftharpoons 2F^{\cdot}_{Me}+2X_X+\tfrac{1}{2}X_2 \quad (1.3.206)$$

and

$$F_2X+V''_{Me} \rightleftharpoons 2F'_{Me}+X_X \quad (1.3.207)$$

or

$$F_2X+\tfrac{1}{2}X_2 \rightleftharpoons 2F'_{Me}+2h^{\cdot}+2X_X \quad (1.3.208)$$

As before, equations (1.3.205)–(1.3.208) represent partial processes which may be expressed by two combined equations. They are obtained by adding equation (1.3.205) to (1.3.206):

$$2F_2X_3+2h^{\cdot} \rightleftharpoons 4F^{\cdot}_{Me}+V''_{Me}+5X_X+\tfrac{1}{2}X_2 \quad (1.3.209)$$

By adding equation (1.3.207) to (1.3.208), we obtain the following expression:

$$2F_2X+V''_{Me}+\tfrac{1}{2}X_2 \rightleftharpoons 4F'_{Me}+2h^{\cdot}+3X_X \quad (1.3.210)$$

The electroneutrality conditions for the lattice in the two cases are as follows:

$$[F^{\cdot}_{Me}]+[h^{\cdot}] = 2[V''_{Me}] \quad (1.3.211)$$

and

$$[F'_{Me}]+2[V''_{Me}] = [h^{\cdot}] \quad (1.3.212)$$

Both conditions can be simplified using the method of Kröger and Vink. When $[F^{\cdot}_{Me}] \gg [h^{\cdot}]$ we have:

$$[V''_{Me}] = \tfrac{1}{2}[F^{\cdot}_{Me}] \quad (1.3.213)$$

and when $[F'_{Me}] \gg [V''_{Me}]$:

$$[h^{\cdot}] = [F'_{Me}] \quad (1.3.214)$$

By utilising equation (1.3.128) and the electroneutrality conditions of the lattice we can express the concentrations of electron holes and cation vacancies in solid solutions F_2X_3–$Me_{1-y}X$ and F_2X–$Me_{1-y}X$ by the following equations:

$$[h^{\cdot}] = \left(\frac{2K_{124}}{[F^{\cdot}_{Me}]}\right)^{1/2} p_{X_2}^{1/4} \quad (1.3.215)$$

$$[V''_{Me}] = \frac{K_{124}}{[F'_{Me}]^2} p_{X_2}^{1/2} \quad (1.3.216)$$

It follows from these equations that the concentration of electron holes in solid solution F_2X_3–$Me_{1-y}X$ decreases with increasing concentration of the dopant, whereas the concentration of cation vacancies in solution F_2X–$Me_{1-y}X$ decreases with increasing concentration of F^{1+} ions. For this reason, at large concentrations of the dopant, the intrinsic electronic or ionic disorder should be taken into account. If we assume that the defects are present only in the cationic sublattice, we can take them into account using equations (1.3.130) and (1.3.131) by means of which equations (1.3.215) and (1.3.216) can be transformed into the following expressions giving the dependence of the concentration of quasi-free electrons or interstitial cations on the concentration of the dopant:

$$[e^-] = \frac{K_{127}[F_{Me}^{\bullet}]^{1/2}}{(2K_{124})^{1/2}} p_{X_2}^{-1/4} \qquad (1.3.217)$$

and

$$[Me_i^{\bullet\bullet}] = \frac{K_{126}[F'_{Me}]^2}{K_{124}} p_{X_2}^{-1/2} \qquad (1.3.218)$$

It follows from these equations that the addition of cations having a valency higher than that of the lattice cations causes an increase of the concentration of quasi-free electrons, whereas addition of F_2X causes an increase of the concentration of interstitial cations. For this reason the addition of a large amount of compound F_2X_3 can cause inversion of the electronic defects (from *p* type to *n* type), whereas the addition of F_2X can cause the inversion of the type of ionic defect. In the latter case interstitial cations become predominant (the solution changes its character and becomes a solution of the F_2X–$Me_{1+y}X$ type).

However, it should be pointed out that, as in the case of addition of dopants to compounds of the $Me_{1+y}X$ type, the disorder in crystal lattices of solid solutions F_2X_3–$Me_{1-y}X$ and F_2X–$Me_{1-y}X$ depends not only on the concentration of the dopant, but also on the equilibrium pressure of the oxidant. However, in this case this effect is reversed. The concentration of intrinsic defects (cationic vacancies and electron holes) increases with increasing oxidant pressure and, for this reason, small amounts of dopant can be expected to have an effect in the region of low partial pressures of the oxidant. In the high pressure region the dopant may have no effect on the disorder in the solid solution. This conclusion is illustrated schematically in Fig. 59 and 60. It follows from the graphs that the low pressure range corresponds to the extrinsic region and the high pressure range to the intrinsic region.

It should be emphasised that the above equation apply not only to defects in the cationic sublattice but also to those occurring in the anionic sublattice. Thus equations (1.3.195) and (1.3.196) as well as (1.3.202) and (1.3.204) describe the effect of various dopants on the character of the

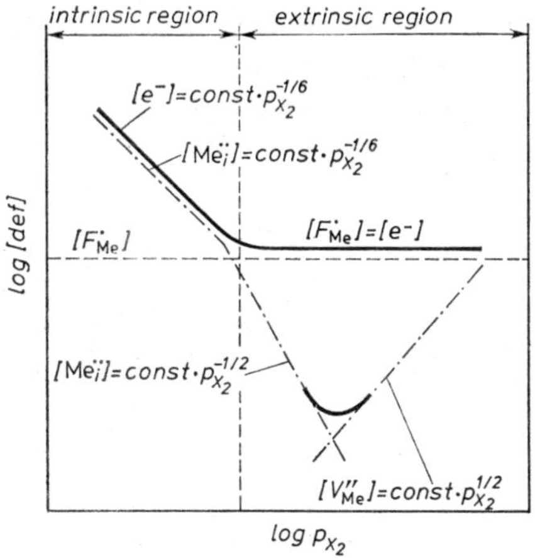

Fig. 59. The effect of a trivalent dopant on the ionic and electronic disorder in a $Me_{1-y}X$ type crystal as a function of equilibrium oxidant pressure.

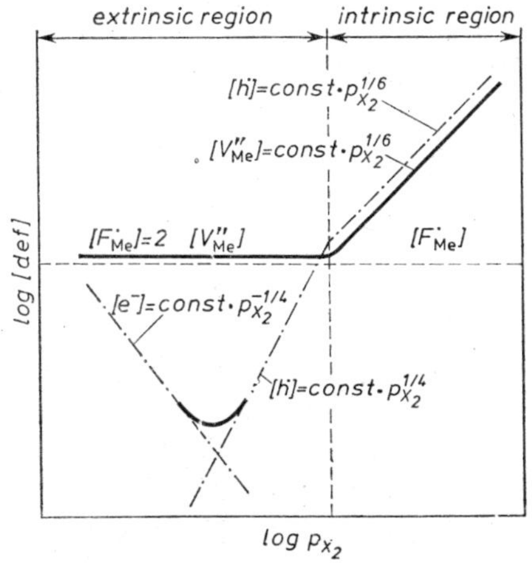

Fig. 60. The effect of a monovalent dopant on the ionic and electronic disorder in a $Me_{1-y}X$ type crystal as a function of equilibrium oxidant pressure.

defects in crystal lattices of compounds with a real excess of metal and a real deficit of the oxidant, since in both cases the excess positive charge in the crystal is compensated by a corresponding number of quasi-free electrons. The only difference consists in the fact that, in the former case, the excess positive charge is due to the presence of interstitial cations and, in the latter case, to the existence of anion vacancies. A similar situation exists in the case of compounds with a real deficit of metal or a real excess of the oxidant.

In conclusion, we can say that addition of ions of a metal having a valency higher than that of the lattice cations of compounds of the $Me_{1+y}X$ or MeX_{1-y} types causes a decrease of the concentration of interstitial cations or of anion vacancies, which is accompanied by an increase of the concentration of quasi-free electrons, whereas addition of cations having a valency lower than that of the lattice cations has the reverse effect. In the case of compounds of the $Me_{1-y}X$ or MeX_{1+y} types addition of ions having a valency higher than that of the lattice ions, causes an increase of the concentration of cation vacancies or of interstitial anions, which is accompanied by a decrease of the concentration of electron holes. Addition of cations having a valency lower than that of lattice cations has the opposite effect. These conclusions are correct only in conditions where the dopant forms a solid solution in the crystal, the ions of the dopant occupy lattice sites and the pressure of the oxidant is suitably chosen.

Finally it should be added that the concentrations of ionic and electronic defects in the compounds discussed of the MeX type can similarly be modified by introducing dopants of the oxidant in the anionic sublattice. However, these cases are rather rare, since the solubility of substances such as nitrogen or halogens in oxides and sulphides of metals and the solubility of sulphur and oxygen in metal halides are very low.

1.3.8 Empirical correlations

The above considerations lead to the conclusion that in order to calculate the concentration of a given kind of defect under specified thermodynamic conditions it is necessary to know the values of the equilibrium constants for the defect reactions and their temperature and pressure dependences, i.e. the entropy and enthalpy of formation of defects. For this reason knowledge of numerical values of enthalpy and entropy of formation for various kinds of point defects is of great theoretical and practical importance.

Unfortunately, in spite of many attempts, a satisfactory method of calculation of these functions has not yet been developed. For this reason there have been many attempts to find empirical correlations between the entropy and enthalpy of formation of defects and other properties of compounds MeX, which can be calculated or measured. It has been found that such correlations exist but only for the enthalpy of formation of defects. The problem of calculation of the entropy of formation of defects remains completely unsolved.

There are several empirical methods for the calculation of the enthalpy of formation of defects. We will only mention here two of them which deserve particular attention. The first method is limited to halides of heavy metals and alkali metals although there are reasons for believing that it is of more general application. It has been shown that the relation between the melting points of these compounds, expressed on the absolute scale, and the enthalpy of formation of defects can be represented by the following linear equation [40]:

$$\Delta H_f^e = 2.14 \times 10^{-3} T_m \tag{1.3.219}$$

where ΔH_f^e is the enthalpy of formation of defects in eV and T_m is the melting point on the Kelvin scale.

It has been shown that the enthalpies of formation of defects of the Schottky type for many halides of heavy metals and alkali metals, calculated

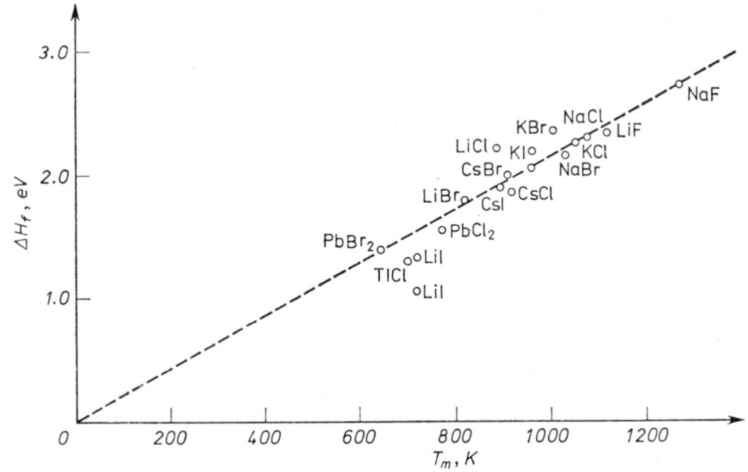

Fig. 61. Enthalpies of formation of defects in metal halides as a function of their melting points (after [40]).

by means of equation (1.3.219), are in good agreement with the data obtained experimentally (the differences do not exceed 10%). The relationship is shown in Fig. 61.

The second method, which also applies to metal oxides, is based on the relation between the energy for dissociation into atoms E_{at} of a compound MeX and the enthalpy of formation of defects in its crystal lattice. E_{at} is the energy required for the decomposition of the compound into atoms in the gas state. It can be determined from the heat of formation of the compound, the sublimation energy of the metal and the dissociation energy of the oxidant molecules. In the case of metal oxides with the general formula Me_aO_b the following relation holds between the above physical quatities [41]:

$$E_{at} = \Delta H_{Me_aO_b} + aL_{Me} + \frac{b}{2} D_{O_2} \tag{1.3.220}$$

where $\Delta H_{Me_aO_b}$ is the heat of formation of the oxide, L_{Me} is the energy of sublimation of the metal and D_{O_2} is the energy of dissociation of oxygen molecules. D_{O_2} is equal to 498.23 kJ/mole (119 kcal/mole).

Equation (1.3.220) is also applicable to sulphides and halides of metals.

It might be expected that in the case of compounds MeX there would be a definite relation between the energy E_{at} and the enthalpy of formation of defects, since this energy is a measure of stability of these compounds and as a result the energies of formation of point defects in crystal lattices of MeX compounds increase with increasing energy for dissociation into

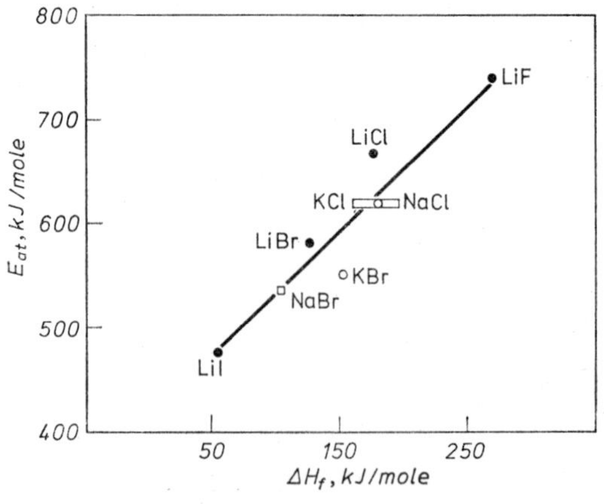

Fig. 62. Enthalpies of formation of defects in metal halides as a function of their atomisation energy (after [41]).

atoms. The correctness of this conclusion has been checked for alkali halides of which the enthalpies of formation of defects have been determined experimentally. The results of these studies are shown graphically in Fig. 62. It can be seen that the relation between the enthalpy of formation of defects and the energy for dissociation into atoms is approximately linear.

Kofstad [42] has shown that an analogous linear relationship exists in the case of refractory metal oxides. The results of his calculations are shown in Fig. 63. In the present state of our knowledge it is not possible to say

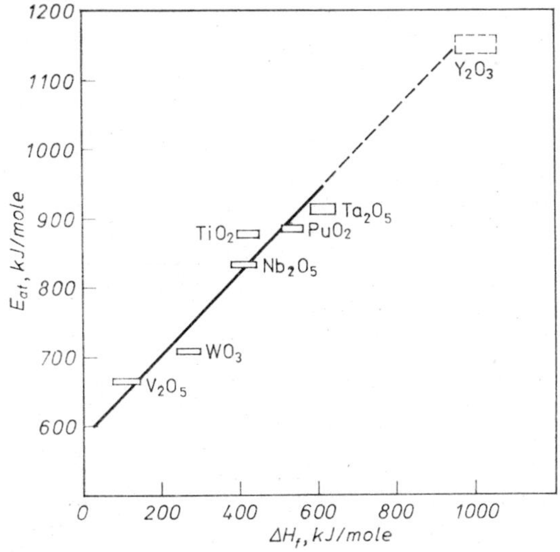

Fig. 63. The dependence of the enthalpy of formation of anion vacancies in metal oxides on their atomisation energy (after [42]).

whether this relation between the enthalpy of defect formation and the energy for dissociation into atoms is of general application and can be used for all compounds of the MeX type as the data available are insufficient to check this interesting empirical relationship.

1.3.9 Empirical relationships between thermodynamic properties of MeX *compounds and thermodynamic properties of defects*

It follows from the above considerations that deviations in MeX crystals from stoichiometry are dependent on the predominant type of disorder in one of the two sublattices of these compounds. The concentration of the predominant defects and consequently the extent of deviation from stoichiometry can be changed reversibly by changing the external conditions, in particular the temperature and the pressure of the component X in the surrounding atmosphere. Thus MeX compounds have definite homogeneity regions and can be regarded as dilute solid solutions. Thermodynamic properties of these solutions must be related to those of the defects. It can be readily shown that this relation in fact exists and be utilised in many calculations which are of importance in the thermodynamics of defects [43].

Let us first consider the relations between thermodynamic properties of compounds of the MeX type and the enthalpy and entropy of defect formation. We will discuss this problem for the example of a compound with a real deficit of metal $Me_{1-y}X$. From the thermodynamic point of view the formation reaction of this type of solid solution can be expressed by the following equation:

$$(1-y)\mathrm{MeX}+\frac{y}{2}\mathrm{X}_2 = \mathrm{Me}_{1-y}\mathrm{X} \qquad (1.3.221)$$

The change of partial thermodynamic potential of the oxidant in the crystal caused by incorporation of one gram-atom of the oxidant is given by the following expression:

$$\Delta\overline{G}_{\mathrm{X}_2} = \Delta\overline{H}_{\mathrm{X}_2} - T\Delta\overline{S}_{\mathrm{X}_2} \qquad (1.3.222)$$

where $\Delta\overline{H}_{\mathrm{X}_2}$ and $\Delta\overline{S}_{\mathrm{X}_2}$ are partial enthalpy and partial entropy of the oxidant respectively.

Applying the law of mass action to reaction (1.3.221) we obtain:

$$K_1 = p_{\mathrm{X}_2}^{-1/2} \qquad (1.3.223)$$

Since the equilibrium constant of any chemical reaction is related to its thermodynamic potential by equation (1.3.24) we can write:

$$\Delta\overline{H}_{\mathrm{X}_2} - T\Delta\overline{S}_{\mathrm{X}_2} = RT\ln p_{\mathrm{X}_2}^{1/2} \qquad (1.3.224)$$

The process leading to deviation from the stoichiometry can be represented by the following defect reaction (where it is assumed that the cation vacancies are completely ionised):

$$\tfrac{1}{2}\mathrm{X}_2 = \mathrm{V}_{\mathrm{Me}}'' + 2h^{\bullet} + \mathrm{X}_{\mathrm{X}} \qquad (1.3.225)$$

Applying the law of mass action and the corresponding electroneutrality condition of the lattice to the above equation we can express the equilibrium constant as follows:

$$K_{\mathrm{V}_{\mathrm{Me}}''} = \frac{4[\mathrm{V}_{\mathrm{Me}}'']^3}{p_{\mathrm{X}_2}^{1/2}} \qquad (1.3.226)$$

Since $y = [V''_{Me}]$ equation (1.3.226) can be expressed in the following form:

$$p_{X_2}^{1/2} = \frac{4y^3}{K_{V''_{Me}}} \qquad (1.3.227)$$

Introducing this expression in equation (1.3.224) we obtain:

$$\Delta \bar{H}_{X_2} - T\Delta \bar{S}_{X_2} = RT\ln\left(\frac{4y^3}{K_{V''_{Me}}}\right) \qquad (1.3.228)$$

The equilibrium constant $K_{V''_{Me}}$ can be expressed in terms of the enthalpy and entropy of formation of defects:

$$\ln K_{V''_{Me}} = \frac{\Delta S''_f}{R} - \frac{\Delta H''_f}{RT} \qquad (1.3.229)$$

where $\Delta H''_f$ and $\Delta S''_f$ are the enthalpy and the entropy of formation of doubly ionised cation vacancies, respectively.

Using this relationship we can eliminate $K_{V''_{Me}}$ from equation (1.3.228) and after further transformations we obtain:

$$\Delta \bar{H}_{X_2} - T\Delta \bar{S}_{X_2} = \Delta H''_f - T\Delta S''_f - R\ln 4 - 3R\ln y \qquad (1.3.230)$$

It follows from this equation that:

$$\Delta \bar{H}_{X_2} = \Delta H''_f \qquad (1.3.231)$$

and

$$\Delta \bar{S}_{X_2} = \Delta S''_f - 3R\ln y - \mathrm{R}\ln 4 \qquad (1.3.232)$$

The above relationships lead to two important conclusions. There is a possibility of determining the enthalpy and entropy of formation of defects indirectly on the basis of numerical values for the partial molar enthalpy and entropy of dissolution of the oxidant in the MeX compound. However, this is possible only when the value of $\Delta \bar{H}_{X_2}$ is independent of the concentration of defects, i.e. on y. On the other hand when $\Delta \bar{H}_{X_2}$ is dependent on the deviation of the compound from stoichiometry, this may mean that there is either a gradual change of the predominant type of disorder or a deviation from the laws of ideal solutions, i.e. the formation of complexes and defect clusters. In any case the dependence of $\Delta \bar{H}_{X_2}$ on y means that in this case the disorder in the crystal cannot be satisfactorily described by a simple defect equation, such as (1.3.225).

It should be mentioned that the above limitation applies only to $\Delta \bar{H}_{X_2}$, since it follows from equation (1.3.232) that $\Delta \bar{S}_{X_2}$ increases with increasing deviation from stoichiometry in spite of the constant value of $\Delta S''_f$.

The above considerations lead to the conclusion that measurements of partial molar thermodynamic properties of MeX compounds can provide valuable information of both qualitative and quantitative character on the structure and thermodynamics of point defects in crystals of compounds MeX.

Kofstad [43, 44] has pointed out that the enthalpy and entropy of formation of point defects in crystals of MeX compounds can also be calculated on the basis of the enthalpy and entropy of formation of these compounds from the elements. In order to illustrate the Kofstad method we shall consider a hypothetical system consisting of two compounds of a metal Me, namely Me_2X and MeX. In order to simplify the problem we shall assume that the first compound shows practically no deviations from stoichiometry, whereas the second contains a real excess of the metal ($Me_{1+y}X$). The deviation of this compound from stoichiometry increases

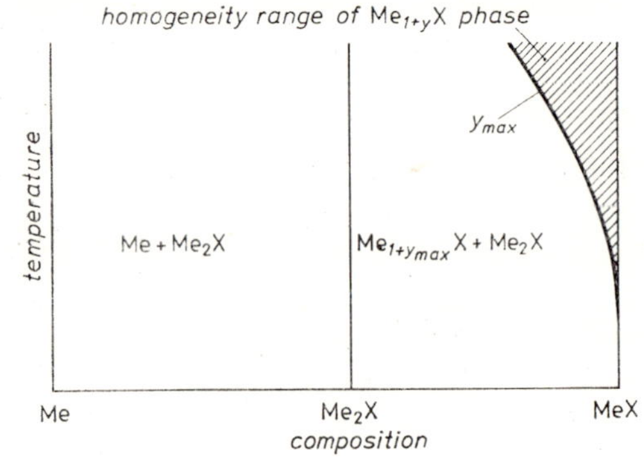

Fig. 64. The scheme of the phase diagram in a hypothetical Me–Me_2X–$Me_{1+y}X$–X_2 system.

with increasing temperature and the maximum value of this deviation at any temperature determines the lower limit of homogeneity of the system. This situation is illustrated by the phase diagram shown in Fig. 64.

The process of formation of defects in the compound $Me_{1+y}X$ can be represented by the following general relation:

$$Me_{Me}+X_X = Me_i^p+pe^- +\tfrac{1}{2}X_2 \qquad (1.3.233)$$

where p is the degree of ionisation of interstitial cations. Applying the law of mass action to this equation we obtain:

$$K_{Me_i} = [Me_i^p][e^-]^p p_{X_2}^{1/2} \qquad (1.3.234)$$

In this case the electroneutrality condition for the lattice has the following form:

$$[Me_i^{p\cdot}]p = [e^-] \tag{1.3.235}$$

Assuming that $y = Me_i^{p\cdot}$ and using condition (1.3.235) we can present equation (1.3.234) in the following form:

$$(p+1)\ln y = \ln K_{Me_i p} - \tfrac{1}{2}\ln p_{X_2} - p\ln p \tag{1.3.236}$$

When the deviation from stoichiometry reaches the maximum value, y_{max}, for a given temperature, the equilibrium pressure of the oxidant p_{X_2} over the compound $Me_{1+y_{max}}X$ is equal to the decomposition pressure of this compound $p^x_{X_2}$. In this extreme case (the lower limit of homogeneity), expression (1.3.236) takes the following form:

$$(p+1)\ln y_{max} = \ln K_{Me_i p} - \tfrac{1}{2}\ln p^x_{X_2} - p\ln p \tag{1.3.237}$$

Below the limiting pressure $p^x_{X_2}$, the decomposition of compound $Me_{1+y_{max}}X$ takes place. Assuming that the maximum deviation from stoichiometry is small, i.e. that $y_{max} \ll 1$, we can formulate the following expression first approximation:

$$2MeX = Me_2X + \tfrac{1}{2}X_2 \tag{1.3.238}$$

The change of the thermodynamic potential of this reaction, which is equal to the decomposition pressure $p^x_{X_2}$, is given by the following equation:

$$\Delta G = \Delta G_1 - 2\Delta G_2 = -\tfrac{1}{2}RT\ln p^x_{X_2} \tag{1.3.239}$$

where ΔG_1 and ΔG_2 are respectively the changes of the thermodynamic potentials in the reactions of formation of compounds Me_2X and MeX from the elements:

$$2Me + \tfrac{1}{2}X_2 = Me_2X \tag{1.3.240}$$

and

$$Me + \tfrac{1}{2}X_2 = MeX \tag{1.3.241}$$

Equation (1.3.239) can be writen in the following form:

$$\frac{1}{2}\ln p^x_{X_2} = \frac{2\Delta G_2 - \Delta G_1}{RT} \tag{1.3.242}$$

ΔG_1, ΔG_2 and $K_{Me_i p}$ are related to the enthalpy and entropy of the corresponding processes according to the following equations:

$$\Delta G_1 = \Delta H_1 - T\Delta S_1 \tag{1.3.243}$$

$$\Delta G_2 = \Delta H_2 - T\Delta S_2 \tag{1.3.244}$$

$$\ln K_{Me_i} = -\frac{\Delta H_{Me_i p}}{RT} + \frac{\Delta S_{Me_i p}}{R} \tag{1.3.245}$$

where ΔH_1, ΔS_1, ΔH_2 and ΔS_2 are the enthalpies and entropies of processes (1.3.240) and (1.3.241) respectively and $\Delta H_{Me_1 p}$, $\Delta S_{Me_1 p}$ are the enthalpy and the entropy of formation of defects according to reaction (1.3.233).

Using equations (1.3.242)–(1.3.245) we can write expresion (1.3.237) in the following form:

$$\ln y_{max} = \frac{\Delta H_{1,2} - \Delta H_{Me_1 p}}{(p+1)RT} + \frac{\Delta S_{Me_1 p} - \Delta S_{1,2}}{(p+1)R} - \frac{p}{(p+1)}\ln p \qquad (1.3.246)$$

where $\Delta H_{1,2} = \Delta H_1 - 2\Delta H_2$ and $\Delta S_{1,2} = \Delta S_1 - 2\Delta S_2$.

The concentration of defects is, as a rule, an exponential function of temperature. Therefore the maximum concentration of defects, equal to the lower limit of homogeneity of compound $Me_{1+y_{max}}X$, is a similar function of the temperature. This relationship can be expressed by the following equation:

$$\ln y_{max} = A - \frac{\Delta H_{max}}{RT} \qquad (1.3.247)$$

where ΔH_{max} is the enthalpy of formation of defects at their maximum concentration.

Comparing expressions (1.3.246) and (1.3.247) we observe that:

$$-\frac{\Delta H_{max}}{RT} = \frac{\Delta H_{1,2} - \Delta H_{Me_1 p}}{(p+1)RT} \qquad (1.3.248)$$

and

$$A = \frac{\Delta S_{Me_1 p} - \Delta S_{1,2}}{(p+1)R} - \frac{p}{p+1}\ln p \qquad (1.3.249)$$

Hence

$$\Delta H_{Me_1 p} = \Delta H_{1,2} + (p+1)\Delta H_{max} \qquad (1.3.250)$$

and

$$\Delta S_{Me_1 p} = \Delta S_{1,2} + pR\ln p + (p+1)RA \qquad (1.3.251)$$

The same procedure can be applied to any type of defect, but in the case of large deviations from stoichiometry corrections for possible interactions of the defects must be introduced and changes in the concentration of occupied lattice sites must be taken into account. In addition, in the equation for the decomposition process of the compound MeX, the coefficients resulting from the deviation from stoichiometry must be taken into account and, in calculations of the decomposition pressures $p^x_{X_2}$, the values of

ΔH and ΔS corresponding to the maximum deviation from stoichiometry must then be used. Obviously this is not always possible, since thermodynamic data concerning non-stoichiometric compounds are still very scarce. Nevertheless the Kofstad method is very valuable, since it indicates that there are new possibilities for the indirect determination of the thermodynamic properties of point defects and it emphasises the close relationship between these properties and the thermodynamics of the crystal.

1.4 Extended defects

The thermodynamic theory of point defects outlined in the previous chapter is limited to cases in which a real crystal can be regarded as an ideal solution of defects in the lattice. This condition is fulfilled when the point defects do not interact with each other and the structure of the crystal is free from major local deformations, i.e. the defects must be distributed in the crystal at random and consequently they must obey the laws of classical statistics. This is possible when the concentration of defects is low. Taking into account the fact that, as a rule, point defects have a definite electrical charge with respect to the lattice, an analogy between aqueous solutions of electrolytes and real ionic crystals is readily drawn. It follows from the theory of solutions that they behave like ideal systems when their concentration do not exceed 0.1 mole per dm^3. In the case of ionic crystals of the MeX type this means that the concentration of defects must not exceed 0.1 at. %.

In practice this condition is completely fulfilled in many cases. For example the concentration of cation vacancies in nickel oxide is lower than the limiting value even at very high temperatures ($\gg$ 1300°C). An analogous situation exists in the case of cuprous oxide and many other oxides. In particular the concentration of defects is extremely low ($\simeq 10^{-4}$ at. %) in oxides and sulphides of the main group metals in the periodic table. However, there is also a relatively large group of transition metal oxides and sulphides with large deviations from stoichiometry, which indicates that in this case the concentration of defects is several times higher than the conventional limiting concentration. For example ferrous oxide (wustite) has deviations from stoichiometry in the direction of deficit of the metal at 1000°C of up to 0.12 moles of iron per mole of FeO

($Fe_{0.88}O$). Hence the concentration of cation vacancies in the crystal lattice of this oxide may be as high as 12 at.%. Later in this chapter we shall show that in fact this concentration is much higher. Sulphides of iron and copper ($Fe_{0.85}S$ and $Cu_{1.75}S$) display even greater deviations from stoichiometry. Oxides of refractory metals, such as TiO_2, Nb_2O_5, MoO_3, WO_3 etc., show considerable deviations from stoichiometry in the direction of deficit of oxygen, which indicates that the concentrations of anion vacancies in their lattices are high.

The substances mentioned above cannot be regarded as very dilute solutions of defects in the lattice, i.e. as ideal solutions obeying the laws of statistical thermodynamics. For this reason the application of the classical theory of point defects to these cases is not justified. However, the results of early studies on these problems seemed to show that there was no interaction between the defects even at concentrations many times higher that 0.1 at.%. For example the results of thermogravimetric studies carried out by Hauffe and Pfeiffer [45] led to the conclusion that the disorder in the crystal lattice of ferrous oxide $Fe_{1-y}O$ is due to the presence of completely ionised cation vacancies and the equivalent number of electron holes which do not interact with each other. These authors have found that in the case of this oxide the deviation from stoichiometry increases with increasing equilibrium oxygen pressure according to the following equation:

$$y = \text{const}\, p_{O_2}^{1/6} \tag{1.4.1}$$

This relationship follows directly from the application of the law of mass action to the following simple defect equilibrium

$$\tfrac{1}{2}O_2 \rightleftharpoons V''_{Fe} + 2h^{\cdot} + O_O \tag{1.4.2}$$

Kofstad [46] and Forland [47], in their studies on defect equilibria in oxides of titanium TiO_{2-y} and niobium Nb_2O_{5-y}, have found that the deviation from stoichiometry in this case decreases with increasing oxygen pressure according to the following equation:

$$y = \text{const}\, p_{O_2}^{-1/6} \tag{1.4.3}$$

This leads to the conclusion that these oxides contain completely ionised anion vacancies which do not interact with each other and which are formed according to the following scheme:

$$O_O \rightleftharpoons V''_O + 2e^- + \tfrac{1}{2}O_2^g \tag{1.4.4}$$

In both the cases mentioned above the concentrations of defects were several times higher than the conventional limiting value of 0.1 at.%. In $Fe_{1-y}O$ the concentration of cation vacancies was 12 at.% and in the oxides of titanium and niobium the concentration of anion vacancies was higher than 1 at.%.

In view of the results of these experiments, the suggestions of certain authors, who were already questioning in the early nineteen fifties the possibility of occurrence of simple defect structures in oxides and sulphides of transition metals showing large deviations from stoichiometry, did not attract the attention of other authors working in this field. Only during the last decade has this situation radically changed. The rapid development of traditional methods and the appearance of new methods of investigation of structures of solids and of kinetics and thermodynamics of defects, made it possible to obtain direct experimental proof for the existence of extended defects in these substances. In particular the use of neutron diffraction, Mössbauer spectroscopy and high resolution electron microscopy made it possible to show that point defects interact with each other even at concentrations below 0.1 at.%. These interactions can be very varied and depend on the kind and concentration of defects and the nature of crystal. The interactions lead to the formation of various complexes and clusters of defects (extended defects) which later may possess short and long range order. This means that defect superstructures may appear and consequently homologous series of intermediate phases. These phases can be formed as a result of assimilation or elimination of defects. The diversity of forms of extended defects is mainly due to the fact that their formation can be accompanied by a local reconstruction of the crystal structure or it can take place without significant change in the atomic positions in the crystal.

The first attempt to systematise these phenomena was made by J. S. Anderson [48–40], the principal authority in this field. He has shown that there are five limiting types of disorder in crystal lattices of compounds of the MeX type, depending on the concentration of defects, temperature, the kind of lattice and the nature of the substance. This classification is shown in Table 2. It can be seen from this table that the thermodynamic theory of point defects can only be applied without limitations to a definite concentration of defects, which largely depends on the kind of oxide and the type of its crystalline structure.

We shall return to this problem later during the discussion of the individual types of disorder.

TABLE 2

The types of disorder in crystal lattices of transition metal oxides

No	Type of disorder	Concentration of point defects at. %	Examples
1	Statistically distributed point defects	$0 < C_d < 10^{-1}$	$Ni_{1-y}O$, $Cu_{2-y}O$, CdO_{1-y}, $Zn_{1+y}O$
2	Quasi-random point defects	$10^{-1} < C_d < 10^1$	$Co_{1-y}O$, $Ti_{1-y}O$ TiO_{1+y}, $V_{1-y}O$, VO_{1+y}
3	Defect complexes, microdomains	$10^{-1} < C_d < 10^1$	$Fe_{1-y}O$, UO_{2+y}
4	The formation of intermediate phases resulting from the defect superstructure (assimilation of point defects)	$10^{-1} < C_d < 10^1$	CeO_{2-y}, PrO_{2-y}
5	The formation of intermediate phases resulting from shear structures (elimination of point defects)	$0 < C_d < 10^{-1}$	WO_{3-y}, MoO_{3-y} Nb_2O_{5-y}, TiO_{2-y}

Table 3 gives Anderson's complete system, which clearly illustrates the relation of the classical theory of defects to modern views on the types of disorder in crystalline structures. This scheme also shows the successive steps in the formation of new intermediate phases resulting from the process of ordering of defects. It follows from Table 3 that, depending on the kind of oxide and the type of defects in its crystal lattice, there are three mechanisms of formation of extended defects that lead to the formation of intermediate phases. Two of these mechanisms consist of a gradual arrangement of point defects leading to the formation of a superstructure in which the defects are assimilated as integral elements of the crystal lattice. The third mechanism consists in elimination of point defects by the process called crystallographic shear. This proces leads to the formation of new intermediate phases in which the concentration of point defects is extremely low ($< 10^{-2}$ at.%) and is not connected with the deviation from stoichiometry of the parent phase. We shall follow this classification and we shall discuss separately these three mechanisms for the formation of extended defects. However, before starting this discussion we should mention the theoretical background to the discovery of extended defects, as their

TABLE 3

Models of disorder in crystal lattices of metal oxides

Interaction negligible	Thermodynamic theory of point defects			Point defects	Non-stoichiometric phases
Non-difusive interaction	Local reconstruction resulting from the relaxation of repulsive forces UO_{2+x}			Randomly distributed complexes of defects	
	Local reconstruction; new coordination polyhedra; Roth complexes $Fe_{1-x}O$				
Diffusive interaction	The formation of clusters from complexes of defects; Koch–Cohen clusters in $Fe_{1-x}O$	Two-dimensional reconstruction; isolated planar defects; shear planes and stracking faults WO_{3-x}	Association or correlation of defects PrO_{2-x}	Short range ordered arrangement; microdomains	
	Partly ordered arrangement of clusters in the form of multiclusters $Fe_{1-x}O$	Ordered aggregation of elements of lattice planes in small regions	Regular association of correlated defects in small regions		
	The formation of ordered superstructure assimilating point defects; new structures	Ordered shear structures eliminating the defects	The formation of ordered superstructure assimilating the defects; new structures	Long range ordered arrangement	Stoichiometric phases
	Successive intermediate phases — homologous series				

existence, like that of dislocations, had been predicted long before the first experimental proofs of their existence were presented.

In view of the ambiguity in the structure of non-stoichiometric compounds the occurrence of large deviations from stoichiometry was causing concern to crystallographers and chemists for many years. If a crystal contains a relatively large number of randomly distributed point defects (> 10 at.%), it is difficult to say which atoms are situated in the normal lattice sites and which are in the interstitial positions. For this reason it is difficult to determine the symmetry of the crystal structure. For example, Fig. 65

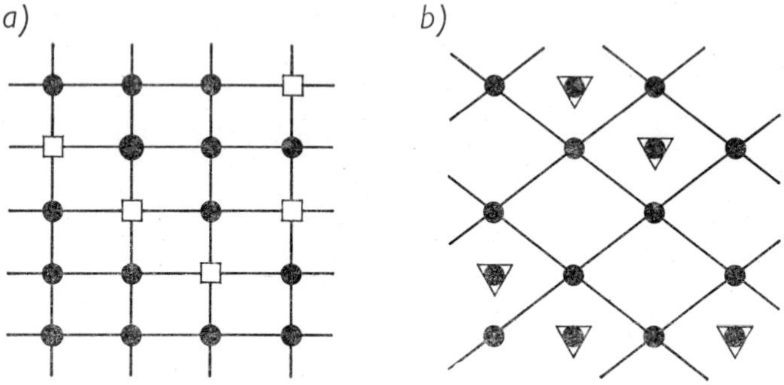

Fig. 65. The scheme of a hypothetical two-dimensional lattice containing 25% defects (after [51]); *a*)—a simple regular lattice containing 25% vacancies; *b*)—a simple diamond lattice containing 25% interstices.

shows a scheme of a hypothetical two-dimensional lattice, which can be regarded as either a primitive regular structure containing 25% vacancies or a simple diamond lattice containing 25% of atoms in the interstitial positions. Problems of this kind are encountered in the identification of the structures of crystals of halides of certain transition metals, e.g. nickel telluride, which exists as a single phase over a wide range of chemical composition between cubic NiTe and hexagonal $NiTe_2$. The former forms crystals having a structure of the nickel arsenide (NiAs) type. These crystals can accomodate increasing amounts of tellurium and when the composition of the substance corresponds to the formula $NiTe_2$, its crystals have a structure of the CdI_2 type.

Theoretical investigation of the above problems by many authors led to the conclusion that an ordered distribution of defects represents a structure which is thermodynamically more stable than that in which the defects are distributed at random [52–54]. By considering the disorder in the crystal lattice of ferrous sulphide Fe_7S_8, and taking into account only the electrostatic energy of the defected crystal lattice, Bertaut [52] has shown that an ordered arrangement of defects represents a much more stable structure for the crystal than that corresponding to their random distribution. On this basis he explained the formation of a superstructure in compounds of the NiAs type, which in the case of ferrous sulphide ($Fe_{1-y}S$) leads to the formation of the compound Fe_7S_8. However, Saunten, Atlas, Anderson [48, 49] and other authors have pointed out that an ordered arrangement of defects does not necessarily lead to the formation of a superstructure, i.e. new phases, which implies long range ordering of defects. On the contrary, at elevated temperatures the interaction between defects is mainly short range, which is almost as favourable as the formation of a superstructure from the energetic point of view. Theoretical considerations by the above mentioned authors have shown that an ordered arrangement of defects over a short range, i.e. the formation of small clusters of defects (complexes), causes a considerable decrease of crystal energy, which is similar to that obtained in the case of a long range ordering [55]. On the other hand, the configurational entropy decrease caused by the formation of these defects is relatively small, which means that the crystal structure containing extended defects is thermodynamically more stable than that containing random point defects. It has been shown that in some cases, e.g. in the case of $LiFe_5O_8$ spinel, the energy difference between crystals with long range and short range order is only about 4.1868 kJ/mole (one kcal/mole), whereas the energy difference between the completely ordered and the random configurations [56] is about 502.42 kJ/mole (120 kcal/mole). At the same time the configurational entropy decrease corresponding to the two kinds of ordering is relatively small (in this case it decreases by only 50%). Westburn and Gronwold [57] came to analogous conclusions in their studies on the order–disorder transformations in the structure of magnetite.

On the basis of these and other results of experimental studies, as well as theoretical considerations, Greenwood and O'Keefe [55] have stated that the phenomenom is quite general. The formation of complexes and clusters of defects, i.e. the short range ordering of defects, decreases the energy of real crystals to a much greater extent than the entropy. Since the energy and the entropy are connected by the following equation:

$$\Delta G = \Delta H - T\Delta S \tag{1.4.5}$$

the decrease of enthalpy ΔH of the process is greater than $T\Delta S$ and as a result $\Delta G < 0$. For this reason point defects in the crystal have a tendency to become ordered spontaneously, which increases with decreasing temperature since the value of $T\Delta S$ decreases with decreasing temperature. Thus a quasi-disordered distribution of defects in crystals can be expected at very high temperatures even at high concentrations of defects and in spite of their interactions, whilst a decrease of the temperature is accompanied by the formation of complexes and defect clusters, which ultimately leads to long range ordering, i.e. to the formation of new intermediate phases.

The plot, shown in Fig. 66, illustrates the dependence of the thermodynamic potential G of the crystal on the concentration of ordered and

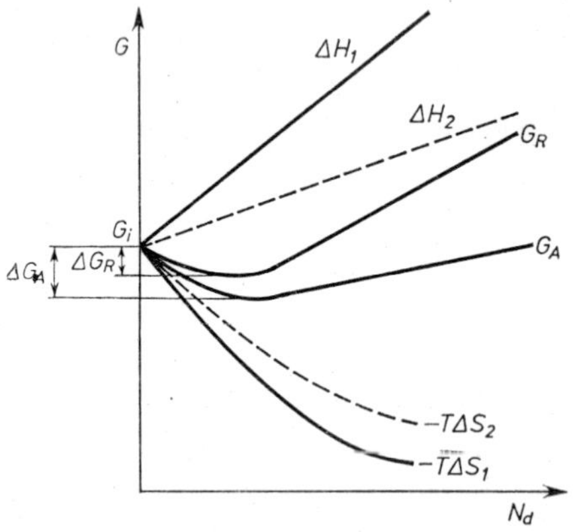

Fig. 66. The dependence of the thermodynamic potential of crystal on the concentration of point defects and extended defects.

disordered point defects. In this plot ΔH_1 and ΔS_1 are the enthalpy and the entropy of formation for random defects and ΔH_2 and ΔS_2 are the enthalpy and the entropy of formation for defect complexes. The plot shows that the absolute value of the free energy of formation of complexes is greater than that of point defects, i.e. that there is always the tendency to spontaneous ordering of these defects in the crystal.

1.4.1 The formation of complexes and defect clusters without reconstruction of the lattice

Following our classification in Table 3, we shall discuss in the present section the first of the three mechanisms for formation of extended defects, which does not involve local reconstruction of the crystal structure.

An example of the extreme type of compounds in which disordered point defects can occur in considerable concentrations are oxides of divalent titanium and vanadium TiO and VO [45, 59–61]. In these unusual chemical compounds defects occur in both the cationic and the anionic sublattices and the concentration of cation and anion vacancies is equal to about 30 at.%, when the compounds have the stoichiometric composition. The deviations of these oxides from stoichiometry can vary within wide limits ($0.8 < y < 1.25$: $Ti_{1\pm y}O$; $V_{1\pm y}O$). When this deviation is very pronounced the concentration of defects is even higher than 30 at.%. Figure 67 shows the dependence of the concentration of cation and anion vacancies in these compounds on the deviation from stoichiometry. It follows from the graph in Fig. 67 that the concentration of defects in these oxides can

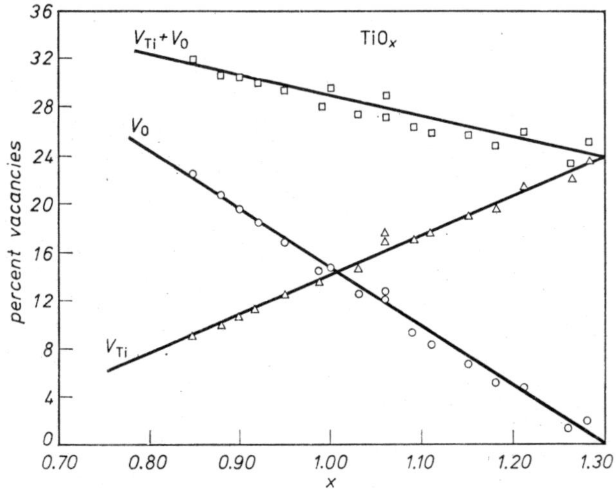

Fig. 67. Concentration of cation and anion vacancies in titanium monoxide as a function of composition (after [48]).

be unusually high ($\ll 40$ at.%). The results of calculations carried out by Van Gool [62], which are a good illustration of the dependence of the concentration of defects on the composition of titanium oxide, have been collected in Table 4.

The possibility of unusually high concentrations of point defects occurring in the oxides of titanium and vanadium is due to their peculiar electronic strustures. The *d*-orbitals of titanium and vanadium ions, in the crystal lattices of these oxides, are so diffuse that they overlap and as a result a strong metallic bond is formed between the cations. This means that the excess positive or negative charge resulting from the deviation from stoi-

TABLE 4

Concentration of point defects in titanium monoxide

Stoichiometric formula	Concentration of vacancies in cation sublattice %	Concentration of vacancies in anion sublattice %	Total concentration of defects in the crystal lattice of TiO_x %
$TiO_{1-0.284}$	6.1	32.8	19.5
$TiO_{1-0.140}$	10.5	23.1	16.8
$TiO_{1-0.108}$	10.8	21.3	16.0
$TiO_{1-0,030}$	14.3	16.8	15.6
$TiO_{1-0.005}$	14.6	15.0	14.8
$TiO_{1+0.060}$	17.2	12.2	14.7
$TiO_{1+0.116}$	18.5	9.1	13.8
$TiO_{1+0.150}$	20.1	8.2	14.2
$TiO_{1+0.195}$	21.2	5.8	13.5
$TiO_{1+0.200}$	22.1	6.5	14.3
$TiO_{1+0.250}$	23.4	4.2	13.8

chiometry (quasi-free electrons or electron holes) is completely delocalised and is not trapped in the form of cations in the $+1$ or $+3$ state. Thus the changes in the stoichiometric composition only cause a rise or a fall of the Fermi level in the forbidden energy band. For this reason these oxides are sometimes called quasi-metals.

The absence of localisation of quasi-free electrons and electron holes in the crystal structures of TiO and VO causes a strong screening of the electric charges of cation and anion vacancies by the electron defects. For this reason the Coulombic interaction forces between the defects are considerably weakened and as a result the existence of a large number of disordered defects is favoured. However, it has been found that even at temperatures close to 1000°C the interactions between defects in TiO and VO are very considerable [48, 49]. It has been shown on the basis of electron diffraction studies that at temperatures below 1000°C non-stoichiometric oxides TiO and VO with apparently disordered distribution of point defects are transformed into ordered phases having definite stoichiometric compositions. The non-stoichiometric titanium oxide TiO_x separates into two ordered phases having different compositions, corresponding to the formulae $TiO_{1.00}$ and $TiO_{1.25}$. Their structures are shown schematically in Fig. 68. It should be noted that both substances form superstructures on the basis of the lattice of the NaCl type in which all the sites of the parent lattice are filled. In this case the cation and anion vacancies are not defects, but are integral elements of the lattice. They constitute the third component of the completely ordered crystal structure. As a result of the similarity to the crystal lattice of NaCl the unit cells of the two kinds

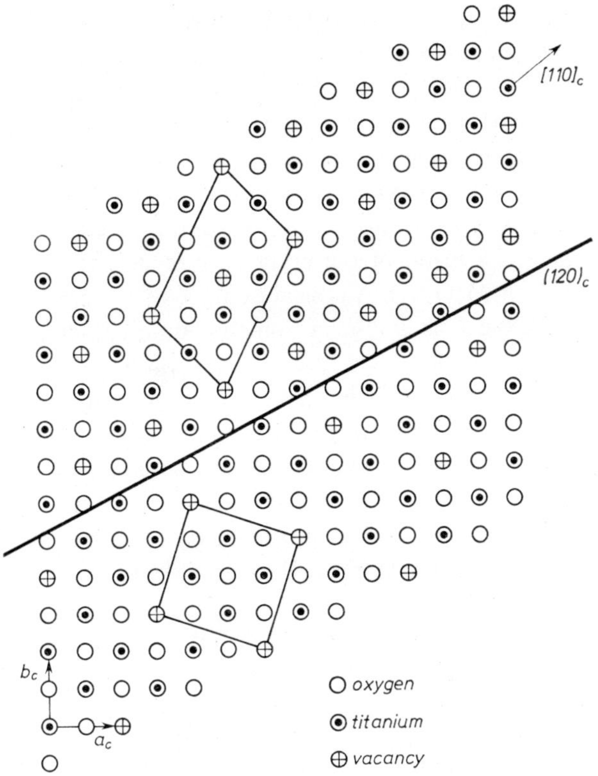

Fig. 68. The scheme of coherently intergrown TiO ($Ti_{0.8}O_{0.8}$) and $TiO_{1.25}$ ($Ti_{0.8}O$) structures (after [48]).

of titanium oxide have similar dimensions and for this reason they can grow coherently (see Fig. 68).

As was mentioned above, the formation of superstructures takes place at temperatures below 1000°C. However, Anderson [48] has pointed out that even in high-temperature non-stoichiometric TiO_x and VO_x phases, above 1000°C, point defects are not distributed completely at random. The fact that these two oxides are completely immiscible in the solid state and other observations indicate that under these conditions short range ordering exists, i.e. that defect complexes are formed. If the defects in both phases were distributed at random, the phases would be able to form solid solutions in the whole range of concentrations, since the structures of TiO_x and VO_x are similar, and they contain the same type of defect at the same concentration.

The formation of defect clusters at high temperatures has also been observed in non-stoichiometric niobium carbide NbC_{1-y} [48]. In this case a strong interaction between defects results in preferential localisation of anion vacancies in third nearest neighbour positions.

Another kind of complex or defect cluster, without reconstruction of the lattice, occurs in oxides of the MeO_2 type with the fluorite structure. These oxides show deviations from stoichiometry in the direction of excess or deficit of oxygen as a result of the occurrence of defects in the anionic sublattice (MeO_{2+y} or MeO_{2-y}). A typical example of a compound showing a real excess of oxygen is non-stoichiometric uranium dioxide UO_{2+y}.

It follows from the theory of point defects that the predominant defects in the crystal structure of this compound are interstitial oxygen anions. Their concentration expressed in mole fractions should be equal to one half the deviation from stoichiometry and, for this reason, the determination of the dependence of the deviation from stoichiometry on temperature and equilibrium oxygen pressure should provide direct information concerning the concentration and the degree of ionisation of interstitial oxygen anions as well as the enthalpy and the entropy of their formation. In agreement with this conclusion, the results of early studies seemed to indicate that a high temperatures the disorder in the uranium dioxide lattice consists of statistically distributed interstitial oxygen anions. However, Willis [63, 64] has shown that the defect structure in this oxide is much more complex than that expected from the classical model. The excess interstitial oxygen ion causes deformations in the anionic sublattice in its neighbourhood, which result in the displacement of the two nearest oxygen ions from their lattice sites to interstitial positions. This leads to the formation of two anion vacancies, since the anionic sublattice in the fluorite structure is so closely packed that oxygen ions cannot occupy interstitial positions without influencing their neighbours. Thus, as a result of the strong repulsive forces between the anions, the presence of one excess oxygen anion leads to the formation of at least four additional defects, thus forming the Willis complex. In the simplest case this consists of three oxygen ions in interstitial positions and two anion vacancies. A diagram of this complex, also decribed by the formula 2:1:2 (two anion vacancies,

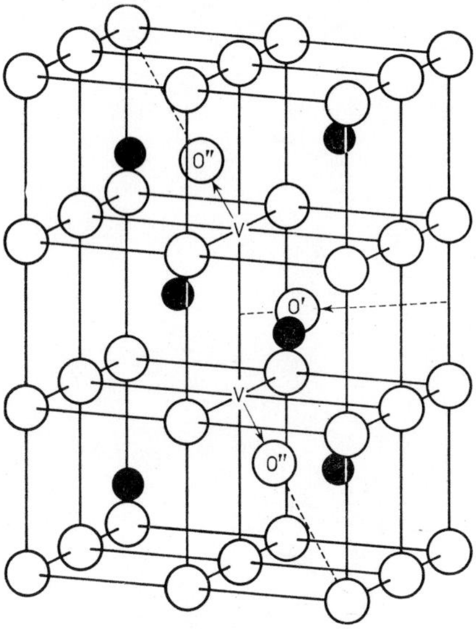

Fig. 69. The scheme of Willis complex (2:1:2 type) (after [48]).

an excess interstitial anion and two interstitial oxygen anions displaced from their lattice sites) is shown in Fig. 69.

The Willis complexes are not always so simple and clusters of the 2:2:2 and 3:2:3 types are frequently encountered [65]. A complex of the 2:2:2 type is shown in Fig. 70. Depending on temperature and oxygen pressure these complexes can interact and form more or less ordered assemblies (multiclusters) which lead to the formation of a superstructure and a new phase of the U_4O_9 type. It is noteworthy that each excess oxygen ion causes the formation of one Willis complex of the 2:1:2 type, whereas two additional oxygen ions cause the formation of larger complexes of the 2:2:2 or 3:2:3 type. For this reason the number of point defects is many times greater than that corresponding to the deviation from stoichiometry and in addition there are not only interstitial anions but also anion vacancies. Finally all these defects can form complexes and hence the possibility of application of the classical theory of point defects to these cases is completely excluded.

The type of disorder discussed above is not limited to oxides having the fluorite structure, but also occurs in metal fluorides and hydrides forming the same type of crystal structure. In particular Willis complexes occur

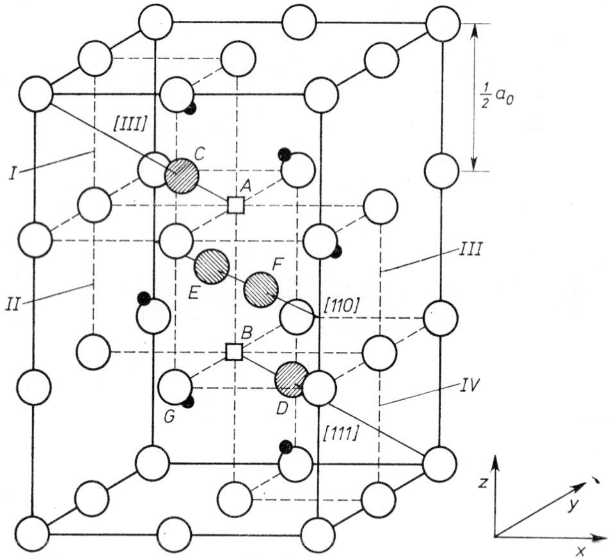

Fig. 70. The scheme of Willis complex (2:2:2 type) (after [48]).

in the crystal lattices of the mixed fluoride CaF_2–YF_3 and in hydrides of rare earth metals of the MH_2–MH_3 type [48].

As was mentioned above, the second group of oxides having the fluorite structure consists of compounds having a real deficit of oxygen MeO_{2-y}. They are mainly oxides of rare earth metals, such as CeO_{2-y}, PrO_{2-y} etc. According to the classical point defect model, this type of deviation from stoichiometry indicates that there are statistically distributed vacancies in the anionic sublattice and that their concentration, expressed in mole fractions, is equal to one half of the oxygen deficit. However, Bevan [66, 67] has shown that, as in the case of uranium dioxide containing an excess of oxygen, the concentration of quasi-free point defects in oxides of the MeO_{2-y} type is not related to the degree of deviation from stoichiometry. In this case the anion vacancies in the cubic coordination polyhedra in these substances are paired across the body diagonal of the cube. Thus the polyhedron contains two paired anion vacancies and 6 oxygen ions, which can be expressed by the formula MeO_6V_2. Since each anion site in the fluorite structure is coordinated to four cations, each of the two anion vacancies is common to three other polyhedra. Thus the whole constitutes a dumb-bell shaped cluster of 7 structural elements, in which the central cube contains two paired anion vacancies which are also components of the remaining 6 polyhedra of complex. This complex, called a Bevan cluster, can be expressed as MeO_6V_2 $(MeO_7)_6$. It is shown schematically in Fig. 71. Each lattice site in this cluster corresponds to a site in the fluorite

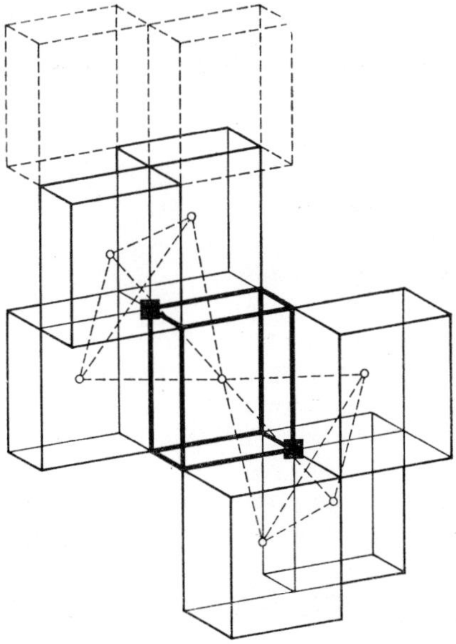

Fig. 71. The scheme of Bevan cluster (after [48]).

structure and the dimensions of the cluster are in agreement with those in the fluorite structure. The infinite repetition of the cluster, considered as the structural element, leads to the formation of a non-stoichiometric crystal of the MeO_{2-y} type.

The Bevan cluster is a kind of "lattice molecule", which may be "plugged into" the fluorite structure in place of the corresponding groups of MeO_8 polyhedra. Depending on temperature the clusters can be randomly distributed in the crystal lattice or they can form "strings", in which they are joined together along their trigonal axis. At high temperatures the presence of these clusters in the structures of hypostoichiometric MeO_{2-y} oxides is the reason for deviations from stoichiometry and at low temperatures it leads to the formation of a series of intermediate phases (e.g. in the case of Pr–O system) [68]. This is illustrated by Table 5. In the latter case we have a defect superstructure resulting from long range ordering of defects. It should be stressed that the anion vacancies paired in the central polyhedron of a Bevan cluster can no longer be regarded as point defects in the classical meaning of this term, as they are immobilised in the lattice and are an integral element of the crystal structure.

TABLE 5

The homologous series of intermediate phases formed in Pr–O system as a result of ordered superstructure of defects

Stoichiometric formulae of Pr_nO_{2n-2} series	n	x in PrO_x	The homogeneity range	Temperature °C
Pr_2O_3	4	1.500	1.500–1.503	1000
Pr_7O_{12}	7	1.714	1.713–1.719	700
Pr_9O_{16}	9	1.778	1.776–1.778	500
Pr_5O_9	10	1.800	1.799–1.801	450
$Pr_{11}O_{20}$	11	1.818	1.817–1.820	430
Pr_6O_{11}	12	1.833	1.831–1.836	400
PrO_2		2.000	1.999–2.000	400
PrO_2		2.000	1.750–2.000	1000

1.4.2 The formation of complexes and defect clusters with reconstruction of the lattice

The formation of extended defects in metal oxides, with large deviations from stoichiometry in the direction of a deficit of metal ($Me_{1-y}O$), causes local reconstruction of the crystal lattice. This reconstruction, which does not take place in the case of the oxides discussed above with the fluorite structure, is caused by strong interactions between the cation vacancies and the metal cations which surround this defect. So far, this phenomenon has only been investigated in the case of ferrous oxide $Fe_{1-y}O$ (wustite) [69–76], but there are reasons for believing that in many other oxides of the $Me_{1-y}O$ type and particularly in manganous oxide the formation of complexes and defect clusters takes place in a similar way.

The range of homogeneity of the wustite phase is unusually wide ($Fe_{0.95}O$–$Fe_{0.85}O$) and, for this reason, this oxide has for many years been the subject of extensive experimental studies and theoretical considerations.

On the basis of X-ray studies [77], it has been known for a long time that deviations from stoichiometry are caused by the presence of defects in the cation sublattice, whereas the anion sublattice is almost perfectly ordered. However, it is believed that the very large deficit of metal observed in wustite at high temperatures is due to cation vacancies, which do not interact with each other and are statistically distributed in the crystal lattice. This conclusion has been drawn from the earlier studies on electrical conductivity of wustite and from the deviations from stoichiometry, since the dependence of both these properties on the partial pressure of oxygen can be expressed by the following empirical equations:

$$\sigma = \text{const}\, p_{O_2}^{1/6} \tag{1.4.5}$$

and

$$y = \text{const}\, p_{O_2}^{1/6} \tag{1.4.6}$$

These relationships are the direct result of application of the law of mass action to the following defect equilibrium:

$$\tfrac{1}{2}O_2 \rightleftharpoons V_{Fe}'' + 2h^{\cdot} + O_O \tag{1.4.7}$$

More recent investigations [71, 76, 78, 79], in which more exact methods were used and which were carried out in a wide range of temperature and oxygen pressure, have shown that the dependence of σ and y on oxygen pressure is more complex and cannot be interpreted on the basis of the classical model of point defects. The results of studies on chemical diffusion coefficients has led to the same conclusions [80–82].

However, the elucidation of the real structure of defects in ferrous oxide only became possible on the basis of the results of modern X-ray and neutronographic studies [69–72, 76] and the utilisation of the Mössbauer effect [83]. The pioneering work carried out by Roth is of particular importance, since its results caused a radical change in the established views on the defect structure in the wustite phase. These studies have shown that not only cation vacancies but also interstitial cations occur in the cation sublattice of wustite and that the concentration of the vacancies is about twice that resulting from the deviation from stoichiometry. Roth was able to prove that a certain number of trivalent iron cations representing electron holes leave their octahedral positions in the lattice and migrate to tetrahedral interstitial positions. As a result additional cation vacancies are formed, but their presence in the crystal lattice is not related to the deviation from stoichiometry. An interstitial Fe^{3+} cation is related to the vacancy resulting from its displacement and, in combination with another cation vacancy resulting from chemical defect formation described by equation (1.4.7), it forms a complex called the Roth complex. This can be expressed by the formula Fe_TV_2, where Fe_T stands for trivalent

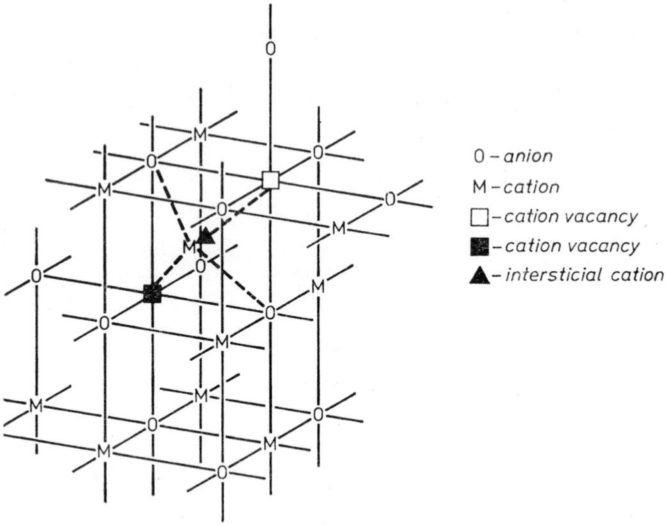

Fig. 72. The scheme of Roth complex (after [48]).

iron cation situated in a tetrahedral position and V_2 for two octahedral cation vacancies. A diagram of the complex is shown in Fig. 72.

It follows from the diagram that the structure element containing the Roth complex consists of four octahedral oxygen anions, two octahedral cation vacancies, two iron ions in the same environment and one trivalent cation in a tetrahedral position. This is similar to an element of the spinel structure of magnetite Fe_3O_4 and for this reason the complex can be formally regarded as a submicrodomain of magnetite in the wustite phase [48].

In the high temperature region Roth complexes only interact with each other slightly and thus are randomly distributed in the crystal structure. However, Koch and Cohen [70] have shown that as the temperature decreases, the complexes approach each other and together with additional cation vacancies begin to form larger assemblies called Koch–Cohen clusters. A diagram of such a cluster is shown in Fig. 73. The diagram shows that the cluster consists of four interstitial Fe^{3+} ions situated in tetrahedral positions and 13 octahedral cation vacancies ($Fe_{T4}V_{13}$). This indicates that during the process of formation of Koch–Cohen clusters considerable aggregation of cation vacancies takes place. In fact, within the cluster there are no iron ions in octahedral positions, since all 13 cation sites of the wustite lattice in this fragment of the structure are empty.

An structure element of a Koch–Cohen cluster (Fe_TV_4) is shown in Fig. 74 in order to demonstrate its similarity to Roth complexes (Fe_TV_2).

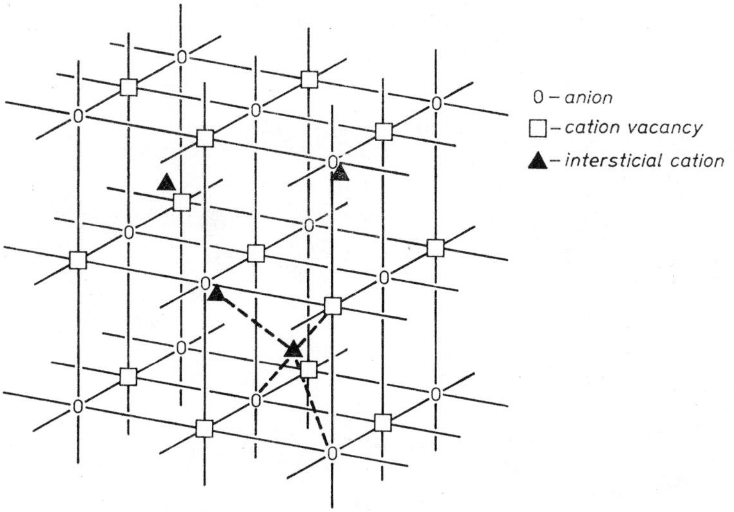

Fig. 73. The scheme of Koch–Cohen cluster (after [48]).

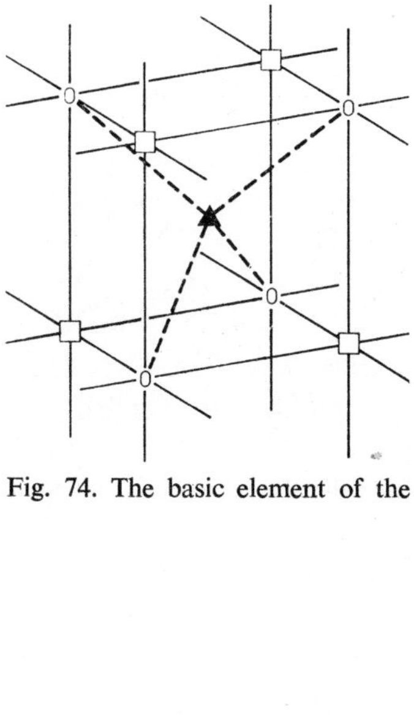

Fig. 74. The basic element of the Koch–Cohen cluster.

It can be seen, by comparing Figs. 72 and 74, that Koch–Cohen clusters are formed from Roth complexes as a result of aggregation of cation vacancies followed by the combination of four elements.

It should be pointed out that, due to the high concentration of cation vacancies, Koch–Cohen clusters have an effective negative charge with respect to the lattice (within the cluster there are 14 oxygen anions and only four ferric ions). For this reason the clusters are surrounded by electron holes, i.e. ferric ions situated in octahedral positions. ThusKoch–Cohen clusters are surrounded by an "atmosphere" of electron holes, which compensates their negative electrical charge. Figure 75 shows a Koch–

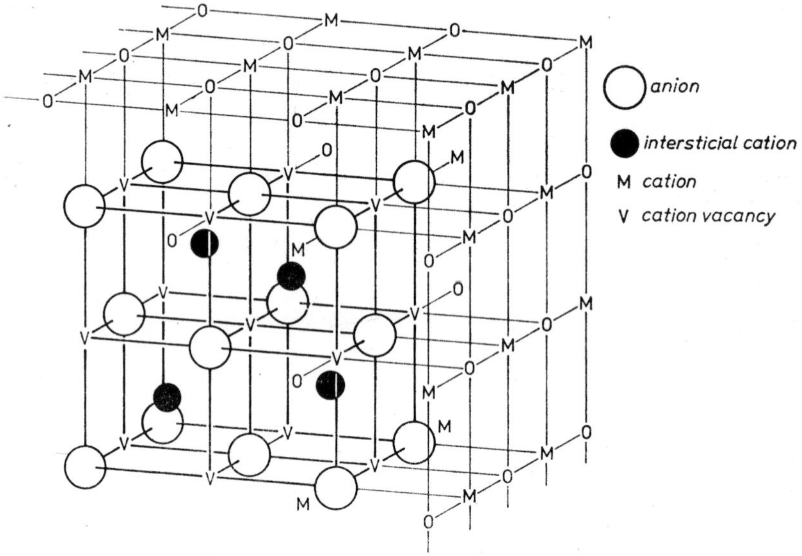

Fig. 75. The scheme of a Koch–Cohen cluster with a fragment of the lattice containing the compensating electron holes " atmosphere" (after [50]).

Cohen cluster and its neighbourhood containing the "atmosphere" of electron holes. As the temperature decreases the clusters associate to form multiclusters, which leads to the formation of a superstructure. Depending on the concentration of cation vacancies, i.e. on the extent of the deviation from stoichiometry and also depending on temperature, the association of clusters can take place in various ways, but the distance between them can never become smaller than the thickness of the "atmosphere" of electron holes. Figure 76 schematically shows one of the ways of forming multiclusters from Koch–Cohen clusters. The smaller cubes shown by the thick lines are Koch–Cohen clusters and the larger ones are the clusters together with their compensating atmosphere of electron holes.

The above discussion leads to the conclusion that above a certain critical concentration the interactions between defects results in the formation

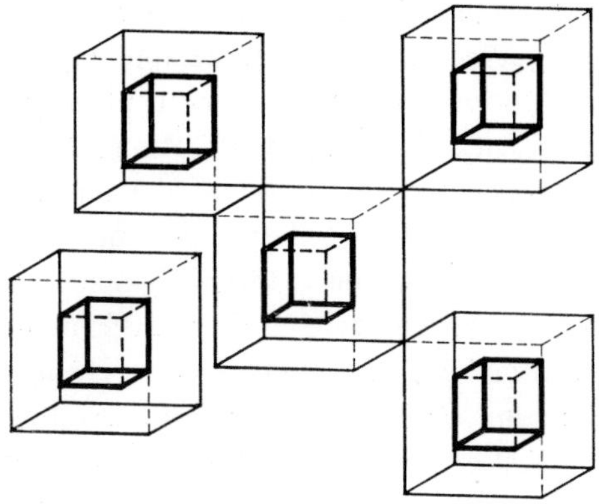

Fig. 76. The scheme of linkage of Koch–Cohen clusters in the wustite structure (after 48]).

of less numerous but often very complex defect assemblies. The tendency to spontaneous agglomeration of point defects is due to the fact that the decrease of energy of the crystal resulting from this process is much larger than the corresponding decrease of entropy, and, for this reason, crystals containing extended defects are thermodynamically more stable. However, the tendency to form extended defects decreases with increasing temperature as a result of the rapid increase of the numerical value of the product $T\Delta S$ ($\Delta G = \Delta H - T\Delta S$). For this reason at high temperatures crystals can contain a considerable number of point defects, with a quasi-disordered distribution. As a rule, the concentration of such defects is not very high, with the exception of titanium and vanadium oxides TiO_x and VO_x, which were discussed above. Thus, even at very high temperatures point defects in crystals with considerable deviations from stoichiometry form complexes which at lower temperatures combine to form less numerous but larger clusters and multiclusters, so that the overall process results in the formation of a superstructure and a homologous series of intermediate phases.

The formation of complexes, clusters and multiclusters of defects leads to submicro-heterogeneity of the solid, since submicro-domains having an atomic configuration (local symmetry) corresponding to a different phase appear in the crystal structure. For example, a Willis complex in the non-stoichiometric uranium oxide UO_{2+y} can be regarded as an element of U_4O_9 phase and a Bevan cluster in PrO_{2-y} is a submicro-domain of the Pr_7O_{12} phase. In non-stoichiometric ferrous oxide $Fe_{1-y}O$, Roth complexes can be regarded as elements of the magnetite phase, Fe_3O_4, in the wustite phase etc. However these submicro-domains cannot be regarded as crystal nuclei of a foreign phase within a definite crystal structure. Although they represent "inclusions" of a different phase in the structure of a given crystal, their existence does not imply the existence of an intermediate state before the formation of the thermodynamically more stable system of another phase. In other words, the presence of Willis complexes in UO_{2+y} does not mean that the oxide is thermodynamically unstable under the given conditions and that it tends to transform to U_4O_9. The same applies to Roth complexes and Koch–Cohen clusters in the wustite phase. Submicro-domains of the foreign phase, resulting from the agglomeration of point defects, remain in a state of dynamic equilibrium with the parent lattice and continuously change their size, positions and orientation. Thus the phase boundary, in the classical meaning of this term, between the parent lattice and the micro-domains does not exist. This has caused conceptional and semantic difficulties in defining these systems, particularly in defining the concept of a phase, since two different crystal structures—the micro-domains and the parent lattice—grow together coherently and continuously change their sizes and positions. The boundary between the two structures is then—owing to the incessant dynamic fluctuation of the domains—completely fluid. The system, although at dynamic equilibrium, is at the same time neither a homogeneous solid solution on the atomic scale nor a heterophase mixture in the usual meaning of this term. For this reason, Greenwood believes that the submicro-heterogeneity of crystals, caused by the presence of extended defects, should be regarded as a stable intermediate state between classical solid solutions and heterogeneous mixtures.

1.4.3 The formation of extended defects by crystallographic shearing

The processes of formation of extended defects, discussed so far, consists of the accumulation of point defects caused by mutual interactions or interactions between the defects and atoms or ions of the parent lattice. The formation of complexes and defect clusters taking place without the reconstruction of the lattice (Willis and Bevan complexes) or with local changes in the lattice (Roth complexes, Koch–Cohen clusters) does not results in the elimination of point defects but in their assimilation in the crystal structure. In some cases the number of point defects is even considerably increased during the formation of complexes or clusters, e.g. Willis and Roth complexes. It should be stressed that the point defects assimilated in complexes and clusters do not show the features asrcibed to them by the classical theory of defects since they are inherent elements of extended defects. However there is a large group of transition metal oxides in which extended defects are formed in an entirely different way. This group consists mainly of higher oxides of titanium, vanadium, niobium, chromium, molybdenum and tungsten. Until recently it was believed that the large deviations in these oxides (TiO_2, V_2O_5, Nb_2O_5, MoO_3, WO_3) from stoichiometry in the direction of deficit of oxygen were due to the presence of anion vacancies (TiO_{2-y}, MoO_{3-y} etc.).

Early studies of the deviations of titanium oxide, niobium oxide and other oxides from stoichiometry were thought to lead to the conclusion that the defect structure in these substances could be interpreted satisfactorily by means of the simple thermodynamic point defect model [84]. However, in the nineteen fifties Magnali [85–87] and then Wadsley [88–90], Anderson [91–94] and other authors had already suggested that defects present in crystals of higher oxides of metals of groups IV *B*, V *B* and VI *B* of the periodic table have an entirely different character and that the large deviations from stoichiometry are not real. Crystallochemical considerations led these authors to the conclusion that the open (loose) structure of the above oxides does not favour a large number of anion

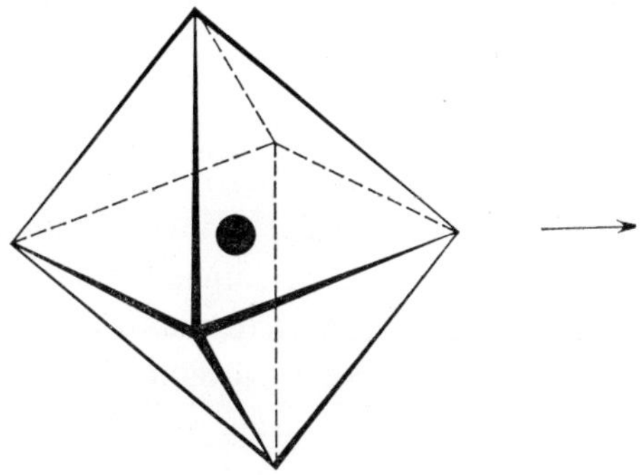

Fig. 77. The scheme of an oxygen octahedron—the basic unit of the crystal lattice of many refractory metal oxides.

vacancies. The basic element of crystal lattices in such oxides is a MeO_6 octahedron, typical of the rhenium oxide structure ReO_3 (Fig. 77). In WO_3 and MoO_3 these octahedra are joined by their apices and form three-dimensional lattices, whilst in TiO_2 and VO_2 they are joined by their edges and form chains, connected by corners of the octahedra. In

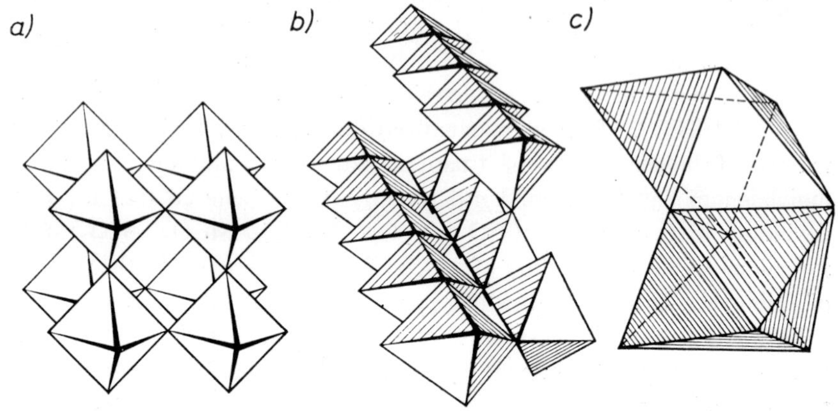

Fig. 78. The scheme of structures of the type of ReO_3, TiO_2 and Ti_2O_3, in which the basic element is the oxygen octahedron shown in Fig. 77.

Ti_2O_3 and Cr_2O_3 the octahedra are connected by common faces. These lattices are shown schematically in Fig. 78.

The characteristic feature of these structures is the octahedral coordination of cations, which is the basic unit in crystal chemistry for oxides of metals of groups IV *B*–VI *B*. The only difference between the crystal structures of higher and lower oxides is the coordination number of anions, which changes from 2 to 3 on passing from oxides of the MeO_3 type to those of MeO_2 type and from 3 to 4 on passing from structures MeO_2 to Me_2O_3. The lattices of the rhenium oxide type are the most open. The presence of anion vacancies in such lattices (WO_3, MoO_3) is unfavourable from the energy point of view and causes a decrease of their stability. Hence, it could be expected that point defects would be eliminated during the process of local reconstruction of the lattice leading to the formation of planar defects. Formally, this process can be regarded as a result of removal of a whole sheet of anion sites from the crystal, which causes mutual approach of the coordination polyhedra along a definite crystallographic plane. This process and the resulting planar defects are called crystallographic shear and crystallographic shear plane defects respectively [48, 95].

It should be stressed that crystal lo chemists had predicted the existence of this type of defect long before they could observe it experimentally. Only in recent years has high resolution electron microscopy made it possible to obtain direct pictures of this type of imperfection in crystal structures. A similar situation existed in the nineteen thirties when Taylor, Orowan [97] and Polanyi [98] predicted the existence of dislocations in metal crystals.

The mechanism of formation of shear structures has not yet been completely elucidated and is the subject of intensive study. However, there is no doubt that the mechanism proposed by Wadsley and Anderson [99] is one of the ways that crystallographic shear planes in the crystals of these oxides are formed. Anion vacancies introduced in a crystal, with for instance the rhenium oxide structure, have a tendency to become ordered as shown in Fig. 79*a*). The diagram represents a cross-section of the ReO_3 structure in which the empty circles represent the positions of anion vacancies. In this configuration of anion vacancies the crystal structure breaks along

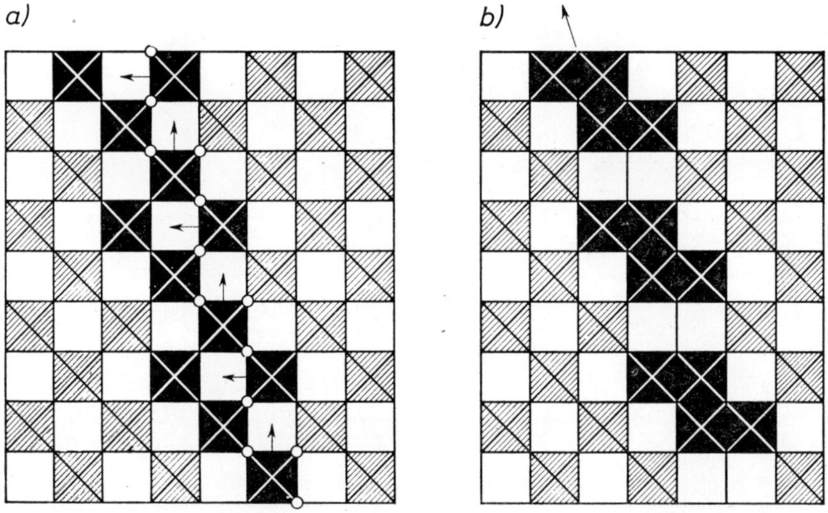

Fig. 79. The scheme of crystallographic shear plane formation (Wadsley defect) (after [99]); *a*)—an ordered arrangement of anion vacancies (o) in ReO_3 structure. (The shaded squares stand for octahedra sharing apexes); *b*)—after the formation of the Wadsley defect.

the octahedra marked in black. The octahedra approach each other and, as a result, the anion vacancies are eliminated and a crystallographic shear plane is formed (Fig. 79*b*). It can be seen from Fig. 79*b*) that such a plane consists of a system of octahedra joined by their edges along a definite crystallographic direction. The ratio of the number of cations to that of anions within the shear plane is obviously greater than in a stoichiometric oxide of the MeO_3 type, since the above crystallographic shear plane structure is of the MeO_2 type.

The crystallographic shear process in crystals of oxides with the rutile structure takes place similarly. However, in this case the coordination octahedra approach each other by changing from edge-sharing to face-sharing octahedra ($TiO_2 \rightarrow Ti_2O_3$).

The above considerations lead to the conclusion that the generation of a shear plane in the crystal causes a deviation from stoichiometry in the direction of a deficit in oxygen (or excess of metal), which is not related to the presence of point defects. As the number of shear planes increases,

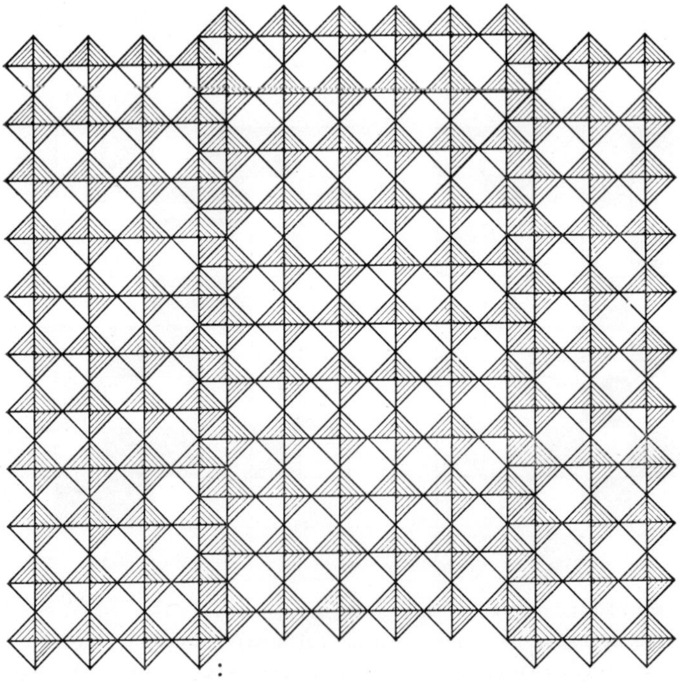

Fig. 80. The scheme of ReO_3 structure containing ordered shear planes.

they begin to interact with each other, which leads to their ordering along definite crystallographic directions. The layers of crystal remaining between the shear planes retain their original structure. This is illustrated by the diagrams in Figs. 80 and 81. When the thickness of the layers between the successive shear planes is exactly the same, a new crystallographic

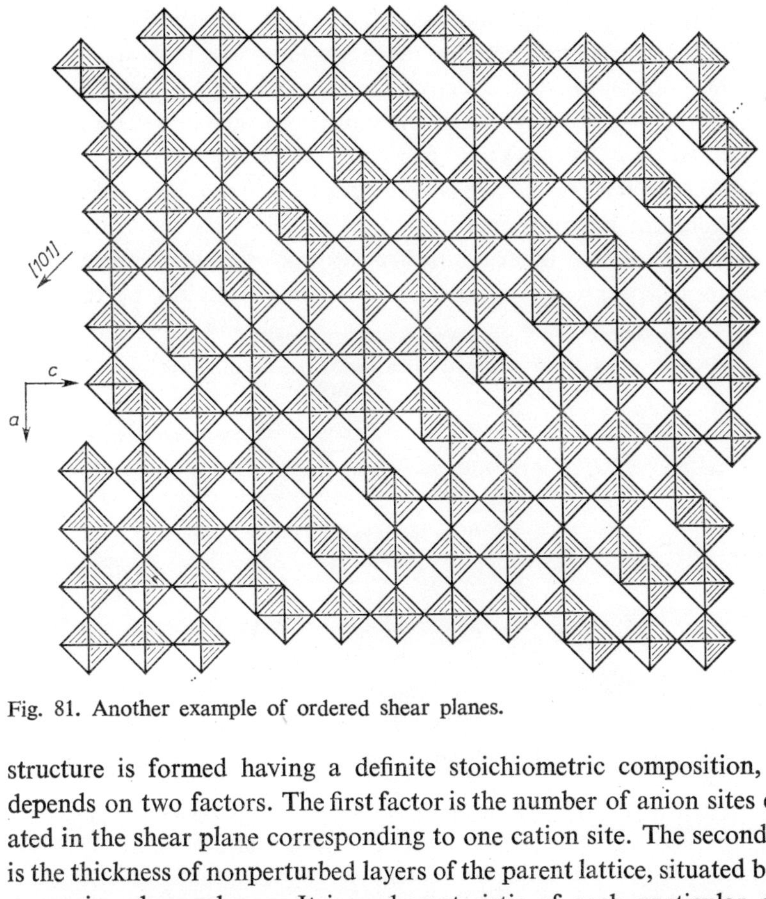

Fig. 81. Another example of ordered shear planes.

structure is formed having a definite stoichiometric composition, which depends on two factors. The first factor is the number of anion sites eliminated in the shear plane corresponding to one cation site. The second factor is the thickness of nonperturbed layers of the parent lattice, situated between successive shear planes. It is a characteristic of each particular crystallographic orientation and as a result of a whole series of intermediate phases, differing very slightly in their chemical composition, can be formed. In view of the formal similarity between these intermediate phases and the homologous series of hydrocarbons (C_nH_{2n+2}, C_nH_{2n}) they are called the homologous series of intermediate phases. They are also often called

Magneli phases, after the Scandinavian scientist, who first predicted their existence. Thus, there are two homologous series of intermediate phases derived from the rhanium oxide structure, corresponding to the formulae Me_nO_{3n-1} and Me_nO_{3n-2}, which are the series of the intermediate phases of molybdenum and tungsten oxides. An analogous series derived from the rutile structure has the general formula Me_nO_{2n-1}. In these formulae the symbol n stands for the number of octahedra in the layers of the parent structure, contained between neighbouring shear planes.

When the thickness of the layers of the parent lattice situated between successive shear planes is not exactly the same, a two-dimensional disorder, causing a new kind of deviation from the stoichiometry of the parent phase, appears in the crystal. We shall return to this problem later.

The existence of homologous series of intermediate phases in transition metal oxides, predicted by crystallochemists and crystallographers in the early nineteen fifties, was fully confirmed experimentally during the nine-

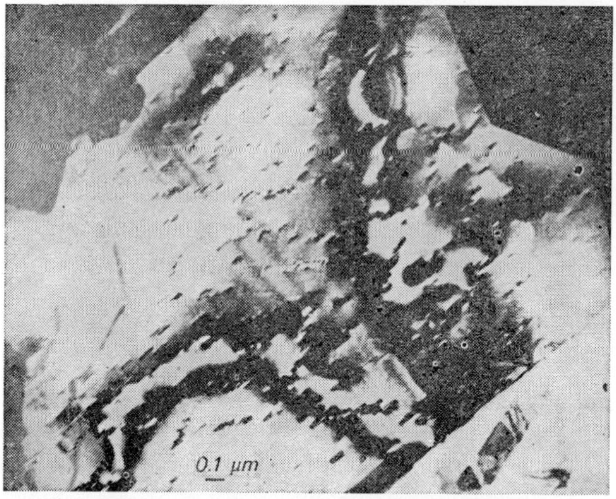

Fig. 82. The crystallographic shear planes in molybdenum oxide (MoO_{3-y}). The picture obtained in a high resolution electron microscope (after [177]).

teen sixties, thanks mainly to the fundamental work carried out by Anderson and his collaborators [100, 101], Allpress [102, 106] and Amelinck [106–108]. By using modern high resolution electron microscopy, they were able not only to prove the existence of definite intermediate phases but also to follow the processes of their formation. Selected photographs illustrating the results of these studies are shown in Figs. 82–85.

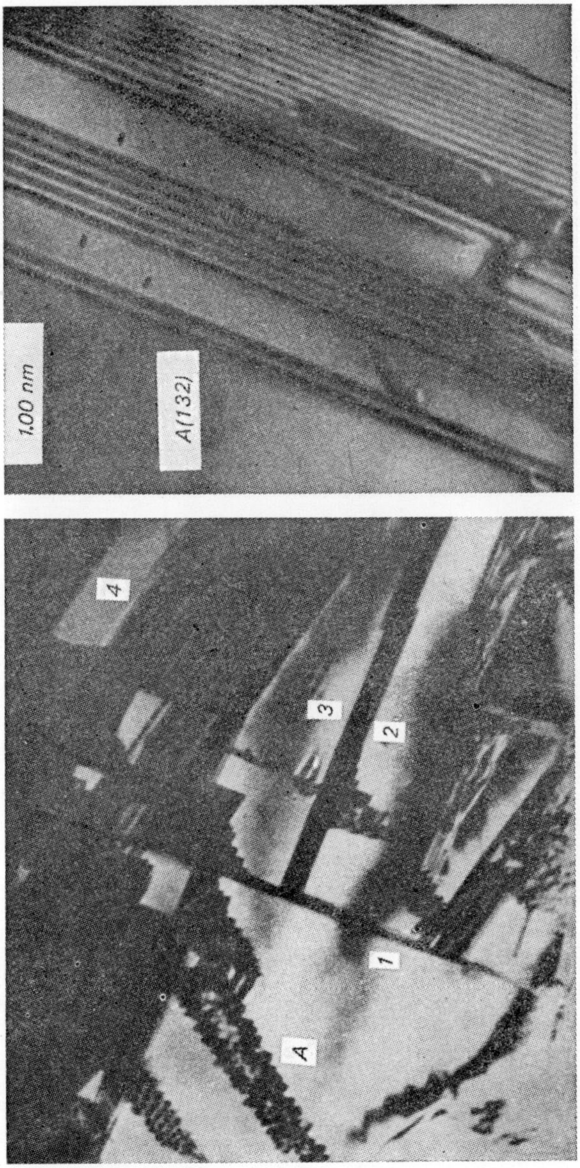

Fig. 83. The crystallographic shear planes in titanium oxide (TiO_{2-y}) (after [762]); *a*)—random Wadsley defects; *b*)—ordered shear planes in $TiO_{1.9986}$.

The homologous series of intermediate phases of titanium oxides Ti_nO_{2n-1} has been the subject of the most extensive studies. The results of this work are collected in Table 6, which shows that the shear planes in two different regions of chemical composition of the intermediate phases

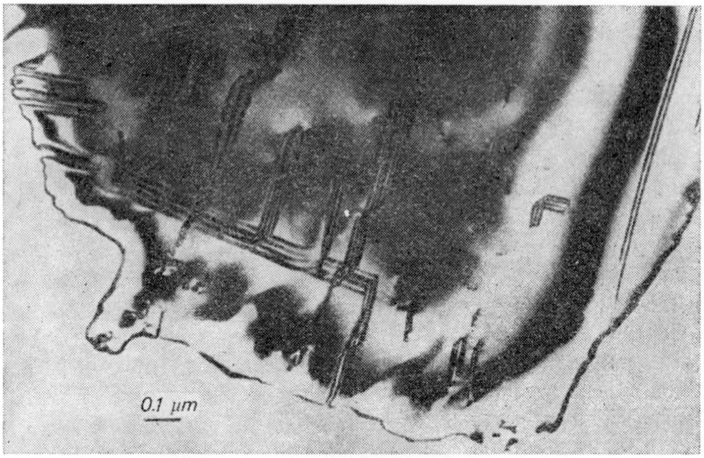

Fig. 84. Crystallographic shear planes in mixed oxide $W_3Nb_{14}A_{48}$ (after [49]).

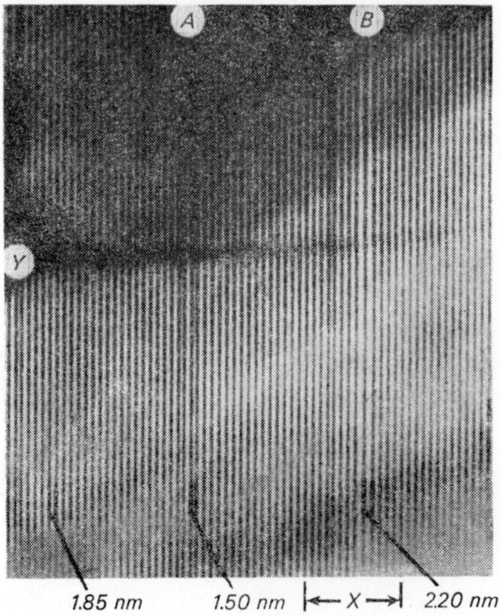

Fig. 85. Single shear planes in weakly reduced rutile (after [177]).

TABLE 6

Intermediate phases in Ti–O system

Stoichiometric composition	n in Ti_nO_{2n-1}	Type of disorder
$TiO_{1.750}$–$TiO_{1.900}$	$4 < n < 10$	Shear planes oriented along $(1\bar{2}1)$ direction
$TiO_{1.90}$–$TiO_{1.937}$	$10 < n < 16$	Shear plane orientation changes from $(1\bar{2}1)$ to $(1\bar{3}2)$
$TiO_{1.957}$–$TiO_{1.973}$	$16 < n < 37$	Shear planes oriented along $(1\bar{3}2)$ direction
$TiO_{1.973}$–$TiO_{1.999}$	$37 < n < 1000$	$(1\bar{3}2)$ shear planes grouped in layers
$TiO_{1.999}$–TiO_{2-x}	$n < 1000$	Wadsley defects in all $(1\bar{3}2)$ orientations
TiO_{2-x}–TiO_2	∞	Point defects in rutile structure at concentrations lower than $10^{-2}\%$

have different crystallographic orientations. In strongly reduced rutile ($TiO_{1.75}$–$TiO_{1.90}$), intermediate phases are formed as a result of the ordering of shear planes having the $[1\bar{2}1]$ orientation. In the intermediate concentration range ($TiO_{1.90}$–$TiO_{1.937}$) the orientation of the shear planes changes from $[1\bar{2}1]$ to $[1\bar{3}2]$. In this range of composition, successive intermediate phases are formed as a result of both the change of the orientation of the shear planes and the decrease in their concentration. In the composition range from $TiO_{1.937}$ to $TiO_{1.973}$ the shear planes are oriented exclusively in the $[1\bar{3}2]$ direction. The distance between them gradually increases and at the composition approximately corresponding to $TiO_{1.973}$, i.e. at n equal to about 37, it is approximately 4 nm. At this distance interactions between the shear planes are practically absent and as a result their ordered arrangement begins to disappear. For this reason in the next composition range, from $TiO_{1.973}$ to $TiO_{1.999}$, domains of rutile structure with randomly distributed shear planes having the $[1\bar{3}2]$ orientation appear. These grow coherently with domains containing assemblies of partly ordered shear planes of the same orientation. At an even smaller degree of reduction of rutile ($TiO_{1.999}$–TiO_{2-y}) only one phase, TiO_2, containing shear planes in all possible $[1\bar{3}2]$ orientations is present. In this case, the shear planes are so far apart that the interaction between them ceases and the defects are distributed in the crystal in complete disorder. Such an isolated shear plane is called a Wadsley defect. It is a two-dimensional defect complex (extended defect) extending through the whole crystal or involving only a fragment of it. The presence of Wadsley defects causes deviations from stoichiometry of the oxide phases by the mechanism entirely different from that resulting from the presence of point defects. Hyde and Bursill [109] have shown that point defects (anion vacancies) in slightly reduced titanium oxide are still eliminated by the crystallographic shear

TABLE 7

Homologous series of intermediate phases in Mo–O

Stoichiometirc formula in Mo_nO_{3n-1} series	n	x in MoO_x	The type of shear structure
MoO_2	1	2.000	Rutile structure; octahedra sharing edges
Mo_4O_{11}	4	2.750	Layers of ReO_3-type octahedra cells thick connected by tetrahedra
Mo_8O_{23}	8	2.875	8-octahedra thick connected to adjacent slobs by sharing edges
Mo_9O_{26}	9	2.889	9-octahedra thick connected to adjacent slobs by sharing edges
$Mo_{10}O_{29}$	10	2.900	10-octahedra thick connected to adjacent slobs by sharing edges
$Mo_{11}O_{32}$	11	2.909	11-octahedra thick connected to adjacent slobs by sharing edges
$Mo_{12}O_{35}$	12	2.916	12-octahedra thick connected to adjacent slobs by sharing edges
$Mo_{14}O_{41}$	14	2.928	14-octahedra thick connected to adjacent slobs by sharing edges
MoO_3		3.000	Layer structure of octahedra connected by shared corners and edges

process when the composition of rutile reaches $TiO_{1.9999}$. This leads to the conclusion that point defects are relatively unimportant in the disorder of crystal lattices of these transition metal oxides. According to Anderson [48] the concentration of anion vacancies in TiO_{2-y} is lower than 10^{-2} at.% ($y < 10^{-4}$).

A similar situation exists in the case of molybdenum, tungsten and niobium oxides as well as many mixed oxides of refractory metals (e.g. homologous series $Cr_2Ti_{n-2}O_{2n-1}$). This is illustrated by Table 7, which shows intermediate phases appearing as a result of crystallographic shear in a molybdenum–oxygen system. It can be seen that in the Mo_nO_{3n-1} homologous series there are at least 7 intermediate phases, which are formed during gradual reduction of molybdenum trioxide.

Tilley, Allpress and Sienko [102] have shown that in the case of tungsten trioxide the intermediate phases are already formed by heating this oxide in vacuo, i.e. even under weakly reducing conditions. A slight loss of oxygen caused by the vacuum leads to the formation of random shear planes (Wadsley defects) even when the composition corresponds to the formula $WO_{2.9999}$, i.e. with the concentration of anion vacancies lower than 10^{-2} at.%. The formation of intermediate phases of the homologous series W_nO_{3n-1} caused by ordering of Wadsley defects begins at the stoichiometric composition corresponding to the formula $WO_{2.9930}$. The concentration of Wadsley defects is then so high that the distance between neighbouring shear planes does not exceed 4–5 nm. In these conditions the interactions between the shear planes lead to their ordering which is equivalent to the formation of new intermediate phases.

In each of the above cases, the orientation of the shear planes (Wadsley defects) can vary and in the successive intermediate phases of a homologous series the orientation of ordered shear planes can change. However, the crystallographic shear vector remains constant and it is hence characteristic of each family of intermediate phases. Thus, in the titanium–oxygen system the disordered shear planes in the structure of rutile (TiO_{2-y}) have the $[1\bar{3}2]$ orientation. In the homologous series of intermediate phases appearing in this structure the orientation of the ordered shear planes gradually changes from [132] to $[1\bar{2}1]$ during the reduction process, but the shear vector remains unchanged and equal to $\frac{1}{2}$ [001]. Similarly in the molybdenum–oxygen system the random Wadsley defects in MoO_{3-y} phase have the [120] orientation, whereas in the first homologous series of intermediate phases of the general formula Mo_nO_{3n-1} the shear planes are oriented in the [211] direction. On passing to the second series, Mo_nO_{3n-2}, the orientation of the shear planes changes from [211] to [311] and in the third series, Mo_nO_{3n-3}, they are oriented in the [411] direction. The shear vector remains unchanged and is equal to $\frac{a}{2}, \frac{b}{7}, 0$. In the homologous series of tungsten oxides W_nO_{3n-2} the crystallographic shear vector is always equal to $\frac{1}{2}$[110], whereas the orientation of the shear planes changes from [120] to [130].

It should be stressed that the shear structures are not always so simple. In many cases, e.g. in the case of one of the intermediate phases in the

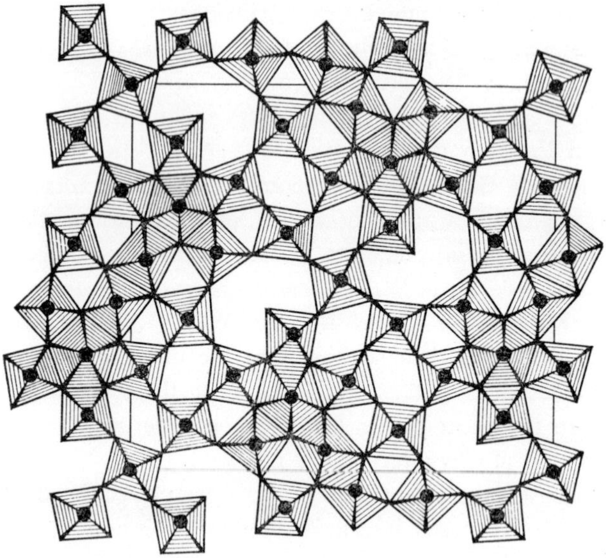

Fig. 86. The scheme of a complex shear structure in molybdenum oxide having the formula $Mo_{17}O_{47}$ (after [68]).

molybdenum–oxygen system (Fig. 86), the crystallographic shear process can give complex products.

A specific kind of shear structure, slightly different from that discussed above, often occurs in the binary niobium–oxygen system and in ternary systems based on niobium oxide, in particular in the Nb–Ti–O and Nb–W–O systems. The crystal lattices of these oxides are based on the rhenium oxide ReO_3 structure. In contrast to the crystallographic shear processes described above, the formation of intermediate phases in the Nb–Ti–O and Nb–W–O systems takes place as a result of simultaneous formation of two perpendicular systems of shear planes along the [100] and [001] directions. This makes it possible to form an enormous number of intermediate phases, which only differ slightly in their chemical composition. This process, called double shearing, leads to the formation of a so-called

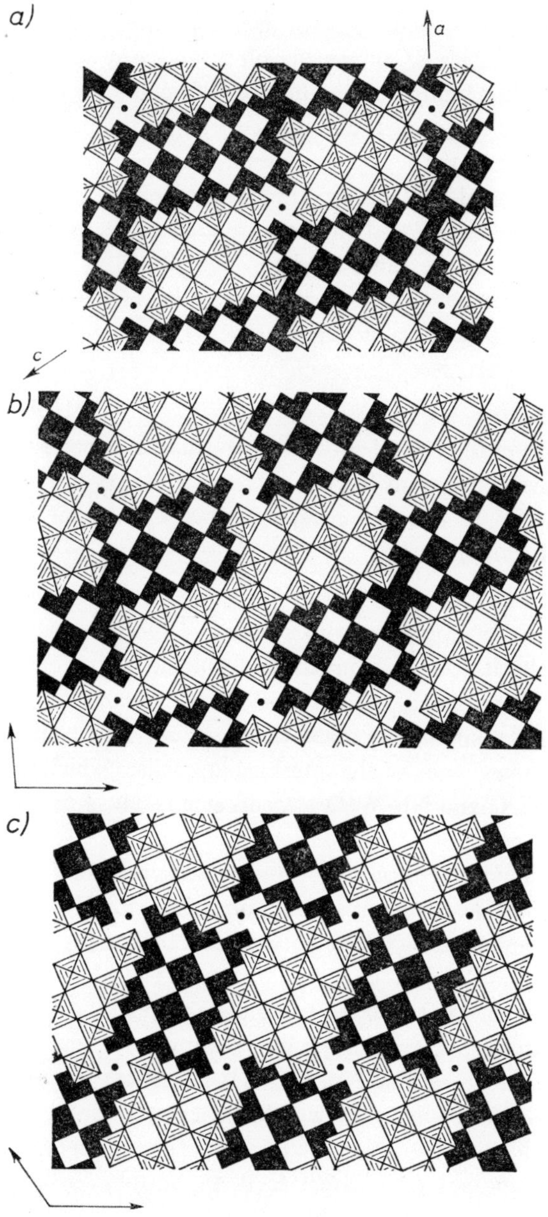

Fig. 87. The scheme of double shear (after [49]).

block structure, since, in structures of the ReO_3 type, rectangular blocks of octahedra appear. They are joined by their edges to the analogous blocks of the parent lattice. Figure 87 shows the block structure schematically. The size of each block is limited by the number of octahedra which it contains. The width and the height of the blocks depend on the numbers n and m of octahedra, whereas the length is unlimited, when the block extends through the whole crystal or it can be limited to a definite number of octahedra. The blocks can be isolated in the parent structure or they can be joined to other blocks at the same level through the edges of the octahedra. In this way groups containing p blocks can be formed. The chemical composition of a block (or a group of blocks) is given by the following formula:

$$Me_{nmp}O_{3nmp-pn+m+4} \tag{1.4.8}$$

The chemical composition of the crystal containing the defects is obtained by adding pairs of the above expressions for two systems of blocks situated in the structure at levels 0 and $\frac{1}{2}$. Thus the composition of the crystal can be calculated from the following formula:

$$n \cdot m_p + n' \cdot m'_{p'} \tag{1.4.9}$$

where n, m, p and n', m' and p' correspond to systems of blocks situated at levels 0 and $\frac{1}{2}$ respectively. It follows from these formulae that double shearing can lead to the formation of an extremely large number of intermediate phases having very similar chemical composition. This is illustrated in Table 8, which shows some of the intermediate phases which are formed in the Nb–O, Nb–Ti–O and Nb–W–O systems as a result of double shearing.

To end this section, we should mention the still unsolved problem of the phase composition of crystalline structures containing random Wadsley defects. As in the case of submicro-heterogeneity caused by the presence of non-ordered complexes and defect clusters, the presence of two-dimensional Wadsley defects makes it difficult to define the phase composition. Also the structure in this case is neither a mixture of two separate phases, nor can it be regarded as a single phase in the classical meaning of this term. This creates a problem as to whether a crystal containing random Wadsley defects at high temperatures is in a state of thermodynamic equilibrium with the surroundings or whether is it an unstable system. Allpress and Roth [105] have shown that the Nb–W–O system is in a state of apparent equilibrium. Heating of this double oxide for several months at a temperature close to its melting point caused an ordering of Wadsley defects in the crystal. On the basis of the results of these studies, Roth [110] suggested that crystals containing Wadsley defects cannot be regarded as thermodynamically stable systems. The change to the stable state takes

TABLE 8

Block structures in niobium oxides and in some ternary oxides

Compound	Structure	Anion to cation ratio
Niobium oxides		
$Nb_{12}O_{29}$	$(3\times4)_\infty$	2.417
$Nb_{22}O_{54}$	$(3\times4)_1+(3\times3)_\infty$	2.454
$Nb_{15}O_{37}$	$(3\times5)_\infty$	2.467
$Nb_{47}O_{116}$	$\{(3\times4)_1+(3\times3)_\infty\}+\{(3\times4)_2\}$	2.468
$Nb_{32}O_{79}$	$(3\times4)_2+(3\times3)_1$	2.469
$Nb_{25}O_{62}$	$(3\times4)_2$	2.480
$Nb_{32}O_{87}$	$(3\times4)_2+(3\times3)_1$	2.486
$Nb_{53}O_{132}$	$\{(3\times4)_1+(3\times5)_\infty\}+\{(3\times4)_2\}$	2.490
$Nb_{28}O_{70}$	$(3\times4)_1+(3\times5)_\infty$	2.500 H–Nb_2O_5
$Nb_{16}O_{40}$	$(4\times4)_\infty$	2.500 N–Nb_2O_5
Ternary oxides		
$TiNb_2O_7$	$(3\times3)_\infty$	2.333
$Ti_2Nb_{10}O_{29}$	$(3\times4)_\infty$	2.417
$TiNb_{24}O_{62}$	$(3\times4)_2$	2.480
$WNb_{12}O_{33}$	$(3\times4)_1$	2.538
$W_4Nb_{26}O_{77}$	$\{(3\times4)_1\}+\{(4\times4)_1\}$	2.563
$W_3Nb_{14}O_{44}$	$(4\times4)_1$	2.588
$W_5Nb_{16}O_{65}$	$(5\times5)_1$	2.635
Oxyfluorides		
Nb_3O_7F	$(3\times\infty)_\infty$	2.667
$Nb_{17}O_{42}F$	$(3\times5)_1+(3\times6)_\infty$	2.530
$Nb_{31}O_{77}F$	$(3\times5)_2$	2.516

place extremely slowly even at temperatures close to the melting point and for this reason the system appears to be at equilibrium. It has not yet been completely explained why the thermodynamic equilibrium in these systems is established so slowly, but it is believed that this is due to the fact that the process of ordering of Wadsley defects is accompanied by extremely small changes in the free energy of the crystal. Thus the crystals containing

Wadsley defects are very near the state of thermodynamic equilibrium, which can be reached only after tedious difficult operations.

It is noteworthy that the exact definition of the equilibrium state becomes more difficult as the sensitivity of the method used for its determination increases. In this connection, Roth quoted Wadsley's statement that the definition of the equilibrium state depends on the wavelength used in its determination. It should be stressed that this statement is equally valid for crystals containing two-dimensional planar defects (Wadsley defects) as for those containing three-dimensional complexes and defect clusters (submicro-heterogeneity).

1.5 Electrical conductivity of ionic crystals

It has already been mentioned that the presence of point defects in ionic crystals of the MeX type is responsible for many of the physico-chemical properties of these compounds. Also their electrical conductance is related to the presence of point defects, since under the influence of thermal vibrations ionic and electronic defects move continuously (self-diffusion), and, under the influence of an external electrical field, this motion can become ordered. In contrast to metallic conductors, the conductivity of ionic crystals increases with increasing temperature, since it is directly proportional to the concentration of the charge carriers, i.e. the lattice defects. Such crystals are called semiconductors.

It was shown in Chapter 1.3.6 that the concentration of defects depends not only on temperature but also on the equilibrium pressure of the X component in the gas phase. Thus the electrical conductivity of ionic crystals is a function of both these parameters. Investigations of the dependence of electrical conductivity of ionic crystals on temperature and partial pressure of the X component in the surrounding atmosphere can provide valuable information about ionic and electronic defects as well as their concentration and mobility.

The electrical conductivity of crystals can be expressed by the following formula:

$$\sigma = c \cdot z \cdot q \cdot B \tag{1.5.1}$$

where σ is the specific conductivity ($\Omega^{-1} \cdot \text{cm}^{-1}$), c is the concentration of charge carriers (number of carriers/cm^3), z is the valency of the carriers, q is the elementary charge and B is the mobility (average velocity of the carrier in cm/s in an electric field with its intensity 1 V/cm) expressed in cm^2/sV.

The observed electrical conductivity of crystals is always the sum of the ionic and the electronic conductances:

$$\sigma_c = \sigma_j + \sigma_e = \sigma_k + \sigma_a + \sigma_{h^\bullet} \cdot \sigma_{e^-} \tag{1.5.2}$$

where σ_c is the total conductivity, σ_j and σ_e are the component ionic and electronic conductivities, respectively and σ_k, σ_a, $\sigma_{h^\bullet}$ and σ_{e^-} are partial cationic, anionic, hole and electronic conductivities, respectively.

Equation (1.5.2) can be also expressed in the following form:

$$\sigma = \sigma(t_j + t_e) = \sigma(t_k + t_a + t_{h^\bullet} + t_{e^-}) \tag{1.5.3}$$

where t_j and t_e are the transference numbers of ions and electrons and t_k, t_a, $t_{h^\bullet}$ and t_{e^-} are the transference numbers of cations, anions, electron holes and quasi-free electrons, respectively. It should be stressed that:

$$t_j + t_e = t_k + t_a + t_{h^\bullet} + t_{e^-} = 1 \tag{1.5.4}$$

This means that the partial electrical conductivity of the given substance is always the product of the total conductivity and the corresponding transport number:

$$\sigma_i = \sigma t_i \tag{1.5.5}$$

Semiconductors can be divided into two main groups: ionic semiconductors and electronic semiconductors. The first group consists of compounds having a stoichiometric composition, in which nearly all the defects are ionic. In this case, in spite of the fact that the mobility of electronic defects is many times higher than that of the ionic defects, the electric current is almost exclusively conducted by ionic defects ($t_j = 1$, $t_e = 0$) and for this reason semiconductors of this type are often called solid electrolytes. Alkali halides, halides of certain heavy metals and a few solid solutions of metal oxides have this type of conductivities under certain conditions.

The second group consists of all the metal oxides and sulphides with deviations from stoichiometry. In the crystals of this type the concentration of ionic defects is comparable to that of electronic defects. Since the mobility of electronic defects is several orders higher than that of ionic defects, the crystals of this type possess practically pure electronic conductivity. This group of semiconductors can be further divided into two subgroups of electronic semiconductors, in which the current carriers are different. One of these subgroups contains compounds with a deficit of metal or an excess of the oxidant ($Me_{1-y}X$ or MeX_{1+y}), in which the negative charge of the ionic defects is compensated by a corresponding number of positively charged electron holes. In these compounds the electric current is conducted almost exclusively by the positive charge carriers, i.e. the electron holes, and for this reason they are called *p* type (positive) semiconductors or hole semiconductors. The compounds with an excess of metal or a deficit of the oxidant ($Me_{1+y}X$ or MeX_{1-y}) are called *n* type (negative) semiconductors, since in this case the electric current is conducted practically exclusively by quasi-free electrons.

A separate group of electronic semiconductors consists of the so-called intrinsic semiconductors in which the electric current is conducted only by electronic defects, which are not connected with the presence of ionic defects in the crystal, but appear as a result of thermal excitation of electrons from the valence band to the conduction band [equation (1.3.91)]. These compounds have either strictly stoichiometric or very nearly stoichiometric compositions and they conduct the electric current by means of either quasi-free electrons or electron holes, depending on the mobility of these two kinds of defects. An example of this type of compound is cupric oxide CuO.

It should be stressed that the above division is largely formal, since it does not take into account the effect of external parameters on the semiconducting properties of the compound. It was shown in Chapter 1.3.6 that the predominant disorder in the crystal lattice can change under the influence of changes of temperature and pressure. For example a *p* type semiconductor can become *n* type or anionic semiconductor (a solid electrolyte). In principle, at a sufficiently high temperature every semiconductor can become an intrinsic electronic semiconductor, since, as the temperature increases, more electrons pass from the valence band to the conduction band (position of equilibrium in eq. (1.3.91) shifts to the right). Nevertheless the above division is very useful, since in each group of compounds one type of disorder predominates within relatively wide limits of temperature and oxidant pressure in the surrounding atmosphere.

Solid electrolytes are intrinsic ionic semiconductors, since the concentration of ionic defects in this type of crystal depends exclusively on temperature (disorder of the Frenkel or Schottky type). Investigations of electrical conductivity of these compounds and determinations of the transference numbers of cations and anions give direct information on the type of disorder in the crystal lattice and the concentration of defects. In this case the transference numbers can be determined by a relatively simple method based on solid state electrolysis in a suitably chosen system. In this way, it has been shown that the Frenkel type disorder occurs almost exclusively in silver halides, whereas in lead halides the disorder is of the anti-Frenkel type. The disorder in crystal lattices of alkali metal halides is of the Schottky type.

The determination of transference numbers is very difficult in the case of *p* and *n* type semiconductors, since in this case the electric current is conducted practically exclusively by electronic defects. For this reason measurements of electrical conductivity of this type of compounds do not provide direct information on the concentration of ionic defects. However, since in *p* and *n* type semiconductors the concentration of electronic defects is a function of that of ionic defects, the results of studies on electrical conductivity of these semiconductors lead to indirect conclusions regarding the character of ionic defects occuring in their crystal lattices. Particularly valuable information can be obtained from determinations of the dependence of the electrical conductivity on the partial pressure of oxidant in the atmosphere surrounding the crystal. If these determinations are carried out in conditions such that thermodynamic equilibrium between the crystal and its surroundings can be reached (at sufficiently high temperatures), then, in the case of *n* type semiconductors, the electrical conductivity will decrease with increasing oxidant pressure and, in the case of *p* type semiconductors, it will increase with increasing pressure of the oxidant. It has already been mentioned that at constant temperature the electrical conductivity is directly proportional to the concentration of charge carriers. Hence using formulae (1.3.97), (1.3.145) and (1.5.1) the following relationship is valid for *n* type semiconductors:

$$\sigma = \text{const}\ [e^-] = \text{const}' p_{X_2}^{-1/n} \qquad (1.5.6)$$

where the numerical value of the exponent $1/n$ depends on the degree of ionisation of the defects.

It should be mentioned that this formula is only an approximation, since the constant depends on the mobility of charge carriers, which may be a function of their concentration. However, the mobility changes are so small in comparison with the changes of concentration of defects caused by changes of equilibrium pressure, p_{X_2}, that they can be neglected.

Using formulae (1.3.107) and (1.3.108) and the general relationship (1.5.1) we obtain the analogous equation for p type semiconductors:

$$\sigma = \text{const}\ [h^{\cdot}] = \text{const}' p_{X_2}^{1/n} \qquad (1.5.7)$$

Thus determination of the dependence of electrical conductivity on the oxidant pressure makes it possible to determine the semiconductor type and determination of the numerical value of the exponent $1/n$ leads to conclusions concerning the degree of ionisation of ionic defects at the temperature of the measurement. However, these determinations do not make it possible to find out whether the defects are present in the cationic or the anionic sublattice, since the p type conduction results from the presence of electron holes, which may appear as a result of either a real deficit of metal (cation vacancies) or a real excess of the oxidant (interstitial anions). Similarly in the case of n type semiconductors the defects can appear in either the cationic sublattice (interstitial cations) or the anionic sublattice (anion vacancies). For this reason the determination of the predominant type of ionic disorder requires additional studies. In this

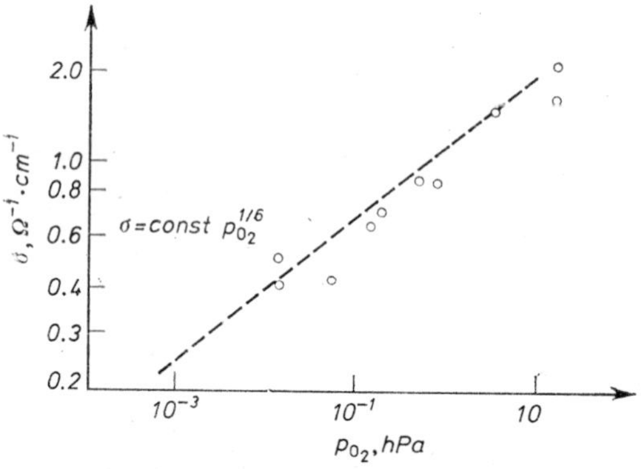

Fig. 88. The dependence of the electrical conductivity of nickel oxide on oxygen pressure at 1300°C (after [111]).

case an important role is played by studies on the mechanism of oxidation of metals (q.v. Chapter 3.1).

Figure 88 illustrates the dependence of electrical conductivity of nickel oxide on the equilibrium oxygen pressure at 1300°C [111]. It can be seen from the graph that in this case p type semiconductor equation (1.3.108) applies, i.e. that completely ionised cation vacancies occur in nickel oxide under these conditions (exponential coefficient $1/n = 1/6$).

Figure 89 shows the results of determinations of the dependence of the electrical conductivity of cobaltous oxide on temperature and oxygen pressure [38, 39]. It is clear from the curves that the degree of ionisation of ionic defects depends on both these parameters. This oxide, like NiO, shows a real deficit of the metal ($Co_{1-y}O$) and hence the results of electrical

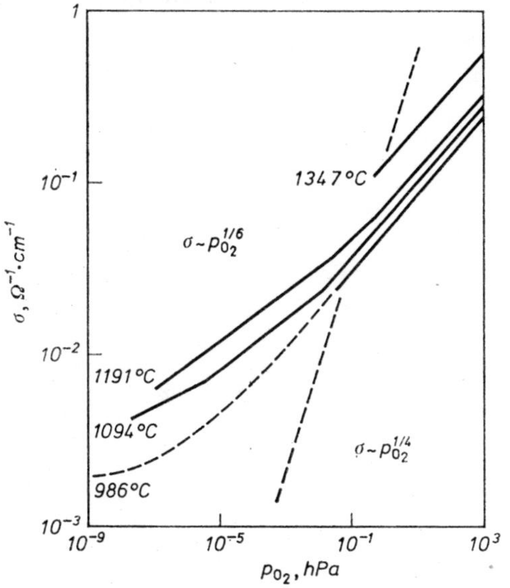

Fig. 89. The dependence of the electrical conductivity of cobaltous oxide on oxygen pressure at several temperatures (after [38]).

conductivity measurement indicate that, at low oxygen pressures, the ionisation of cation vacancies is complete ($1/n = 1/6$), whereas at higher pressures singly charged vacancies predominate (equation (1.3.107)).

In this figure the broken line divides the two regions in which the different types of dependence of σ on p_{O_2} are observed. This division shows that as the temperature increases the region of completly ionised cation vacancies is shifted in the direction of higher oxygen pressure. This is understandable in view of the fact that the degree of ionisation of defects increases with increasing temperature.

The effect of oxygen pressure on the degree of ionisation of defects is also in agreement with theoretical predictions. The concentration of cation vacancies increases with increasing oxygen pressure [equations (1.3.109)–(1.3.111)] and hence the complete ionisation of these defects becomes more difficult to achieve when the equilibrium oxygen pressure in the atmosphere surrounding the crystal increases.

Figure 90 and 91 show the results of electrical conductivity measurements for two oxides containing different kinds of ionic defects ($Me_{1+y}X$ and MeX_{1-y}) [112, 113]. In both cases the electrical conductivity decreases with increasing oxygen pressure (equation (1.5.3)), which indicates that both compounds are *n*-type semiconductors and not *p*-type semiconductors.

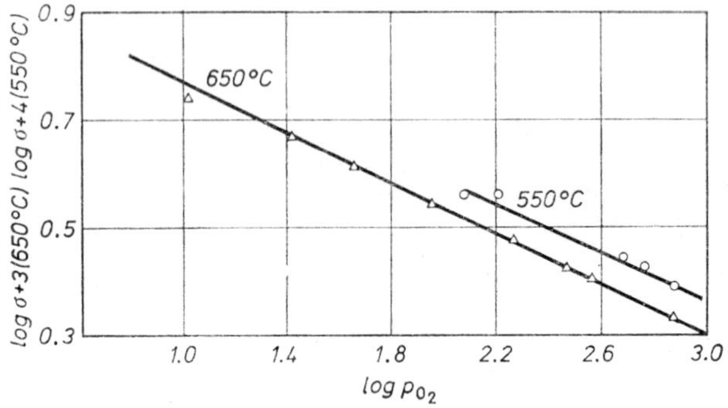

Fig. 90. The dependence of the electrical conductivity of zinc oxide on oxygen pressure at 550°C and 650°C (after [112]).

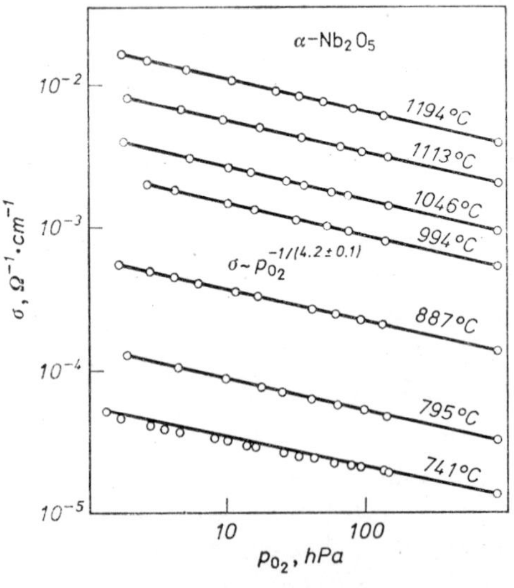

Fig. 91. The dependence of the electrical conductivity of niobium oxide on oxygen pressure at several temperatures (after [113]).

It has been shown by other methods that in zinc oxide the defects are situated in the cationic sublattice (interstitial cations), whereas in niobium oxide they are situated in the anionic sublattice (anion vacancies). The results of the above studies indicate that, in the range of temperature and oxygen pressure investigated, monovalent interstitial cations predominate in zinc oxide [equation (1.3.97)], whereas doubly ionised anion vacancies predominate in niobium oxide (equation (1.3.98)).

As has already been mentioned the electrical conductivity of ionic crystals increases with temperature. In general, the following exponential relationship is valid over a wide temperature range:

$$\sigma = \text{const} \exp\left(-\frac{E}{RT}\right) \tag{1.5.8}$$

This is understandable in view of the fact that the concentration of defects increases exponentially with temperature [compare equations (1.3.112)–(1.3.119)]. The activation energy E of the conductivity contains both the energy required for the formation of defects and the energy required for their migration in the periodic field of the crystal lattice. Thus relationship (1.5.8) is complex and is only completely valid if the mobility of defects, which are the current carriers, is an exponential function of temperature as well. This is the case in solid electrolytes in which ionic defects are the current carriers.

In the case of electronic semiconductors the mobility of charge carriers may also be an exponential function of temperature. This is often observed at high temperatures in the case of oxides and sulphides of transition metals. It is due to the fact that energy levels of electrons and electron holes at high temperatures do not, as a rule, form bands, but are localised on specific atoms of the crystal structure; e.g. in nickel oxide electron holes are localised on cations in the form of Ni^{3+} and hence $h^{\bullet} = Ni^{3+}$. Thus, the current carriers move by jumping from one localised level to another, which requires a well defined activation energy. This type of charge transfer in semiconductors is called the "hopping electron mechanism" or "small polarons mechanism". The latter name is due to the fact that, in oxides and sulphides of transition metals, an electron defect polarises the crystal lattice around it and thus causes a local deformation of the structure. Such a defect together with its closest surroundings (the polarisation field) can be regarded as a quasi-molecule, called a polaron. When the diameter of the polaron is smaller than the lattice constant, the defect is called small polaron. The migration of the polaron in the crystal lattice requires that it passes through the energy barrier during successive jumps, as in the diffusion process for ionic defects. For this reason the mobility of electron defects is a temperature dependent activated process according to the following equation:

$$B = \text{const} \exp\left(-\frac{\Delta H_{\text{m}}}{RT}\right) \tag{1.5.9}$$

where ΔH_{m} is the activation energy of the motion (diffusion) of the current carrier (polaron) in the periodic field of the crystal lattice. Obviously the same relationship is also valid for the mobility of charge carriers in solid electrolytes, where the transport of charge takes place with the participation of ionic defects.

Heikes and Johnston [114] considered the mechanism of motion of small polarons in crystal lattices of compounds of the MeX type and came to the conclusion that it is a diffusion process. Consequently, the mobility depends on the diffusion coefficient according to the Nernst–Einstein equation:

$$B = \frac{q}{RT} D_q \tag{1.5.10}$$

where q is the negative or positive elementary charge and D_q is the diffusion coefficient of the polaron. Since the diffusion coefficient is an exponential function of temperature [see equation (2.1.6)], equation (1.5.10) can be expressed in the following form:

$$B = \frac{\text{const}}{T} \exp\left(-\frac{\Delta H_{\text{m}}}{RT}\right) \tag{1.5.11}$$

Thus, the mobility of small polarons is an exponential function of temperature but, according to Heikes and Johnston, in this case, the preexponential factor is not completely independent of temperature in contrast to equation (1.5.9). This problem was the subject of detailed theoretical studies carried out by Holstein [115], Sewell [116] and Friedman [117]. They came to the conclusion that at low temperatures small polarons can move in a narrow band in the crystal by tunnelling. The width of this band decreases and as a result the mobility of polarons also decreases with increasing temperature. At a certain critical temperature the average free path of polarons becomes comparable with the lattice constant and the band model looses its physical meaning. Thus, at high temperatures the polarons move by jumping from one localised level to another by passing through the energy barrier between the levels, so that their mobility increases exponentially with increasing temperature. Thus, according to the Holstein theory, the mobility of the electronic charge carriers in compounds of the MeX type decreases with increasing temperature but, after reaching a certain minimum value, it increases exponentially with increasing temperature.

Equations (1.5.9) and (1.3.116)–(1.3.119) can be used for the derivation of general relationships between the electrical conductivity of MeX compounds, temperature and oxygen pressure. Introducing equations (1.3.9) and (1.5.117) in the general relationship (1.5.1), we obtain the following expression for p type semiconductors:

$$\sigma = \text{const}\, p_{X_2}^{1/n} \exp\left(-\frac{\frac{2}{n}\Delta H_f + \Delta H_m}{RT}\right) \tag{1.5.12}$$

Introducing equations (1.3.118), (1.3.119) and (1.5.9) in equation (1.5.1), we obtained the corresponding expression for n type semiconductors:

$$\sigma = \text{const}\, p_{X_2}^{-1/n} \exp\left(-\frac{\frac{2}{n}\Delta H_f + \Delta H_m}{RT}\right) \tag{1.5.13}$$

In the case of intrinsic ionic semiconductors, in which the concentration of defects does not depend on the oxidant pressure, equations (1.5.12) and (1.5.13) are reduced to the following form:

$$\sigma = \text{const}\, \exp\left(-\frac{\Delta H_f + \Delta H_m}{RT}\right) \tag{1.5.14}$$

Numerical values of the constants appearing in the above equations depend on several factors including the entropy of formation of the defects.

It can be seen, by comparing equations (1.5.12) and (1.5.13) with empirical equation (1.5.8), that the activation energy E_σ, determined by measuring the electrical conductivity, is related to the enthalpy of formation of defects and the energy of activation for their motion in the crystal by the following expression:

$$E_\sigma = \frac{2}{n}\Delta H_f + \Delta H_m \tag{1.5.15}$$

which is valid for p type and n type semiconductors with deviations from stoichiometry.

For intrinsic ionic semiconductors, we have:

$$E_\sigma = \Delta H_f + \Delta H_m \tag{1.5.16}$$

The above considerations lead to the conclusion that investigations of the dependence of electrical conductivity on temperature and oxidant pressure can provide valuable information on the degree of ionisation of defects and the enthalpy of their formation or the energy of activation for their movement in the crystal lattice, when the numerical value of one of the last two properties is known. However, in a relatively large group of electronic semiconductors, particularly those consisting of elements, the movement of charge carriers (quasi-free electrons or electron holes) is not an activated process even at high temperatures. In such cases the mobility is either completely independent of temperature or slightly increases and sometimes even decreases with increasing temperature. In these semiconductors the energy levels of electrons are not localised on definite elements of the crystal structure, but form wide or narrow bands (broad band or narrow band semiconductors). In such cases there are several kinds of temperature dependence of the mobility of charge carriers (electrons or electron holes), but none of them is exponential. If, in the transport of charge carriers, they are predominately trapped on impuritisites, e.g. ionic defects, the temperature dependence of the mobility can be expressed by the following equation [118–120]:

$$B = \text{const}\, T^{3/2} \tag{1.5.17}$$

i.e. in this case the mobility slightly increases with increasing temperature. On the other hand, when the charge carriers are trapped on atoms in the lattice only (e.g. in semiconductors consisting of elements), the mobility decreases with increasing temperature according to the following equation:

$$B = \text{const}\, T^{-3/2} \tag{1.5.18}$$

This temperature dependence in due to the fact that the amplitude of vibrations of the atoms in the structure increases with temperature and as a result the periodic field of the lattice gradually disappears.

The charge carriers are often trapped on both the lattice atoms and on impurities. In such cases the overlap of the mechanisms of trapping may lead to the lack of any temperature dependence of the mobility and when one of the mechanisms predominates, a weak positive or negative dependence is observed.

The electrical conductance of broad band and narrow band semiconductors is an exponential function of temperature irrespective of the increase or the decrease in the mobility of the quasi-free electrons or electron holes with increasing temperature, since the electrical conductivity principally depends on the concentration of the carriers, which increases exponentially with increasing temperature.

In order to characterise the semiconducting and transport properties of a substance and to elucidate the defect structure it is necessary to know not only the electrical conductance of this substance but also the concentration of charge carriers and their mobility. Both these important properties can be determined by combined measurements of electrical conductivity and of the so-called Hall effect. The principle of these measurements is shown schematically in Fig. 92. The crystal to be examined is placed in a homo-

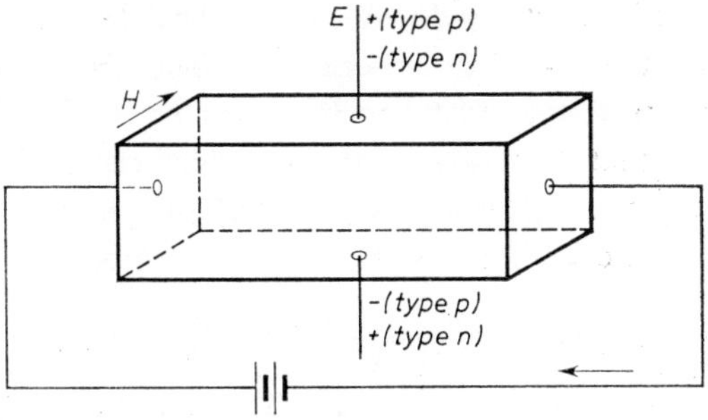

Fig. 92. The scheme of the Hall's method.

geneous magnetic field and an electric current is passed perpendicularly to this field. Under the influence of the magnetic field the paths of the current carriers in the crystal become bent and as a result an electric field, perpendicular to the magnetic field and the applied current, appears in the crystal. The sign of the electric field with respect to the magnetic field depends on the type of current carriers present (Fig. 92).

The intensity of the electric field is directly proportional to the intensity of the magnetic field H and the current density i:

$$E = RHi \tag{1.5.19}$$

where R is the proportionality coefficient called the Hall constant. The numerical value of this constant depends on the concentration and the type of current carrier. This dependence can be simply expressed to good approximation by the following equation:

$$R = \frac{i}{c_i q_i} \tag{1.5.20}$$

where c_i and q_i are the concentration and the charge of the given type of carrier.

It follows from equation (1.5.20) that the sign of the Hall constant depends on the type of current carrier. This constant is positive in the case of *p* type semiconductors (q^+) and negative in the case of *n* type semiconductors (q^-). When the numerical value of the Hall constant is known, the concentration of the carriers can be calculated and the results of these calculations together with the electrical conductivity make it possible to calculate the mobility of the carriers. By introducing equation (1.5.20) in equation (1.5.1) we obtain:

$$B_i = \sigma R \qquad (1.5.21)$$

The sign of the current carriers in a semiconductor can also be determined relatively simply by determining the so-called Seebeck coefficient (the thermoelectric power). For this purpose the crystal to be examined is placed in a temperature gradient such that this is a temperature difference ΔT between the two ends of the crystal. This causes a potential difference ΔV to appear between the ends of the crystal, proportional to the temperature difference:

$$\Delta V = \pm Q \Delta T \qquad (1.5.22)$$

The sign of the thermoelectric coefficient Q is determined by the type of charge carrier. If the colder end of crystal is positive with respect to the hotter end, the current is carried by electron holes (type *p*). If the colder end has the negative sign, the current carriers are quasi-free electrons (type *n*). The numerical value of the coefficient Q, like that of the Hall constant, can be used for the calculation of the concentration of charge carriers but, as a rule, this method is less accurate, since it is difficult to determine the temperature difference ΔT between the two ends of the crystal exactly.

Measurements of electrical conductivity also provide valuable information on the effect of impurities on defect equilibria in ionic crystals. It was shown in a previous chapter that incorporation of higher valency cations in the lattice of compounds with Frenkel type defects causes an increase of concentration of cation vacancies (see equations (1.3.163)–(1.3.169)). For this reason, when for example divalent cations in the form of lead dibromide are added to a crystal of silver bromide with a Frenkel type defects, the electric conductivity of the mixed crystal is expected to be higher than that of pure AgBr and should increase with increasing concentration of the dopant since, as a result of the absence of electronic defects in crystals of this type, the electric current is almost exclusively conducted by ionic defects. The results [121] shown in Fig. 93 fully confirm the correctness of these conclusions.

It follows from equations (1.3.177) and (1.3.178) that incorporation of higher valency cations in a crystal lattice with Schottky type defects should

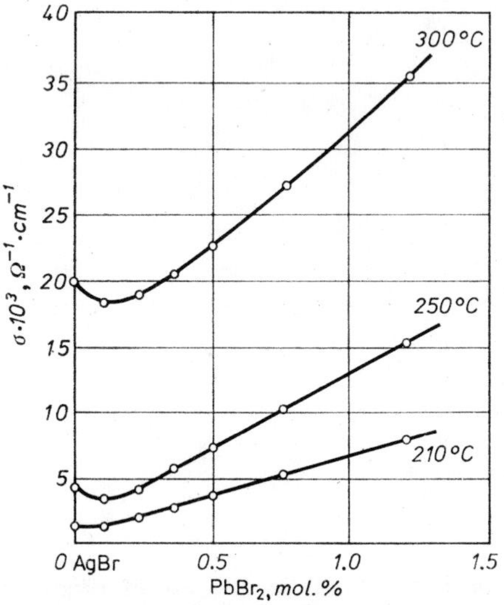

Fig. 93. The effect of a $PbBr_2$ dopant on the electrical conductivity of silver bromide (after [121]).

cause an increase of the concentration of cation vacancies and a simultaneous decrease of the concentration of anion vacancies. In this case cation vacancies would be expected to be predominant in the mixed crystal and as a result the current should be almost exclusively conducted by cations migrating through cation vacancies, whereas in the pure crystal with Schottky type defects the current carriers are cations and anions. In addition the electrical conductivity of the mixed crystal may be higher than that of the parent substance since the concentration of cation vacancies depends on the concentration of the dopant. The results of determinations of transference numbers and electrical conductivity of alkali chlorides containing strontium and other divalent metals fully confirm the correctness of these conclusions (Fig. 94) [122, 123].

It was also shown in a previous chapter that the effect of impurities on crystal defects is most pronounced at not too high temperatures and that at high temperatures this effect may disappear completely (the limiting solutions of equations (1.3.180) and (1.3.182)). Obviously the temperature at which the effect of the impurity on the crystal defects disappears increases

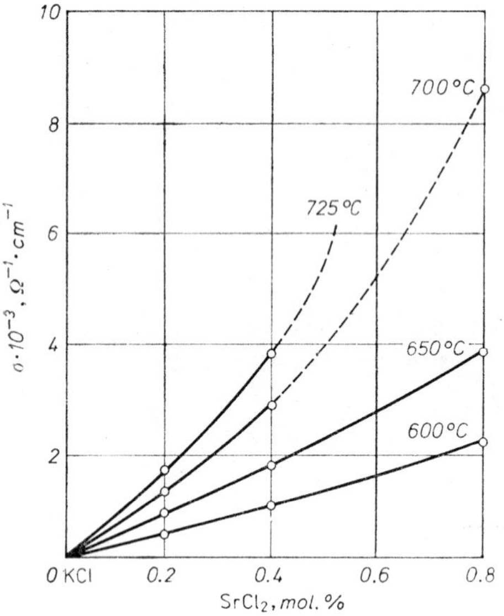

Fig. 94. The effect of a divalent dopant on the electrical conductivity of potassium chloride (after [122]).

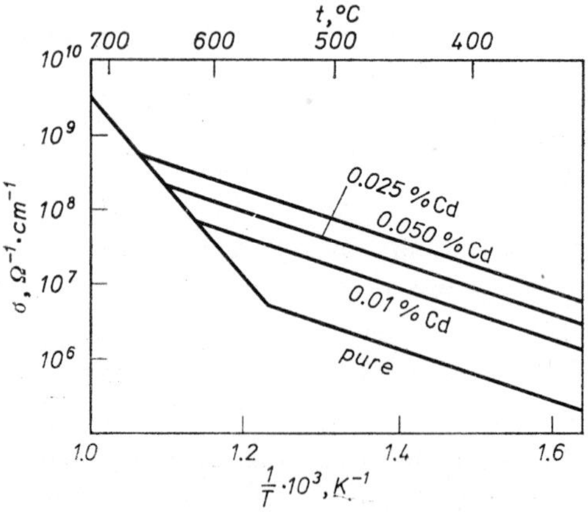

Fig. 95. The effect of divalent dopant on the temperature dependence of the electrical conductivity of sodium chloride.

with increasing concentration of the impurity. Since the electrical conductance is directly proportional to the concentration of defects, the temperature dependence of the conductance of mixed crystals should be analogous.

Figure 95 shows the results of studies on the effect of temperature on the electrical conductivity of NaCl crystals containing additions of cadmium ion. These results fully confirm the conclusions obtained from the theory of defects. It follows from this figure that at temperatures below 550°C the electrical conductivity of mixed crystals of $CdCl_2$–NaCl is higher than that of pure sodium chloride and that the difference increases with increasing concentration of the dopant.

It should be noted that the slopes of the straight lines obtained at the plot $\log\sigma$–$1/T$ for the two temperature regions are different. This is due to the fact that in the low temperatures range the concentration of defects is determined by the concentration of the dopant (equations (1.3.183) and (1.3.185)) and for this reason it does not depend on temperature. Hence in this range the increase of the conductivity with increasing temperature is due exclusively to the increase of the mobility of defects ($\Delta H_t = 0$ in equation (1.5.14) and hence $E_\sigma = \Delta H_m$ in equation (1.5.8)). On the other hand in the high temperatures range the increase of the conductivity with increasing temperature is the sum of the effect of the increase of mobility and effect of the increase in the concentration of defects [$E_\sigma = \Delta H_t + \Delta H_m$ in equation (1.5.8)]. For this reason the line is steeper in this temperature range (the activation energy for conduction is higher).

It is apparent from the plot shown in Fig. 95 that an analogous break in the line is also observed in the case of the pure substance. This is due to the fact that in practice it is difficult to obtain absolutely pure sodium chloride completely free from divalent impurities. These impurities, like cadmium dopant, determine the concentration of defects in the low temperatures region. However, since the concentration of impurities in the "pure" material is lower than in the artificially contaminated (doped) preparations, the low temperatures region is narrowest in the case of the "pure" material. The graph indicates that there is a simple relationship between the concentration of the impurity and the temperature at which the discontinuity in the plot of the function $\log\sigma$–$1/T$ appears. This makes it possible to evaluate the concentration of divalent impurity in samples of "pure" sodium chloride. In the case shown in the graph it is equal to about 3×10^{-3} at.%. This example clearly shows the enormous effect of traces of impurities on semiconducting properties of solids and explains the meaning of the term "pure" in the case of semiconductors. Obviously this applies equally to ionic and electronic semiconductors.

In the case of electronic semiconductors the electric charge is almost exclusively transferred by electronic defects. It was shown in the previous chapter that when the deviations from stoichiometry are large, the concentration of electronic defects in crystal is closely connected with the concentration of ionic defects. Therefore also in these substances the impurities must affect the electrical conductivity. In the case of compounds

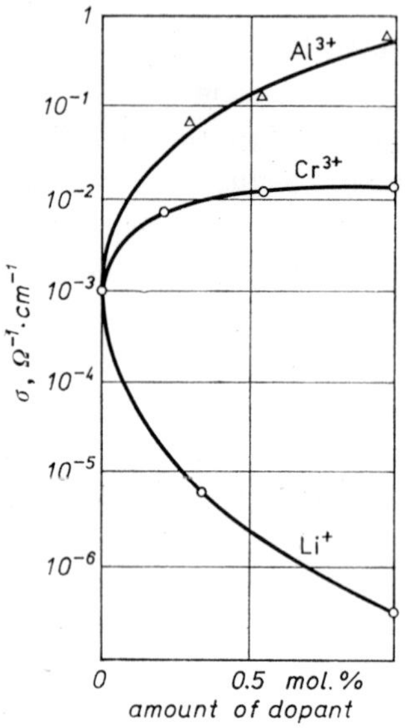

Fig. 96. The effect of mono- and trivalent dopant on the electrical conductivity of zinc oxide (after [124]).

showing an excess of metal, e.g. ZnO, cations having a valency higher than that of the lattice cations should cause an increase of the concentration of quasi-free electrons and consequently an increase of the electrical conductivity, whereas cations having a lower valency should have the opposite effect. On the other hand in the case of compounds with a metal deficit, e.g. NiO, univalent cations should cause an increase of the concentration of electron holes and hence an increase of the electrical conductivity, whereas trivalent cations should have the opposite effect. Figures 96 and

97 show the effects of various kinds of impurities on the electrical conductance of zinc oxide and nickel oxide. They fully confirm the correctness of conclusions derived from the above theoretical considerations.

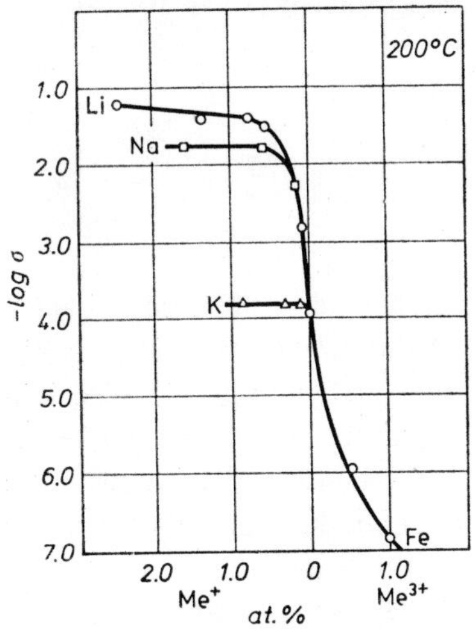

Fig. 97. The effect of mono- and trivalent dopant on the electrical conductivity of nickel oxide (after [125]).

2. Diffusion in the solid state

2.1 The mechanism of lattice diffusion

As a result of thermal vibrations occuring in crystal lattices, point defects are always in motion. The random walk of the defects, which are in the state of thermodynamic equilibrium with the surroundings, is called self-diffusion [126–128]. The diffusion of foreign atoms in crystals is referred to as hetero-diffusion. The self-diffusion process is not connected with the flow of material. It arises when a concentration gradient of defects appears in the crystal and some of the material begins to move in a definite direction in order to eliminate the gradient. It should be stressed that the diffusion process can take place not only as a result of the concentration gradient of defects but also under the influence of other forces constituting the electrochemical potential gradient in the system. In some cases, e.g. under the influence of a strain gradient, the diffusion process can even take place in the direction opposite to the concentration gradient.

In polycrystalline materials the mechanism of diffusion is very complex, since the transport of material is not limited to the migration of point defects, but also takes place along grain boundaries, internal surfaces, dislocations, etc. The contribution of the individual forms of diffusion to the total diffusion in polycrystalline materials largely depends on the temperature of the system. At sufficiently high temperatures (above the Tammann temperature) lattice diffusion is predominant. Its mechanism will be discussed in this chapter.

The diffusion process is governed by two fundamental laws derived by Fick in 1880. Although these laws were not derived for diffusion processes taking place in solid phases, they apply to any state of matter, as result of their perfectly general character regarding macroscopic diffusion processes.

Fick's first law gives the relation between the flux, i.e. the amount of substance flowing in unit time through unit surface area perpendicular to the flux, and the concentration gradient of the diffusing substance:

$$j = -D\frac{\partial c}{\partial x} \tag{2.1.1}$$

where j is the flux of the component diffusing in direction x, c is its concentration perpendicular to the direction of flow and $\partial c/\partial x$ is the concentration gradient in the direction of flow. The minus sign in equation (2.1.1) is due to the fact that the diffusion process takes place in the direction of decreasing concentration of the diffusing component. The proportionality coefficient D appearing in this equation is called the diffusion coefficient. It determines the rate of diffusion at unit concentration gradient and hence it is a measure of the rate at which the system tends to eliminate the concentration difference. It can be also regarded as a measure of the rate of diffusion with a constant concentration gradient.

The diffusion coefficient is usually expressed in cm^2/s. In the temperature range 20–1500°C the values of D for solids are within the limits from 10^{-20} to 10^{-4} cm^2/s.

It should be mentioned that diffusion in solids can take place not only under the influence of a concentration gradient but also under the influence of other factors contributing to the electrochemical potential gradient in the system. The concentration gradient usually plays the principal role, but in some cases, e.g. in the case of diffusion of carbon in steel, this role is played by such factors as the strain field in the crystal and as a result the diffusion takes place in the direction opposite to that of decreasing concentration [129]. The electric field can have a similar effect, since the diffusion in ionic crystals usually involves ions and not electrically neutral atoms or complexes.

The determination of the diffusion coefficient on the basis of Fick's first law would requires measurements at a concentration gradient that does not change with time. It is very difficult to create such experimental conditions, particularly in the case of diffusion in solids. On the other hand concentration changes taking place during diffusion processes can be measured fairly readily. The determination of diffusion coefficient on the basis of results of such measurements is made possible by Fick's second law, which gives the relation between the concentration gradient and the rate of change of concentration caused by diffusion at a given point in the system:

$$\frac{\partial c}{\partial t} = \frac{\partial}{\partial x}\left(D\frac{\partial c}{\partial x}\right) \tag{2.1.2}$$

When the diffusion coefficient is independent of concentration, equation (2.1.2) becomes:

$$\frac{\partial c}{\partial t} = D\frac{\partial^2 c}{\partial x^2} \tag{2.1.3}$$

Equations (2.1.2) and (2.1.3) can be solved for certain limiting conditions and as a result coefficient D can be determined on the basis of various kinds of measurements.

The theoretical description of the diffusion process requires the consideration of the elementary act, which is a jump of an atom or an ion in the crystal from one stable position to another, i.e. it is necessary to pass from the macroscopic phenomen on described by Fick's laws to the microscopic mechanism of the diffusion process.

It has already been mentioned that the movement of atoms in crystals is possible mainly due to the fact that the crystal structure is not perfect. Since various kinds of defects can occur in this structure depending on the kind of crystal and the external conditions, there are many mechanisms of lattice diffusion. In some rare cases the diffusion process can take place without the participation of defects. The following types of mechanisms of lattice diffusion can be distinguished:

(1) the exchange mechanism,
(2) the ring mechanism,
(3) the vacancy mechanism,
(4) the relaxation mechanism,
(5) the intersitial mechanism,
(6) the interstitial mechanism with:
 a—collinear or
 b—non-collinear displacement
(7) the dumbbell interstitial mechanism,
(8) the crowding mechanism,
(9) the ascending diffusion mechanism.

The exchange mechanism consists of the exchange of lattice positions involving two atoms situated next to one another in the crystal structure. This process is shown schematically in Fig. 98. It follows from the diagram that a diffusion process taking place by the double exchange mechanism does not require the presence of defects in the crystal. However, it can be shown that such a process is very improbable in crystals consisting of relatively closely packed atoms, since in this case the double exchange requires a considerable deformation of the lattice and hence an enormous activation energy.

However Zener [130] has shown that this type of diffusion can take place much more readily, when the exchange of positions in the crystal

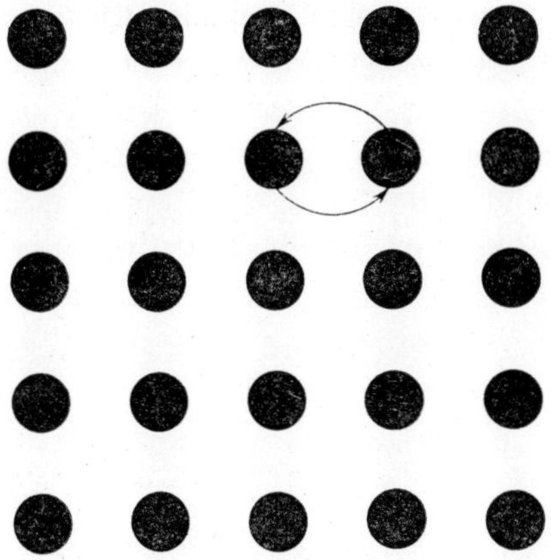

Fig. 98. Double exchange diffusion mechanism.

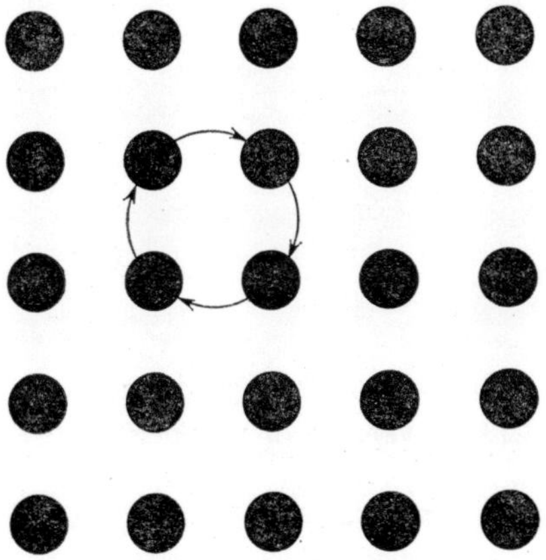

Fig. 99. Ring mechanism of diffusion.

involves not two, but three, four or even five atoms. This kind of mechanism, called the ring mechanism, is shown in Fig. 99. It should be stressed that both the ring mechanism and the exchange mechanism are unimportant, in spite of the fact that the activation energy of the ring diffusion is much lower than that of the double exchange process, since real crystals always contain a certain number of point defects, thanks to which the movement of atoms in the lattice is easier from the energy point of view.

The simplest mechanism of diffusion and most often encountered in practice is the vacancy mechanism. It occurs in both metals and compounds of the MeX type. Defects of the vacancy type predominate over a wide range of temperature and pressure in many metallic structures (oxides, sulphides, spinels). Diffusion processes taking place according to this mechanism consist of successive jumps of atoms or ions from lattice sites to neighbouring vacancies. Hence the condition necessary for the change of the position of an atom in the lattice is the appearance of a vacancy in the neighbourhood of the atom. Obviously the direction of diffusion of the vacancy in this process is opposite to that of the atom

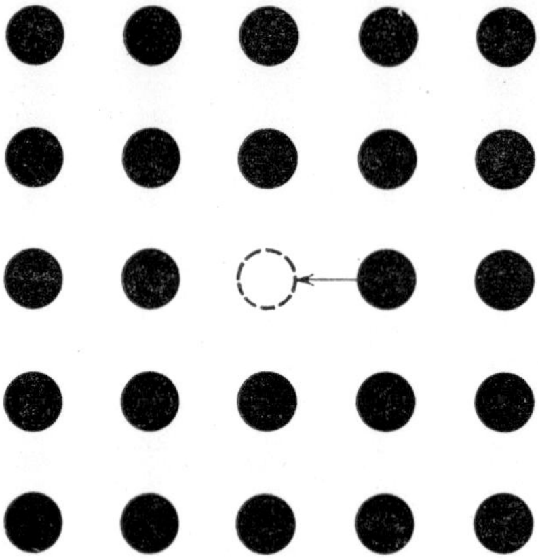

Fig. 100. Vacancy mechanism of diffusion.

and consequently the vacancy type of diffusion can be regarded as either the movement of atoms or the equivalent movement of vacancies. However, it will be shown later that the diffusion coefficient of vacancies is not equal to that of atoms. The vacancy mechanism of diffusion is shown schematically in Fig. 100.

A variant of the vacancy mechanism is called the relaxation mechanism. It takes place when the atoms surrounding the vacancy are shifted towards it and the disturbance of the normal structure of the crystal lattice caused by this shift extends to a distance equal to several lattice constants. This situation is shown schematically in Fig. 101. The disturbance of the structure can be compared to a localised melting of the crystal around the vacancy.

In this region atoms can diffuse as they do in a liquid phase as a result of random thermal movements.

When interstitial atoms or ions are the predominant defects, two mechanisms of lattice diffusion can be distinguished. One of them, called the simple interstitial mechanism or briefly the interstitial mechanism, consists in successive jumps of an atom or an ion from one interstitial position to another. This mechanism is shown schematically in Fig. 102.

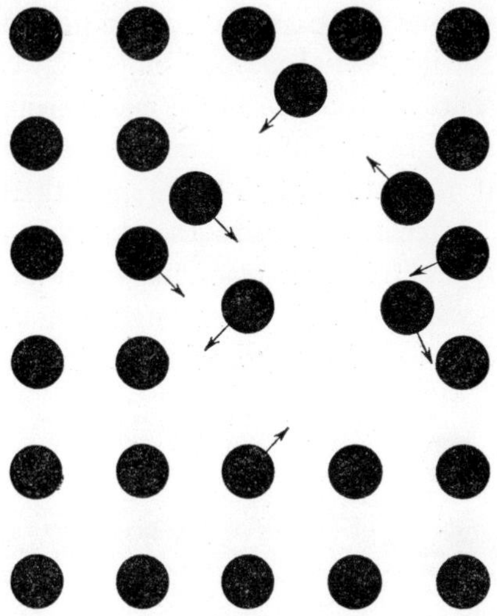

Fig. 101. Relaxation mechanism of diffusion.

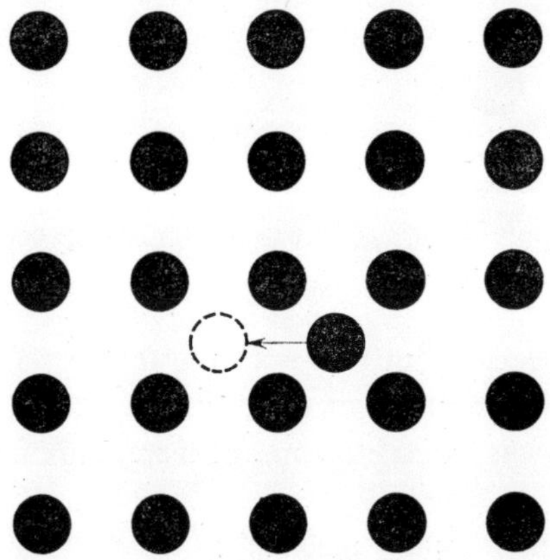

Fig. 102. Interstitial mechanism of diffusion.

The process taking place according to this mechanism requires considerable deformation of the crystal lattice at the moment that the atom jumps. This is possible when the crystal consists of loosely packed elements or when the diameter of the intersitial atoms is much smaller than that of the lattice atoms, i.e. impurities are present. It has been found that this type of mechanism occurs in the case of interstitial solid solutions of hydrogen or carbon in metals [131].

When the diameters of interstitial atoms are comparable with those of the atoms constituting the lattice or one of the sublattices, diffusion usually takes place, by a second kind of interstitial mechanism, which is called the interstitialcy mechanism. There are two variants of this mechanism, which are called collinear and non-collinear. In both of them the elementary act of diffusion consists in displacement of a lattice atom by its interstitial neighbour to an interstitial position. After these jumps the displacing atom occupies the lattice site. This process can take place in two ways. The displacing and the displaced atoms can move along a straight line as is shown schematically in Fig. 103. This is referred to as the collinear

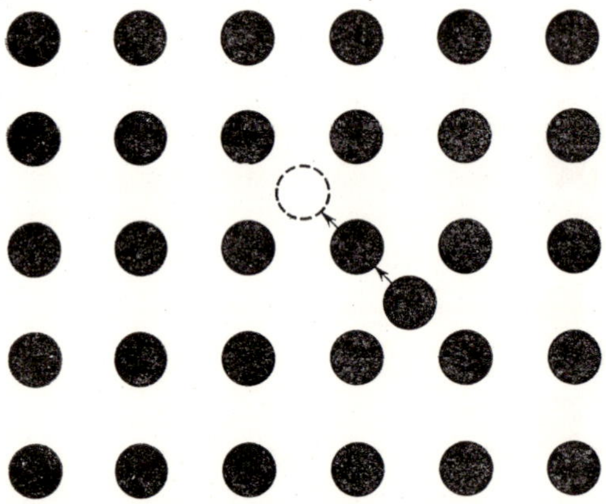

Fig. 103. Interstitial mechanism of diffusion. The collinear variant.

interstitial mechanism. When the displaced atom moves to the interstitial position at an angle to the direction of movement of the displacing atom, the mechanism is called the non-collinear interstitial mechanism. This is shown schematically in Fig. 104.

It is noteworthy that in the case of interstitialcy mechanism the distance between the initial and the final position of the defect in the lattice is greater than that traversed by the atom during the elementary diffusion act. This has a considerable influence on the correlation effect, which will be discussed in Chapter 2.2.

In some cases the displacement can take place according to a third variant, which is called the dumbbell interstitial mechanism [126]. In such

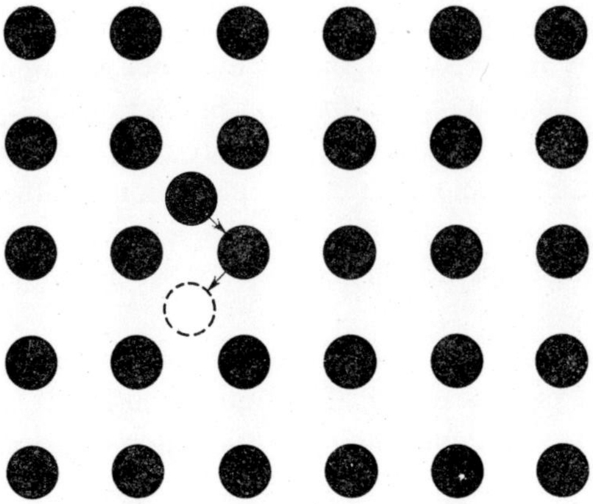

Fig. 104. Interstitial mechanism of diffusion. The non-collinear variant.

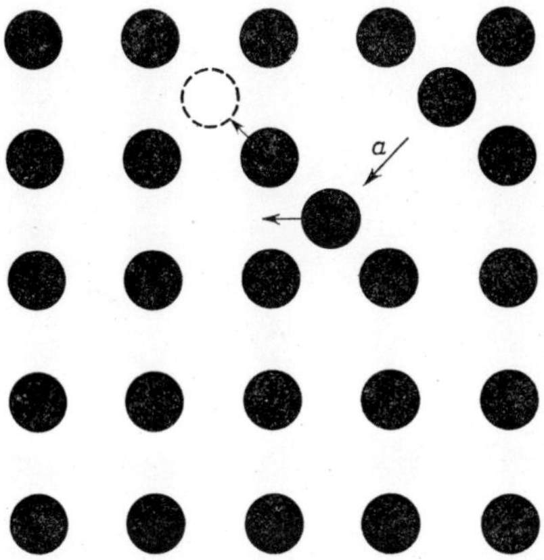

Fig. 105. Dumbbell (interstitial) mechanism of diffusion (the first stage).

cases the most stable configuration of atoms in the crystal lattice is the system in which the presence of an interstitial atom causes a shift of its nearest neighbour in the direction of the nearest intersitial position. This leads to the formation of a system of two interstitial atoms situated symmetrically with respect to the lattice site, as it is shown schematically in Fig. 105.

It should be stressed that in this configuration the distances of the two interstitial atoms from the empty lattice site, which they surround, are considerably smaller than in the case of "normal" interstitial atom. In other words an interstitial atom in the dumbbell position is situated asymmetrically with respect to both lattice planes but symmetrically with respect to the atom displaced from the lattice site to similar dumbbell position.

In the case of this type of configuration the diffusion process obviously also takes place by the interstitial mechanism but in a more complex way. In this case the passage of an atom from one interstitial position to another consists of three elementary acts in contrast to the collinear and non-collinear variants in which two jumps are sufficient (compare Figs. 103 and 104).

In the case of the dumbbell interstitial mechanism the first elementary act takes place at the moment of appearance of an interstitial defect in the neighbourhood of a given lattice atom. The interstitial atom displaces the lattice atom to the dumbbell position. Then it makes the second jump in the direction of its neighbour situated in the nearest lattice site and displaces it to the corresponding dumbbell position. Finally, during the

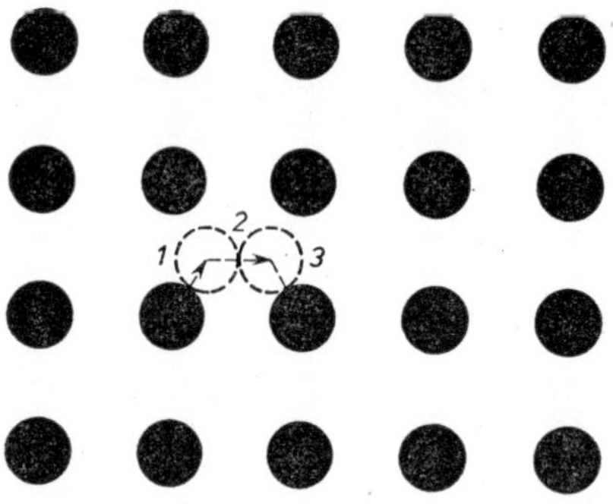

Fig. 106. Dumbbell mechanism of diffusion (the individual stages).

third elementary act the atom displaced from the lattice site to the dumbbell position moves away and occupies the lattice site. The succesive stages of this mechanism of diffusion are shown schematically in Fig. 106.

Another possibility of movement of atoms in crystal lattices of solids is due to the so-called crowding mechanism [126], which is connected with the occurrence of specific interstitial defects in crystals. In contrast to the typical interstitial defects the crowding configuration consists in the presence of an additional atom situated in the lattice plane, but not in the interstitial position. This is shown schematically in Fig. 107.

As a result of the presence of an additional atom in the row of atoms, all the atoms situated at a distance of up to about ten lattice constants

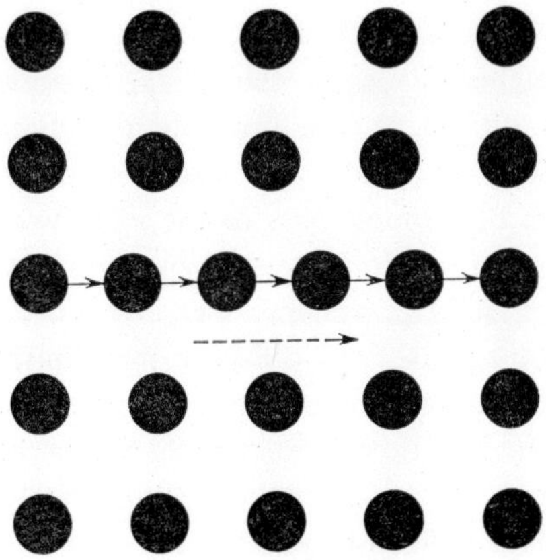

Fig. 107. Crowding mechanism of diffusion.

from the additional atom are shifted from their equilibrium positions in the line of the row (Fig. 107). This leads to the formation of a system of atoms in non-equilibrium positions, which can move in the crystal lattice in both directions along the row. It is very probable that during the annealing process this type of diffusion is responsible for the elimination of the excess of interstitial atoms, which are formed in metals during cold rolling or as a result of high energy irradiation.

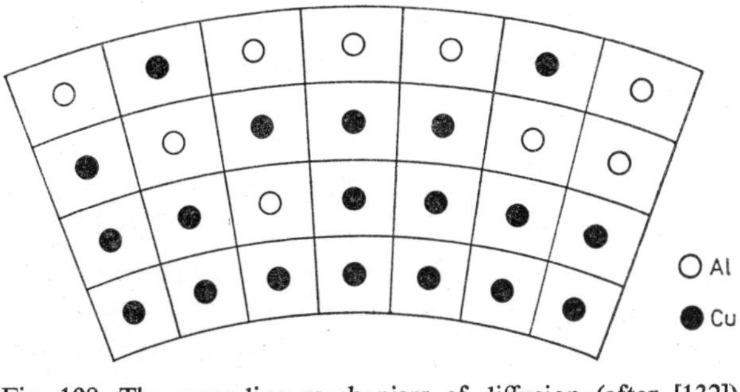

Fig. 108. The ascending mechanism of diffusion (after [132]).

A specific type of diffusion can take place in alloys of metals under the influence of stress gradients. A model for the mechanism of this type of diffusion, called ascending diffusion, was proposed by Konobeevski [132]. The scheme shown in Fig. 108 represents this model in the form of an elastically bent block of a crystal in which the lattice is stretched in the upper part and compressed in the lower one. Both the stress and the degree of plastic deformation change continuously. Since the components of the solid solutions have different diameters, the larger elements have the tendency to shift in the direction of stretched layers of the block, whereas the atoms having smaller diameters diffuse in the opposite direction. Consequently a concentration gradient, opposing the stress gradient, appears in the solid solution, in which the components of the alloy were uniformly distributed before the deformation. It follows from the above discussion that atoms can move in crystals in various ways. However, in practice in the great majority of solid phases the diffusion mechanism only takes place by the vacancy mechanism and the interstitial mechanism.

Let us now consider the energetic and kinetic reasons for the jumps of elements in crystal structures, in order to obtain a mathematical relation between the macroscopic parameters for diffusion, i.e. the self-diffusion coefficients for atoms or ions, and the elementary acts in jumps of defects represented by the coefficient of diffusion for defects, which is a microscopic parameter of the process.

It has already been mentioned that the self-diffusion process consists of a completely random walk of the defects in the crystal. Hence the successive jumps of the defects in the lattice are not correlated. However, it will be shown later that the motion of atoms or ions in crystals resulting from the motion of defects can have a certain degree of correlation [126, 131, 133–135]. For the time being we assume, for the sake of simplicity, that the movements of both defects and atoms are not correlated. Having obtained the diffusion equations we shall introduce the corrections for the correlation effect. This assumption is necessary, since statistical methods are used for the derivation of the diffusion equations.

The jumps of defects and consequently the jumps of atoms in crystals are due to thermal vibrations of very high frequency. In the majority of solids the frequency of vibrations of atoms or ions in the lattice is of the order of 10^{13} s^{-1}. The frequency of jumps of atoms is lower by several orders of magnitude. For example, at 700°C, the jumps frequency of atoms is of the order of 10^{8} s^{-1}. This means that once in every 100 000 vibrations a thermal fluctuation appears sufficiently large for an atom to overcome the energy barrier separating it from the next stable position in the crystal structure.

Figure 109 shows schematically the three stages in the passage of an interstitial atom in a simple regular lattice from one interstitial position to another. It also shows the energy of this atom as a function of the

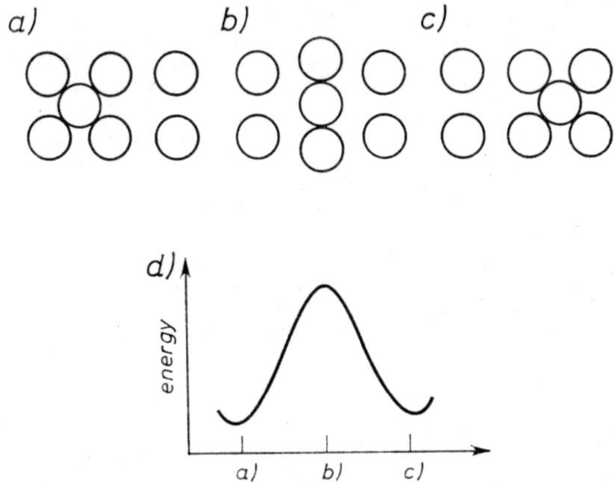

Fig. 109. The scheme illustrating the individual stages of passage of a defect from one interstitial space to another (after [131]).

position coordinate. It can be seen from the plot that the atom must pass through an energy barrier E_D, which has a peak situated half way between the two stable interstitial positions. An analogous energy barrier must be overcome by the atom in the case of other mechanisms of diffusion. This energy is called the energy of activation of diffusion.

If we assume that the distances traversed by the atom during successive jumps are equal and that the jumps are completely random, i.e. uncorrelated, then we can use statistical methods to give the following simple expression for the coefficient of diffusion of defects, which is also called the microscopic diffusion coefficient:

$$D_d = \alpha a_0^2 \omega \tag{2.1.4}$$

where α is the geometrical coefficient. Its numerical value depends on the structure of the crystal and the mechanism of diffusion. For a regular face centred or body centred lattice and for the interstitial diffusion mechanism $\alpha = 1$. ω is the frequency of jumps and a_0 is the distance travelled by the atom during the jump from one position to anothe.. This distance is comparable with the lattice constant.

It has already been mentioned that a jump of a defect in a crystal is an activated process and consequently the frequency of jumps is an exponential function of temperature. On the basis of statistical mechanics and transition state theory, Zener derived the following equation:

$$\omega = \chi\nu\exp\left(-\frac{\Delta G_{\mathrm{m}}}{RT}\right) = \chi\nu\exp\left(\frac{\Delta S_{\mathrm{m}}}{R}\right)\exp\left(-\frac{\Delta H_{\mathrm{m}}}{RT}\right) \quad (2.1.5)$$

where χ is the transmission coefficient determining the probability that an atom having sufficient energy will jump to another position and will not return to its original position; ν is the frequency factor which in the first approximation is equal to the frequency of vibrations of atoms or ions in the lattice sites (the Debye frequency).

According to Zener, χ is equal to one in the case of metals. ΔG_{m} is the change of the free energy of the system during the jump and ΔS_{m} and ΔH_{m} are the entropy and enthalpy changes of this process. Introducing equation (2.1.5) to equation (2.1.4), we obtain:

$$D_{\mathrm{d}} = \alpha a_0^2 \chi\nu\exp\left(\frac{\Delta S_{\mathrm{m}}}{R}\right)\exp\left(-\frac{\Delta H_{\mathrm{m}}}{RT}\right) \quad (2.1.6)$$

It follows from this equation that the energy of activation for diffusion of defects is equal to ΔH_{m}. Equation (2.1.6) is valid for any mechanism of diffusion of point defects.

So far we have been considering the diffusion of defects in crystals, i.e. the diffusion of vacancies and atoms or ions situated in interstitial positions. Although the movement of defects is equivalent to the movement of atoms, the expressions obtained for the diffusion coefficient of defects are not valid for the diffusion of atoms, since the defects represent a small fraction of all the atoms present in the crystal. For this reason the probability for a jump of an atom is proportional to that of finding this atom in the interstitial position or to the probability of finding a vacancy in the immediate neighbourhood of this atom. This probability increases with increasing concentration of defects in the crystal and is equal to the mole fraction of defects. Therefore using equation (2.1.4) we can express the diffusion coefficient of the atoms, i.e. the macroscopic coefficient of self-diffusion, in the following form:

$$D = \alpha a_0^2 \omega N_{\mathrm{d}} \quad (2.1.7)$$

or:

$$D = D_d N_d \tag{2.1.8}$$

where N_d is the mole fraction of defects.

Now let us explain the physical meaning of the geometrical coefficient α. For this purpose we must consider the random walk mechanism. If we assume that the jumps of a diffusing atom can take place in any direction and that they are completely accidental and independent of each other, then the effective displacement of the atom in the crystal after n jumps R_n is equal to the algebraic sum of individual jumps s_1, s_2 etc:

$$R_n = S_1 + S_2 + \ldots = \sum_{j=1}^{n} S_j \tag{2.1.9}$$

When the probability of individual jumps in all the directions is the same and the distances travelled during the jumps do not differ from each other, then the effective displacement of an atom in space after n jumps is equal to zero ($R_n = 0$). This does not mean that the diffusing atom remains in its starting position after making n jumps. It only means that all jumps are equally probable. In fact the effective displacement of the atom caused by the random walk can be anything between zero and $\pm nS$. A measure of the average displacement of an atom in the crystal structure is the square of R_n. $\overline{R_n^2}$ is given by the following equation:

$$\overline{R_n^2} = \sum_{j=1}^{n} S_j^2 + 2\sum_{j=1}^{n-1} \sum_{k=j+1}^{n} S_j S_k \tag{2.1.10}$$

If, as before, the vectors of the individual jumps are equal to each other, i.e. if $|S_1| = |S_2| \ldots = |S_j| = S$, which is true in crystals having regular structures and if the jumps are at random and not correlated with each other, then the second term of the right hand side of equation (2.1.10) is equal to zero. The mean square of the shift of the front of diffusing substance $\overline{R^2}$ in three dimensional space is given by the following relationship:

$$\overline{R_n^2} = 6Dt \tag{2.1.11}$$

where t is the time, in which the shift takes place. Comparison of equations (2.1.10) and (2.1.11) gives the following expression:

$$D = \frac{1}{6}\frac{n}{t}S^2 \tag{2.1.12}$$

where n/t is the number of jumps in unit time. This quantity depends on several factors, including the mechanism of diffusion. For example, in the case of the vacancy mechanism, the number of jumps in unit time n/t depends on the concentration of defects, since the jump of an atom is possible only when a vacancy exists in its close neighbourhood. The value of n/t depends also on the frequency of jumps ω, at which the defects migrate in the crystal structure. Finally, the value of n/t depends on the number Z of nearest neighbour positions in the crystal structure to which the atom can jump. Thus. the value of n/t is given by the following equation:

$$\frac{n}{t} = ZN_{\mathrm{d}}\,\omega \qquad (2.1.13)$$

Hence equation (2.1.12) can be expressed in the following form:

$$D = \tfrac{1}{6}ZS^2\omega N_{\mathrm{d}} \qquad (2.1.14)$$

The jump distance of an atom in the crystal is a function of the structure of the crystal and is related to the lattice constant as follows:

$$S = Aa_0 \qquad (2.1.15)$$

where the proportionality coefficient, A, can have various values depending on the mechanism of diffusion and the structure of the crystal. Introducing equation (2.1.15) in equation (2.1.14) we finally obtain:

$$D = \tfrac{1}{6}ZA^2a_0^2\,\omega N_{\mathrm{d}} \qquad (2.1.16)$$

It can be seen by comparison equations (2.1.7) and (2.1.16) that the geometrical coefficient α in expressions (2.1.4) and (2.1.6) is given by the following equation:

$$\alpha = \tfrac{1}{6}ZA^2 \qquad (2.1.17)$$

In the case of the vacancy mechanism of diffusion in a regular body centred structure each atom has 8 nearest neighbours or positions, to which it can jump. Thus $Z = 8$. The distance of the jump S is in this case related to the lattice constant a_0 as follows:

$$S = \frac{\sqrt{3}}{2}a_0 \qquad (2.1.18)$$

It follows from this expression that $A = \frac{\sqrt{3}}{2}$. Introducing these values in equation (2.1.17) we find that $\alpha = 1$.

In the case of the simple interstitial diffusion mechanism with the same crystal structure, each atom of an interstitial dopant has 4 nearest octahedral positions, to which it can jump. Thus $Z = 4$. The jump distance is related to the lattice constant as follows:

$$S = \tfrac{1}{2}a_0 \tag{2.1.19}$$

Hence the constant A in equation (2.1.15) is equal to 1/2. Introducing the above numerical values in equation (2.1.17) we find that $\alpha = 1/6$. The above considerations lead to the conclusion that the geometrical coefficient α is a complex quantity but that it can be readily calculated on the basis of the knowledge of the mechanism of diffusion and the lattice constants for the crystal structure.

The diffusion coefficient for defects is, as a rule, independent of their concentration in the crystal, since this concentration is usually very small, whereas the diffusion coefficient of atoms is directly proportional to the concentration of defects [equations (2.1.7) or (2.1.8)]. The concentration of defects is temperature dependent. It follows from the discussion in the last chapter that this dependence can be expressed by the following equation:

$$N_{\mathrm{d}} = \exp\left(\frac{\Delta S_{\mathrm{f}}}{R}\right)\exp\left(-\frac{\Delta H_{\mathrm{f}}}{RT}\right) \tag{2.1.20}$$

where ΔS_{f} and ΔH_{f} are the entropy and the enthalpy of formation of defects, respectively. Introducing equation (2.1.20) in equation (2.1.8) and utilising equation (2.1.6) we obtain the following general relationship:

$$D = \alpha v a_0^2 \chi \exp\left(\frac{\Delta S_{\mathrm{f}}+\Delta S_{\mathrm{m}}}{R}\right)\exp\left(-\frac{\Delta H_{\mathrm{f}}+\Delta H_{\mathrm{m}}}{RT}\right) \tag{2.1.21}$$

It follows from equation (2.1.21) that the activation energy for diffusion of atoms in crystals depends on both the energy of formation of defects and the energy required for their migration in the periodic field of the crystal lattice. Experimental studies have confirmed the theoretical predictions that the temperature dependence of the diffusion coefficient is, as a rule, an exponential function, which can be expressed by the following general empirical equation:

$$D = D_0 \exp\left(-\frac{E_{\mathrm{D}}}{RT}\right) \tag{2.1.22}$$

By comparing equations (2.1.22) and (2.1.21) we obtain the following relationships:

$$E_D = \Delta H_f + \Delta H_m \tag{2.1.23}$$

and

$$D_0 = \alpha a_0^2 \chi \nu \exp\left(\frac{\Delta S_f + \Delta S_m}{R}\right) \tag{2.1.24}$$

It was shown in the last chapter that the concentration of defects in a definite temperature range can be strictly controlled by the concentration of a dopant in the crystal. In this case the activation energy of diffusion is equal to the activation energy of migration alone and equation (2.1.21) becomes:

$$D = N_d \alpha a_0^2 \chi \nu \exp\left(\frac{\Delta S_m}{R}\right) \exp\left(-\frac{\Delta H_m}{RT}\right) \tag{2.1.25}$$

It was shown in section 1.3.7 that the effect of dopant is largely temperature dependent. The concentration of defects increases rapidly (exponentially) with increasing temperature and, at a certain critical temperature, it can reach values considerably exceeding the concentration of the additive. This leads to the conclusion that at low temperatures the concentration of defects is controlled by the amount of foreign atoms in the crystal (the extrinsic region), i.e. it is independent of temperature, but at higher

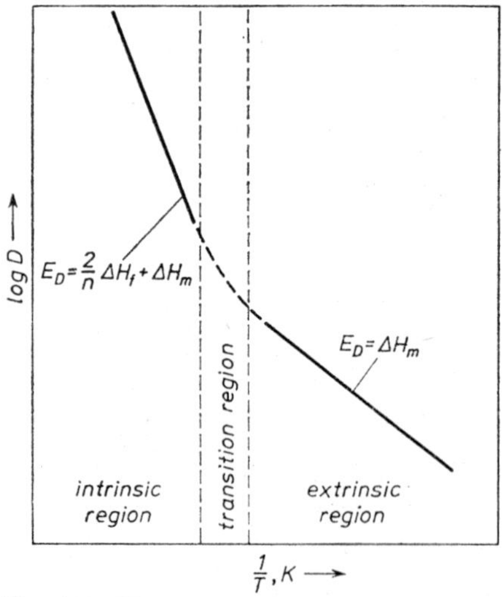

Fig. 110. The scheme of the temperature dependence of the diffusion rate in a doped crystal.

temperatures the dopant has no effect on the disorder in the crystal (the intrinsic region). With respect to diffusion processes this means that on passing from the extrinsic region to the intrinsic region the activation energy for diffusion changes. This is shown schematically in Fig. 110.

It follows from a comparison of equations (2.1.6) and (2.1.25) that, when the concentration of defects is controlled by the dopant, the activation energy for defect diffusion is equal to that of atoms, since both are equal to the activation energy for migration. However, the absolute values of D_d and D are quite different, since ,as it follows from equation (2.1.8), the diffusion coefficient of atoms is several times smaller than that of defects (the values of the mole fractions of defects are, as a rule, lower than 10^{-2}). In order to illustrate this difference, several selected values of D and D_d for metal oxides are compared in Table 9.

Another significant difference between the diffusion coefficient of defects and that of atoms consists in the fact that the former is independent of the concentration of defects in the crystal, whereas the latter is always a definite function of concentration [equation (2.1.8)].

TABLE 9

Diffusion coefficients of defects D_d and self-diffusion coefficients of ions D_{Me} in some MeX-type compounds

Compound	Temperature, °C	Partial pressure of component X_2, hPa	D_d, cm²/s	D_{Me}, cm²/s
Cu_2O	800	1013.25	1.2×10^{-5}	7.2×10^{-9}
Cu_2O	1000	1013.25	9.7×10^{-5}	9.7×10^{-8}
NiO	1000	212.78	1.0×10^{-7}	1.0×10^{-11}
NiO	1100	1013.25	2.5×10^{-7}	3.6×10^{-11}
CoO	1000	435.7	3.3×10^{-7}	2.3×10^{-9}
CoO	1100	1013.25	9.0×10^{-7}	8.3×10^{-9}
FeO	1000	1.013×10^{-10}	1.3×10^{-6}	1.7×10^{-7}
Cu_2S	550	1013.25 (sulphur vapour)	4.0×10^{-4}	1.0×10^{-4}
Cu_2S	350	liquid sulphur	8.7×10^{-5}	2.0×10^{-8}
FeS	670	1013.25	1.0×10^{-6}	1.6×10^{-8}
FeS	800	1013.25	8.2×10^{-6}	1.3×10^{-7}
Ag_2S	200	liquid sulphur	4.7×10^{-1}	5.0×10^{-5}

It was shown in the last chapter that the concentration of defects in compounds of the MeX type with deviations from stoichiometry depends on the equilibrium pressure of the X component in the surrounding atmosphere. Consequently, the diffusion coefficient of atoms or ions in crystals of this type must be an analogous function of the oxidant pressure. Therefore for compounds with a real deficit of metal ($Me_{1-y}X$) the diffusion coefficient of atoms or ions of metal at a constant temperature is given by the following relationship:

$$D_{Me} = \text{const}\, p_{X_2}^{1/n} \tag{2.1.26}$$

The corresponding relationship for compounds with a real excess of metal ($Me_{1+y}X$) is:

$$D_{Me} = \text{const}\, p_{X_2}^{-1/n} \tag{2.1.27}$$

In the case of a real excess of the oxidant (MeX_{1+y} compounds), when the deviations from the stoichiometric composition are caused by defects present in the anionic sublattice, the self-diffusion coefficient of the oxidant D_X depends on its pressure in the surrounding atmosphere, according to equation (2.1.26). In the case of a real deficit of the oxidant (MeX_{1-y}) this dependence is given by equation (2.1.27).

Equations (2.1.26) and (2.1.27) are valid for isothermal conditions. Using the general relationships (1.3.116) and (1.3.118) it is possible to obtain general relationships between the self-diffusion coefficient, temperature and equilibrium oxidant pressure. On introducing equation (1.3.116) in equation (2.1.8) and using equation (2.1.6), we obtain the following general dependence of the self-diffusion coefficient of the metal in compounds of the $Me_{1-y}X$ type (or of the oxidant in compounds of the MeX_{1+y} type) on the equilibrium oxidant pressure and temperature:

$$D = a_0^2(p_{X_2}^{1/n}) \exp\left(\frac{\frac{2}{n}\Delta S_f + \Delta S_m}{R}\right) \exp\left(-\frac{\frac{2}{n}\Delta H_f + \Delta H_m}{RT}\right) \tag{2.1.28}$$

Similarly, on inserting equation (1.3.118) in equation (2.1.8) and using equation (2.1.6), we obtain an analogous equation for compounds of the $Me_{1+y}X$ or MeX_{1-y} type:

$$D = a_0^2(p_{X_2}^{-1/n}) \exp\left(\frac{\frac{2}{n}\Delta S_f + \Delta S_m}{R}\right) \exp\left(-\frac{\frac{2}{n}\Delta H_f + \Delta H_m}{RT}\right) \tag{2.1.29}$$

It follows from these relationships that, by determining the dependence of the self-diffusion coefficient on the equilibrium oxidant pressure and temperature, it is possible to obtain valuable information not only about the mobility of elements of the crystal structure, but also about the structure of defects, the thermodynamics of their formation and the mechanism of their diffusion in the crystal lattice, since the activation energy determined experimentally is related to the heat of formation of defects ΔH_t and the real activation energy of diffusion ΔH_m by the following equation:

$$E_D = \frac{2}{n}\Delta H_t + \Delta H_m \tag{2.1.30}$$

Hence, when the enthalpy of formation of defects and the degree of their ionisation are known, the activation energy of diffusion can be calculated and, conversely, when ΔH_m is known, the heat of formation of the defects can be calculated. However it clearly follows from the above discussion that measurements of self-diffusion coefficients are of value only if they are carried out not only at a definite temperature but also at a closely controlled pressure of component X in the atmosphere surrounding the crystal. This fact is not always attended to in studies of diffusion in solids.

One of the principal purpose of studies of self-diffusion in ionic crystals and metals is the determination of the mechanism of diffusion together with the entropy an the enthalpy (energy) of activation of this process. This problem is very complex, since not only the simple vacancy or interstitial mechanism must be taken into account, but also the degree of ionisation of defects, their interactions etc. In order to reach a correct interpretation of the experimental studies it is necessary to create a suitable theoretical model. In this case the possibility of estimating the numerical value of the pre-exponential factor D_0 in equation (2.1.22) is of particular importance.

It follows from equation (2.1.24) that this factor consists of several terms including the entropy term. The calculation of this term for the assumed mechanism of diffusion presents great difficulties. Weit and Zener [137, 138] proposed the following solution for the simplest case of diffusion of an interstitial dopant in a metal (solutions of gases in metals). In this case the general equation (2.1.21) becomes:

$$D = \alpha a_0^2 \chi \nu \exp\left(\frac{\Delta S_m}{R}\right) \exp\left(-\frac{\Delta H_m}{RT}\right) \tag{2.1.31}$$

In the case of interstitial diffusion through octahedral positions of a body centred cubic crystal lattice of a metal, the geometrical coefficient $\alpha = 1/6$, a_0 is the lattice constant of the metal, ν is the frequency of vibrations of interstitial atoms of the dopant and ΔS_m and ΔH_m are the entropy and the enthalpy of diffusion, respectively. Wert and Zener have shown that the frequency of vibrations of the atoms of an interstitial dopant is related to the activation energy of diffusion by the following equation:

$$\nu = \frac{\Delta H_m}{2m\lambda^2} \tag{2.1.32}$$

where m is the mass of an interstitial atom of the dopant and λ is the distance between two interstitial positions.

In order to calculate the value of ΔS_m, Wert and Zener utilised the following thermodynamic relation:

$$\Delta S_m = -\frac{d(\Delta G_m)}{dT} \tag{2.1.33}$$

Assuming that the change of the free energy of the lattice ΔG_m, accompanying a jump of an atom of the dopant from one interstitial position to another, is approximately equal to the stress energy appearing during this process, Wert and Zener have shown that ΔS_m can be expressed as follows:

$$\Delta S_m = \beta\left(\frac{\Delta H_m}{T_t}\right) \tag{2.1.34}$$

where T_t is the melting point of the metal and β is defined by the following equation:

$$\beta = -\frac{d(u/u_0)}{d(T/T_t)} \tag{2.1.35}$$

where u is the shear modulus of the lattice. At the absolute zero, $u = u_0$.

It has been estimated that the numerical value of β for the majority of metals varies from 0.35 to 0.40. It follows from equation (2.1.34) that ΔS_m is directly proportional to ΔH_m. Since ΔH_m and β are positive, ΔS_m must also always be positive. This important conclusion makes it possible to estimate the lower limit of numerical values of the pre-exponential coefficient D_0 in equation (2.1.22) a priori.

It can be seen by comparing equation (2.1.22) with equation (2.1.31), that, for interstitial diffusion of a dopant in a metal, D_0 is given by the following equation:

$$D_0 = \alpha a_0^2 \nu \exp\left(\frac{\Delta S_m}{R}\right) \tag{2.1.36}$$

Assuming that $\nu \cong 10^{13}$ s^{-1}, $a_0 = 3\times10^{-8}$ cm and $\alpha = 1/6$ and that the values of the entropy of migration ΔS_m are positive we obtain:

$$D_0 > 10^{-3} \text{ cm}^2/\text{s} \tag{2.1.37}$$

Experimental studies have shown that in the case of diffusion of carbon in α-iron, $D_0 = 0.020$ cm^2/s. At the same time it has been found that the activation energy of diffusion $E_D = 82\,736$ kJ/mole (20 000 kcal/mole). If it is assumed that the diffusion process takes place by the interstitial mechanism, $E_D = \Delta H_m$. Replacing ΔH_m in equation (2.1.34) by the measured value of E_D, and putting $\beta = 0.4$, $T_t = 180$ K, we find that $\Delta S_m = 2.2R$. Hence, the entropy factor: $\exp(\Delta S_m/R) = 9$. On inserting this value in equation (2.1.36) and inserting $\nu = 10^{13}$ s^{-1}, $\alpha = 1/6$ and $a_0 = 3\times10^{-8}$ cm we find that $D_0 = 0.015$ cm^2/s. Bearing in mind the simplifications made during the calculations, we observe that this result is in excellent agreement with the experimentally determined value, $D_0 = 0.020$ cm^2/s. This agreement between the calculated and experimentally determined values of D_0 confirms the correctness of the model assumed for the diffusion of carbon in α-iron. A similar agreement between the calculated and experimentally determined values of D_0 was also obtained in the case of diffusion of oxygen and nitrogen in other metals.

The considerable success of the above theory prompted Zener [136] to attempt to extend it to the vacancy mechanism of diffusion. In this case the entropy term in the pre-exponential factor of the coefficient D_0 consists of two parts: $\Delta S = \Delta S_f + \Delta S_m$ [see equation (2.1.24)]. The estimation of the value of ΔS_m is, in this case, the same as in the case previously discussed of interstitial mechanism. A new problem is the estimation of the value of ΔS_f. This is a very complex problem, which can only be solved to a first approximation at the present time. The application of statistical mechanics to the solution of this problem leads to the conclusion that the entropy of formation of vacancies is related to the frequency of vibrations of atoms in the lattice by the following equation:

$$S_f = \text{const} \ln\left(\frac{\nu_0}{\nu_f}\right) \tag{2.1.38}$$

where ν_0 is the mean frequency of vibrations of an atom in the lattice of a crystal free from defects and ν_f is the mean frequency of vibrations of the same atom situated in the neighbourhood of a vacancy.

Since the lattice is less rigid in the direction of the vacancy, ν_f is smaller than ν_0 and, as a result, $\Delta S_f > 0$. For the sake of simplicity it is assumed that ΔS_f is proportional to ΔS_m and, as a result, the total entropy change ΔS can be expressed by the following simplified formula:

$$\Delta S \cong \lambda\beta\left(\frac{\Delta H}{T_t}\right) \tag{2.1.39}$$

where ΔH is the total enthalpy of the process, λ is a new constant, approximately equal to one for all metals having the same crystal structure.

The values of ΔS calculated on the basis of equation (2.1.39) have been checked experimentally. It was found that the calculated values agree with the experimental ones when $\lambda = 0.55$. It was also found that in the case of metals having regular body centred structures, ΔS is always positive, in agreement with Zener's theory. However the results of the experimental studies have shown unambiguously that, in certain cases, ΔS is negative. So far this fact has not been fully explained. Consequently the validity of equation (2.1.39), unlike that of equation (2.1.34), is limited. In view of the importance of this problem in the theory of solids, attempts are being made to develop Zener's concept or to develop new solutions by means of statistical mechanics. The discussion of these studies is outside the range of the present book.

2.2 The correlation effect

In previous considerations of the mechanism of diffusion it was assumed that, irrespective of the kind of defects present in a crystal lattice, the successive jumps of atoms take place in completely random directions, i.e. they are not correlated with each other. This assumption is justified on condition that, after n jumps, all possible directions for the $(n+1)$-th jump are equally probable. Such a situation exists in the case of vacancies or interstitial atoms, since after any number of jumps all the structural elements surrounding the defect are identical. Thus the diffusion of defects is a non-correlated process. This is understandable in view of the fact that the frequency of vibration of atoms in their lattice sites and in interstitial positions is several orders higher than that of the jumps. For this reason equilibrium in the region surrounding the defect is very rapidly established between successive jumps and the next jump takes place without any effect of the previous jump on its direction. This leads to the conclusion that the diffusion of defects in crystals is a random walk process, in which the successive jumps are completely uncorrelated. However not every random walk is a non-correlated movement. In particular, the self-diffusion of radioactive tracers in crystal lattices of metals and compounds of the MeX type is not uncorrelated when the diffusion takes place according to the vacancy mechanism or the interstitialcy mechanism [133–135].

In order to avoid any ambiguity it should be explained that a tracer is a radioactive or a stable isotope of the lattice (or sublattice) element of the crystal, in which we wish to determine the self-diffusion coefficient. In the case of a metal it is an isotope of this metal and in the case of compounds of the MeX type it is a radioactive or a stable isotope of the component for which we wish to determine the mobility in the given sublattice.

In order to explain the correlated character of the self-diffusion of tracers, we will first discuss the vacancy mechanism of diffusion in metals. The results of this discussion will also be completely valid for the vacancy mechanism of diffusion in ionic crystals of the MeX type.

A tracer atom can make a jump, if a vacancy is present in its immediate neighbourhood. The second jump would be uncorrelated with the first, if the probability of occurrence of the second jump were the same in all possible directions. Hence the probability for the appearance of a vacancy should be the same for all the lattice sites surrounding the tracer atom. However this atom arrives from a position which is vacant at the time of its arrival and the time required for the vacancy to move away is of the same order as that between successive jumps. Thus, at the moment when the tracer atom, vibrating in its lattice site, "is preparing" for the next jump, the chance that the position from which it has arrived is unoccupied is greater than that for any other position around the tracer atom. Consequently, the two jumps are correlated, since it is more probable that after making the first jump the tracer atom will return to its former position than that it will jump in another direction.

The above situation is illustrated schematically in Fig. 111, which shows a plane of a close packed two-dimensional lattice containing a vacancy in position 6 and a tracer atom in position 7. It is evident from the diagram that, if the tracer has arrived from position 6 by exchange with the vacancy, the most probable next jump of this atom is the return to position 6. The second in the order of probability are the jumps to positions 1 and 5, since this process can take place only after a jump of the vacancy from position 6 to positions 1 or 5, from which it can jump to position 7. The least probable is the jump of the tracer to position 3, since it requires

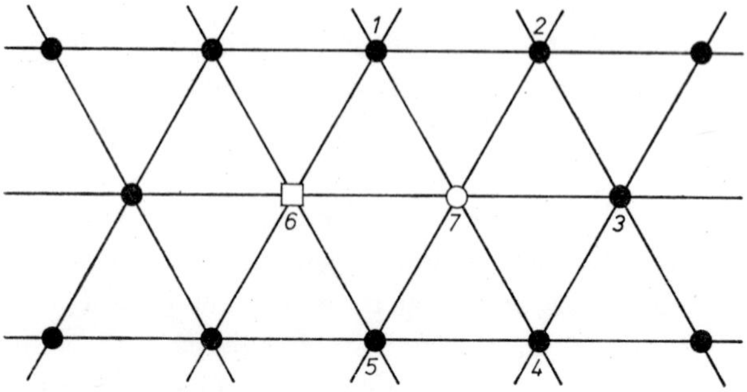

Fig. 111. Two-dimensional close packed lattice with the tracer atom in position 7 and a vacancy in position 6 (after [131]).

successive jumps of the vacancy from position 6 through positions 1 and 2 or 5 and 4 to position 3, from which it can finally jump to position 7.

Since the jumps to position 6 and, to a lesser degree, to positions 1 and 5 are more probable than the jumps to the other positions, the distance travelled by the tracer atom during *n* jumps is shorter than that travelled by the vacancy. For this reason the self-diffusion coefficient of the tracer, which is a measure of the rate of this process, is smaller than that of the

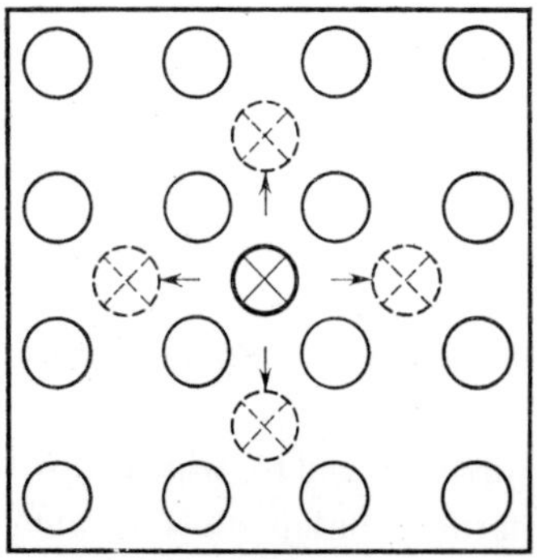

Fig. 112. The scheme of the possible positions of the tracer atom in the case of diffusion taking place according to the simple interstitial mechanism.

atoms constituting the lattice, since the self-diffusion of the tracer is a correlated random walk, whereas the movement of the vacancy and consequently that of the atoms constituting the lattice are not correlated.

An analogous situation exists in the case of diffusion of tracers by the interstitial mechanism.

The correlation effect is absent in the case of the simple interstitial mechanism (Fig. 112), since the successive jumps of atoms from one interstitial position to another are equally probable for all the directions.

It has been explained in a previous chapter that diffusion taking place by the interstitial mechanism may be of the collinear or the non-collinear type (Fig. 103 and 104). In both cases the diffusion of tracer atoms has a definite degree of correlation. This can be explained for one example, e.g. the collinear mechanism, since in the case of non-collinear mechanism the principle of the correlation is the same and the differences are of a quantitative character only.

Figure 113 shows a plane in a two-dimensional lattice containing a tracer atom initially in an interstitial position (Fig. 113*a*) and then, after a jump, in a lattice site (Fig. 113*b*). In the interstitial position the tracer atom is

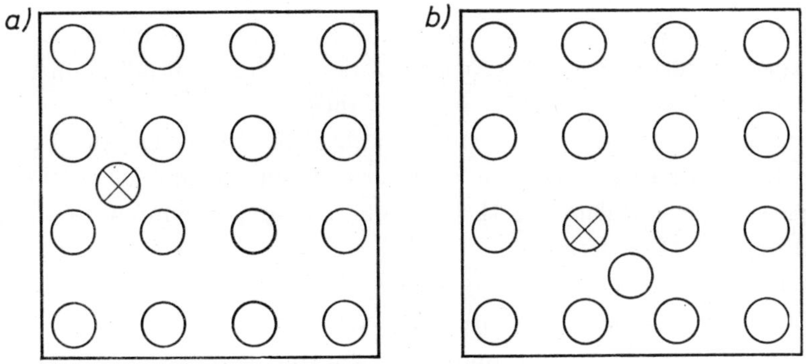

Fig. 113. Two-dimensional, simple regular lattice: *a*) with the tracer atom in the interstitial position; *b*) with the tracer atom in the normal lattice site.

surrounded on all sides with atoms identically situated in space. For this reason a jump may not be correlated with the jump preceding it. However in this case the movement of atoms in crystal consists of the displacement of atoms situated in lattice sites by atoms situated in the interstitial positions. Therefore the tracer atom makes its successive jump to a lattice site by displacing one of its neighbours to the interstitial space (Fig. 113*b*).

In this situation the surroundings of the tracer are no longer uniform, since one of the nearest interstitial positions is occupied by the atom previously displaced to this position by the tracer atom. For this reason the probability that the tracer atom will make its next jump in the direction from which is has come is greater than the probability that it will jump in the opposite direction. According to this mechanism of diffusion there is a definite degree of correlation between the successive jumps of the tracer atom. An analogous situation exists with the non-collinear mechanism.

The above considerations lead to the conclusion that in the case of the two principal mechanisms for diffusion in solids, the vacancy mechanism and the interstitial mechanism, a certain degree of correlation between the successive jumps of the tracer atom always exists. Thus, in these cases, the diffusion coefficient of the tracer must be smaller than that of the lattice atoms. When the diffusional movement of the tracer is uncorrelated, as in the case of the simple interstitial mechanism, these differences are absent. The relation between the diffusion coefficients of the tracer and the lattice atoms can be expressed by the following equation:

$$\frac{D_t}{D} = f \tag{2.2.1}$$

where D_t is the diffusion coefficient of the tracer and f is the so-called correlation coefficient,* which can have various values depending on the mechanism of diffusion and the structure of the crystal.

Obviously in the simple interstitial mechanism, $f = 1$. On the basis of the theory of correlation and by means of fairly complex equations it is possible to calculate the Haven coefficient very accurately for any particular case. The derivation of these equations is outside the range of the present book.

In the case of regular structures, i.e. isotropic crystals, in which the distances of jumps in all directions are identical, the correlation theory makes it possible to calculate the coefficient, f, in a relatively simple way. In this case, the general equation becomes:

$$f = 1+2\sum_i \langle\cos\theta_i\rangle \tag{2.2.2}$$

where $\langle\cos\theta_i\rangle$ is the mean value of the cosine of the angle between the directions of two successive jumps. When the successive jumps are not

* At the International Congress on Diffusion in Solids held in Belgirate in September 1972, the correlation coefficient was given the name "Haven coefficient" [139].

correlated, as in the case of diffusion of defects and lattice atoms, the sum of the mean values of $\cos\theta$ is equal to zero and, as a result, $f = 1$. This result is not valid when a definite degree of correlation between the successive jumps exists. In the case of the vacancy mechanism of diffusion of a tracer in a regular lattice, equation (2.2.2) becomes:

$$f_V = \frac{1 + \langle\cos\theta_i\rangle}{1 - \langle\cos\theta_i\rangle} \tag{2.2.3}$$

and in the case of the interstitial mechanism of diffusion:

$$f_I = 1 + \langle\cos\theta_i\rangle \tag{2.2.4}$$

where f_V and f_I are the correlation coefficients of the vacancy and the interstitial mechanisms of difusion, respectively. Since in both cases $\langle\cos\theta_i\rangle$ is negative, $f < 1$.

Manning [140–141] has shown that, in the case of the vacancy mechanism of diffusion, the correlation coefficient can be calculated in a much simpler way on the basis of the following equation:

$$f = \frac{1-P}{1+P} \tag{2.25}$$

where P is the probability of a jump of a tracer atom to the neighbouring vacancy and $(1-P)$ is the probability that the neighbouring vacancy will move away as a result of jumps of the lattice atoms. Manning has also shown that, to a first approximation, P is equal to the reciprocal of the number of nearest lattice sites around the tracer atom. Hence equation (2.2.5) can be expressed in the following form:

$$f = \frac{1 - \dfrac{1}{Z}}{1 + \dfrac{1}{Z}} = \frac{Z-1}{Z+1} \tag{2.2.6}$$

This equation indicates that in the case of the vacancy mechanism of diffusion in a regular lattice the value of the correlation coefficient can be estimated in a very simple way on the basis of knowledge of the geometrical structure of the crystal alone. In order to illustrate the usefulness of the simplified formula (2.2.6) for a rapid estimation of the numerical value of the correlation coefficient, we will consider the possibility of jumps of a tracer atom in three kinds of regular lattices. The number, Z, of the nearest lattice sites in a simple regular lattice, to which the tracer atom can jump, is equal to 4. In body centred and face centred structures it is equal to 6 and 8 respectively. By introducing these values in equation (2.2.6), we obtain the corresponding numerical values of the correlation coefficient, which are shown in Table 10. The table also shows, for comparison, the exact values of coefficient, f, calculated by the formulae in the

TABLE 10

Correlation factors for self-diffusion of tracer atoms in regular lattices

Structure	The number of the nearest lattice sites Z	Approximate values of f calculated from eqn. (2.2.6)	Accurate values of f	Error of values of f calculated in % from eqn. (2.2.6)
Cubic simple	6	0.71	0.65311	9
Cubic body centred	8	0.78	0.72722	7
Cubic face centred	12	0.85	0.78146	8

correlation theory and the percentage deviations from these values, resulting from the simplifications used in the method based on equation (2.2.6). It can be seen that the error does not exceed 10%.

The above considerations are limited to correlations occurring in self-diffusion processes. It should be mentioned that hetero-diffusion, i.e. diffusion of foreign atoms (dopants) in crystal lattices of compounds of the MeX type, is also a correlated process. In this case the atoms of the dopant can be differentiated from the lattice atoms in the same way as in the case of tracer atoms. However the calculation of the correlation coefficient for hetero-diffusion processes is much more difficult than for self-diffusion as a result of the considerable differences between the masses of atoms of the dopant and of the lattice atoms. These differences causes considerable differences between the frequency of jumps of atoms of the dopant and that of the lattice atoms. This leads to complications, which have so far made it impossible to develop general formulae for the calculation of correlation coefficients in hetero-diffusion processes.

Table 11 gives examples of the correlation coefficients for self-diffusion calculated for several structures using vacancy and interstitial mechanisms.

TABLE 11

Correlation factors for various mechanisms to diffusion referred to various crystallographic structures

The type of crystal lattice	The mechanism of diffusion	Correlation factor
Diamond lattice	vacancy	0.50000
Simple cubic lattice	vacancv	0.65311
Cubic body centred lattice	vacancy	0.72722
Cubic face centred lattice	vacancy	0.78146
Hexagonal close packed lattice	vacancy	0.78121
Simple cubic lattice	interstitial	
	a) collinear type	0.80000
	b) non-collinear type	0.96970
Cubic face centred lattice	interstitial	
	a) collinear type	0.66666
	b) non-collinear type	0.72740

The comparison shown in this table indicates that the correlation effect is in general small and may be neglected in the determination of diffusion coefficients in solids by the tracer methods. However Table 11 also shows that the differences between the correlation coefficients obtained for the same structures, but with different mechanisms of diffusion, are considerable. This offers an unusually valuable possibility of determining the mechanism of diffusion on the basis of the determination of D by several methods. For this reason the correlation theory is of great importance.

Diffusion coefficients in solids can be determined not only by the methods based on the use of tracers but also by other methods, such as methods based on the determination of electrical conductivity, rate of sintering, kinetics of oxidation of metals etc. These methods make it possible to determine the diffusion coefficient of lattice atoms, i.e. the value of D free from the "error" caused by the correlation effect. If the value of D_t can also be determined, the correlation coefficient, f, can be calculated by means of equation (2.2.1). If the values of both D and D_t have been determined with sufficient accuracy, the comparison of the experimentally determined correlation coefficient with that calculated from the correlation theory makes it possible to determine the mechanism of diffusion.

Before giving an example of this kind of study, we will discuss the physical meaning of the correlation coefficient, f. It consists of two additive terms,

f' and f''. The first is related to the difference between the distances travelled during the elementary act by the tracer atom and the defect of structure (f'). The second term is related to the real phenomenon of correlation between the successive jumps of tracer atoms (f''). The terms are related to one other by the equation:

$$f = f'f'' \tag{2.2.7}$$

The physical meaning of the term f'' has already been explained. The term f' is not equal to one, when the distance travelled by the tracer atom during one jump is shorter than the distance between the positions of the defect in the structure before and after the jump. In the case of the vacancy mechanism, $f' = 1$, since the distance travelled by the tracer atom during one jump is the same as that travelled by the vacancy in the opposite direction (Fig. 100). The same applies to the simple interstitial mechanism (Fig. 102). In this case $f' = 1$ and since $f'' = 1$, also (no correlation between successive jumps) $f = 1$.

In the case of the interstitial mechanism $f' < 1$. It can be seen from the diagrams shown in Figs. 101 and 102 that, during the elementary act, the tracer atom moves from an interstitial position to a lattice site, or vice versa. In both cases it travels a distance, a_T, equal to the distance travelled by a lattice atom during the elementary act. However the passage of a lattice atom from an interstitial position to a lattice site is equivalent to the appearance of an identical atom displaced from the lattice site in a neighbouring interstitial position. Thus a jump of a lattice atom from the interstitial position to a lattice site means, in the collinear mechanism, a displacement of the lattice atom by a distance equal to 2 a_T. Hence $f' = 0.5$.

In the case of the interstitial mechanism of the non-collinear type, the difference between the above distances is smaller (Fig. 104) and is equal to 1.6 a_T. Hence in this case $f' = 2/3$. The correlation theory makes it possible to calculated the term f'' for both variants of the interstitial mechanism and consequently to calculate the values of the correlation coefficients. The results of such calculations are shown in Table 11. They indicate that the differences between the values of the correlation coefficients obtained for various mechanisms of diffusion, particularly for the three variants of the interstitial mechanism, are considerable and make it possible to determine the mechanism of diffusion by means of suitable experiments.

Figure 114 gives examples of the results of studies on the mechanism of self-diffusion of silver in α-Ag_2S [142]. This sulphide is an excess metal *n*-type semiconductor ($Ag_{2+y}S$). For this reason, its anionic sublattice is practically free from defects, whereas the cationic sublattice contains defects in the form of interstitial silver cations and quasi-free electrons.

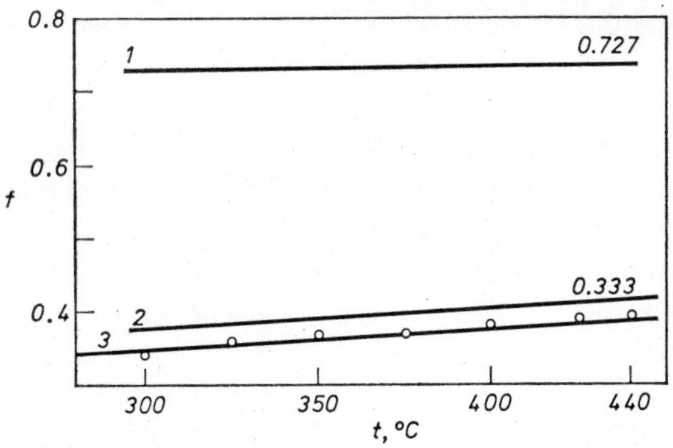

Fig. 114. The temperature dependence of the correlation factor for the diffusion of ions in Ag_2S crystal (after [142]). Curves *1* and *2* illustrate the theoretical dependences calculated by Friauf for the interstitialcy mechanism of the collinear type (curve *2*) and the non-collinear type (curve *1*); curve *3* represents experimental results.

Hence the diffusion of silver in α-Ag_2S can take place by one of the three variants of the interstitial mechanism.

In order to elucidate this problem it is necessary to determine experimentally the correlation coefficient for the process of self-diffusion of radioactive silver ions (tracers) in α-Ag_2S and to compare the result of this determination with the three correlation coefficients calculated for the mechanisms of diffusion which are possible in this case. This requires knowledge of self-diffusion coefficients of the lattice silver ions (D_{Ag}) and of the tracer ($D_{Ag(t)}$). The value of $D_{Ag(t)}$ was determined by Allen and Moore [143] and by Mrowec and Rickert [144] for several temperatures between 200 and 400°C. The values of D_{Ag} for the same temperature range were determined by Bartkowicz and Mrowec [145, 146] on the basis of measurements of the ionic component of the electrical conductivity of silver sulphide and of the kinetics of sulphidation of silver.

The results of these studies were used for calculation of the temperature dependence of the correlation coefficient by means of equation (2.2.1). The results of these calculations are represented by curve *3* in Fig. 114. The plot also shows the temperature dependence of the correlation coefficient calculated by Friauf [147] for the collinear and the non-collinear variants of the interstitial mechanism. It can be seen from the graph that the experimentally determined curve *3* and the theoretical curve *2* almost overlap, which means that the process of diffusion of silver in α-Ag_2S takes place by the collinear mechanism.

The mechanism of self-diffusion of silver in AgCl was investigated in a similar way. It was found that diffusion of silver ions in this compound also takes place by the interstitially mechanism, but it is more complex than in the case of silver sulphide, since in the case of the chloride the diffusion of silver ions takes place by both the collinear and the non-collinear variant but the former is predominant.

It is noteworthy that the studies of this type are still rare, since the determination of exact values of self-diffusion coefficients of lattice atoms or ions presents considerable difficulties.

2.3 The relation between diffusion and ionic conductivity

It has been shown in a previous chapter that the electrical conductivity of ionic crystals of the MeX type is related to the existance of point defects. In the case of compounds with no deviations from stoichiometry the concentration of electronic defects is so small in comparison with ionic defects that the electric current is carried almost exclusively by ionic defects (ionic semiconductors, solid electrolytes). When the concentration of ionic defects is comparable with that of electronic defects, the current is carried almost exclusively by the latter (electronic semiconductors).

In both cases there is a definite relationship between the ionic component of the electrical conductivity of the MeX type crystals and the self-diffusion coefficient of defects. This relationship follows directly from the Nernst–Einstein equation:

$$\frac{\sigma_d}{D_d} = \frac{c_d q_d^2}{kT} \tag{2.3.1}$$

where σ_d is the cationic or the anionic component of the ionic electrical conductivity of the crystal, connected with the given type of defects, D_d is the self-diffusion coefficient of defects corresponding to σ_d, c_d is the concentration of defects, q_d is their charge, k is the Boltzmann constant and T is the absolute temperature.

It follows from equation (2.3.1) that when the concentration of point defects and the ionic conductivity of crystal are known, it is possible to calculate the self-diffusion coefficient of the defects, which cannot be evaluated by means of direct measurements. However, as it has already been mentioned, the exact determination of the concentration of defects in crystals in thermodynamic equilibrium with the surrounding atmosphere and the determination of transport numbers also are very difficult and consequently the determination of ionic conductivity of crystals is a difficult problem.

However the self-diffusion coefficient of defects is related to the diffusion coefficient of atoms by equation (2.1.8). In addition the mole fraction of defects appearing in this equation is related to the concentration of defects as follows:

$$c_{\mathrm{d}} = N_{\mathrm{d}} C \qquad (2.3.2)$$

where C is the concentration of the given kind of ions in the crystal undefected.

Introducing equation (2.3.2) in equation (2.3.1) and using equation (2.1.8) we obtain:

$$\frac{\sigma_{\mathrm{d}}}{D} = \frac{C q_{\mathrm{d}}^2}{kT} \qquad (2.3.3)$$

Equation (2.3.3) is very important, since it connects two important physico-chemical properties, which can be determined experimentally. This makes it possible to draw significant conclusions regarding the mechanism of diffusion and also calculate the ionic conductivity of crystal, which, as a rule, is not readily determinable experimentally. The ionic conductivity can be determined on the basis of equation (2.3.3) with the aid of self-diffusion measurements carried out by means of radioactive tracers. In this case a correction must be introduced in equation (2.3.3) in order to take account of the difference between the self-diffusion coefficient of the lattice atoms (D) and that of the tracer atoms (D_{t}), which is given by the correlation coefficient f. Hence by taking into account equation (2.2.1) we obtain the following general relationship:

$$\frac{\sigma_{\mathrm{d}}}{D_{\mathrm{t}}} = \frac{C q_{\mathrm{d}}^2}{f kT} \qquad (2.3.4)$$

It follows from this equation that the exact determination of the diffusion coefficient, D_{t}, and the ionic conductivity, σ_{d}, makes it possible to calculate the correlation coefficient and hence to determine the mechanism of dif-

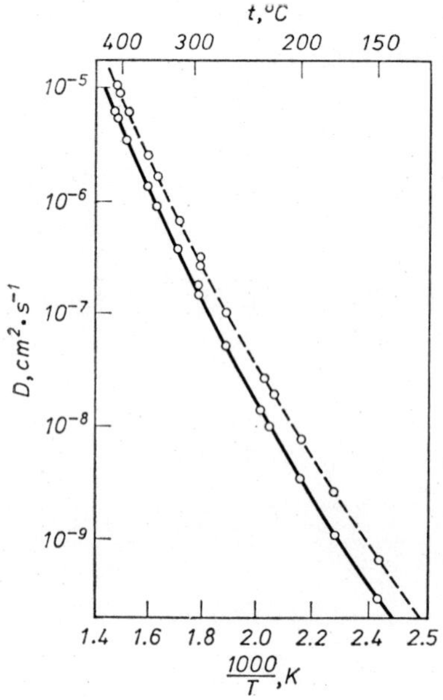

Fig. 115. The temperature dependence of the self-diffusion coefficient of silver in AgBr (after [148]). Continuous line—the values obtained by means of the radioactive silver isotope; the broken line—the values obtained from electrical conductivity measurements.

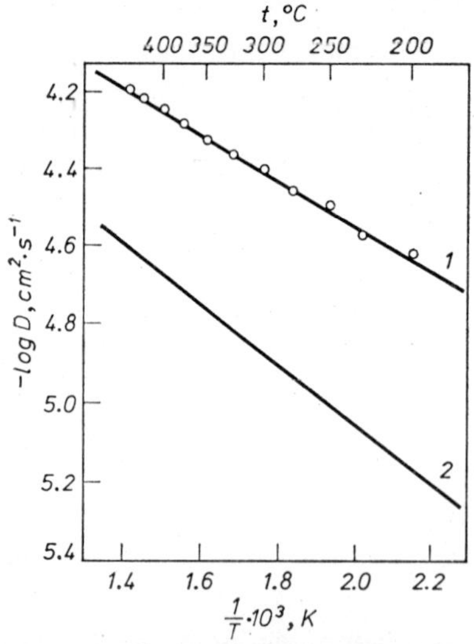

Fig. 116. The temperature dependence of the self-diffusion coefficient of silver ions in Ag_2S crystal (after [142]). Curve *1*—the values calculated from electrical conductivity measurements; curve *2*—the values determined by the tracers method.

fusion:

$$f = \frac{D_t C q_d^2}{\sigma_d k T} \tag{2.3.5}$$

Figures 115 and 116 show, as an example, the temperature dependence of self-diffusion coefficients of silver in silver bromide [148] and silver sulphide [142] calculated on the basis of measurements of electrical conductivity and determined by means of radioactive tracers. It can be seen from the graphs that the values of D_{Ag} calculated by means of equation (2.3.3) are considerably higher than those determined by means of the radioactive silver isotope. The differences are due to the correlation effect.

It should be observed that the results of measurements of the ionic component of electrical conductivity of crystals, over a range of temperatures, make it possible to determine the activation energy of the diffusion process. However this is often erroneously interpreted, since the temperature dependence of σ_d is not differentiated from that of D or D_t and the activation energy of diffusion is calculated from the slope of the plot of $\log \sigma$ against $\frac{1}{T}$. It clearly follows from equations (2.3.3) and (2.3.4) that the ionic component of the electrical conductivity of the crystals is not directly proportional to the self-diffusion coefficient, but to the product of this conductivity and the temperature:

$$\sigma_d T = \frac{C q_d^2}{k} D = \text{const}\, D \tag{2.3.6}$$

or

$$\sigma_d T = \frac{C q_d^2}{f k} D_t = \text{const}\, D_t \tag{2.3.7}$$

Hence the calculation of the correct value of the energy of activation of self-diffusion on the basis of measurements of ionic conductivity can be carried out by plotting $\log \sigma_d T$ against $1/T$, not $\log \sigma_d$ against $1/T$.

2.4 The Kirkendall–Frenkel effect

The fundamental studies carried out by Kirkendall and Smigelskas [149] on the mechanism of diffusion in solids and, in particular, in metal alloys and the theoretical analysis of these processes carried out by Darken [150] have played an important part in the elucidation of the nature of these processes. Kirkendall carried out an experiment in order to find out whether diffusion processes taking place in an alloy–metal system cause a shift of the phase boundary. For this purpose thin molybdenum wires were placed on opposite surfaces of a rectangular block made of α-brass. The distance between the two layers of wires (d) was measured exactly and the block together with the wires was coated electrolytically with a thick layer of copper (Fig. 117). Then the block was subjected to diffusion an-

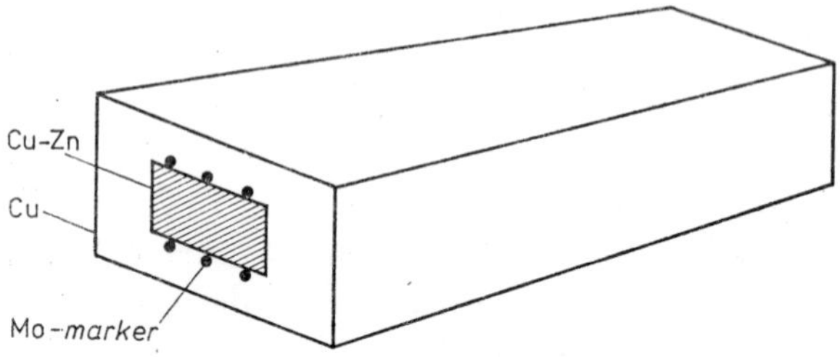

Fig. 117. The scheme of the Kirkendall's experiment.

nealling and after cutting it along a plane perpendicular to the length of the wires the distance between the layers of wires was measured again. It was found that this distance decreased as a result of annealing time and that this decrease was proportional to the square root of the time of annealing.

The shift of the molybdenum wires, which did not participate in the diffusion process, could only be due to inequality of the diffusional fluxes of zinc and copper, which were flowing in opposite directions. In this case the number of atoms of zinc passing in unit time from brass to copper was greater than the number of atoms of copper flowing during the same time in the opposite direction. As a result the number of metal atoms on the copper side increased and consequently on this side new lattice planes were formed. At the same time the number of lattice planes on the side of brass decreased and as a result the geometrical line dividing the copper from the brass marked by the molybdenum wires, retreated. Thus the wires were acting as markers in the diffusion process.

This method was first used in 1929 [151] for the inverstigation of the mechanism of oxidation of iron. The results of these studies caused a radical change of views about the mechanism of oxidant of metals and became the experimental basis on which the ionic–electronic theory of gas corrosion of metals and alloys was founded. During the last twenty years the marker method has been developed widely and it is now used in studies on mechanisms of many solid state reactions, particularly in studies of the types of defects occuring in crystal lattices of compounds of the MeX type [152, 153].

The phenomenon of the shift of the dividing line in the brass–copper system, indicated by means of molybdenum markers, has been called the Kirkendall effect. This effect was later observed in many other metal systems and also in high temperature reactions involving diffusion in the solid state not only in metal phases but also in their absence [152]. This indicates that this phenomenon is of general character.

According to Le Claire [154] the principal importance of the Kirkendall effect in the case of metallic systems consists in the fact that it makes the determination of the mechanism of diffusion possible and at the same time it shows that one diffusion coefficients is insufficient for the complete characterisation of diffusion processes in binary alloys.

When the Kirkendall effect is observed, it is certain that the diffusion process takes place according to the vacancy mechanism. In the cases of ring diffusion and interstitial diffusion this phenomenon cannot take place, since the migration of atoms according to these mechanisms does not lead to the formation of new lattice planes.

Another important result of the Kirkendall's experiment was the proof that in a solid solution the values of diffusion coefficients of the components are not equal to each other. This problem was studied in detail by Darken [155], who introduced the concept of partial diffusion coefficients for the components of binary solid solutions and the so-called chemical diffusion coefficient, D, for the characterisation of diffusion processes in the system as a whole. This coefficient is also called the interdiffusion coefficient.

On the basis of theoretical considerations Darken derived the following simple equation showing the relation between the chemical diffusion coefficient in a binary alloy and the partial diffusion coefficients of the two components:

$$D = D_A N_B + D_B N_A \tag{2.4.1}$$

where N_A and N_B are the mole fractions of components A and B, respectively, in the alloy.

The values of the partial diffusion coefficients depend, as a rule, on the composition of the alloy. For this reason D is a complex, non-linear function of concentration. In the case of very dilute solid solutions (when $N_B \to 0$

and consequently $N_A \to 1$) the chemical diffusion coefficient is approximately equal to the partial diffusion coefficient of the solute.

Assuming that the Kirkendall effect is the only effect of diffusional annealing in a binary system, we can use the following, Darken's, equation together with equation (2.4.1) for the experimental determination of the values of D_A and D_B by measuring the movement of the markers:

$$\frac{x_m}{2t} = (D_A - D_B)\frac{dN_A}{dx} \tag{2.4.2}$$

where x_m is the movement of the markers during time t and dN_A/dx is the concentration gradient in the diffusion zone.

Equation (2.4.2) is valid on condition that the rate of movement of markers is directly proportional to the square root of the annealing time. This condition is fulfilled only when the concentrations of both components at the interfacial drifting together with the markers remain unchanged during the annealing.

However it has been found that the Kirkendall effect is usually accompanied by another phenomenon, called the Frenkel effect, which causes an important limitation to the applicability of Derken's equations. It has been observed that during the annealing of the alloy–metal or metal–metal systems a constriction appears and pores develop near the partition line as shown in Fig. 118. The constriction and the pores appear on the side of the

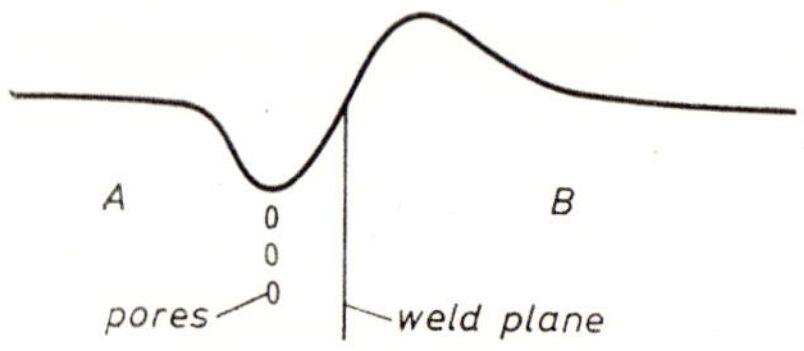

Fig. 118. The scheme of the Frenkel effect.

system, from which more matter flows away in unit time than flows in from the other side. In the scheme shown in Fig. 118 this means that $D_A > D_B$.

Figure 119 shows a microphotograph of a transverse cross-section of a silver–gold system annealed at 900°C for 98 hours [156]. It can be seen that the system is much deformed and that pores are present on the silver side ($D_{Ag} > D_{Au}$). Seith and Kotmann have found that in this case the pores occupy about 40% of the surface of the section.

The appearance of deformations and pores in the system is caused by the non-equivalence of the diffusional fluxes of the components. This provides a further proof of the correctness of the hypothesis that the diffusion takes place by the vacancy mechanism, since the movement of

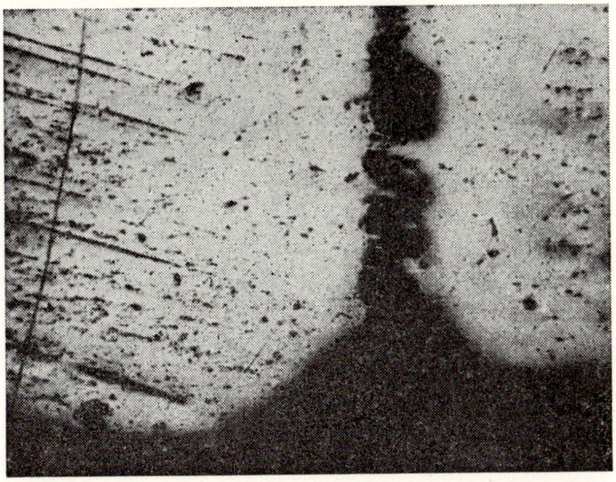

Fig. 119. Microphotograph of the cross-section of the Ag–Au joint after the diffusion annealing (after [156]).

atoms in one direction is equivalent to the movement of empty sites in the opposite direction. In the case of the copper–brass system the flux of vacancies is in the direction of brass, since the flux of zinc atoms flowing out is greater than the flux of copper atoms flowing in the opposite direction. In the places where brass has a very large number of vacancies they form pores, which have regular shapes in many cases (e.g. octahedral or spherical).

The flow of vacancies from copper to brass causes a decrease of their equilibrium concentration in the former phase. Since during annealing, the flow of vacancies is uninterrupted, it must be accompanied by processes leading to the formation of these defects in the lattice. It has been suggested that when the concentration of vacancies becomes lower than that corresponding to the equilibrium state, new vacancies are formed at the grain boundaries and dislocations.

Although the Kirkendall effect accompanied by the Frenkel effect was first observed in metallic systems, recent studies have shown that phenomena of this type in general occur during reactions taking place in solid phases with the participation of lattice diffusion [157, 158]. As a rule the diffusional fluxes of reagents are not equivalent to each other, and this causes changes in the geometry of the system and the formation of porous zones. These changes disturb the original experimental parameters and cause deviations from the course required by the model based on the classical theory of diffusion. For these reasons the experimental determination of the real values of partial diffusion coefficients is very difficult.

2.5 The effect of pressure on diffusion

When a high hydrostatic pressure acts on a crystal, the distances between atoms or ions in the crystal become slightly smaller. Since the repulsive forces between atoms rapidly increase with decreasing distance between them, the external pressure is expected to slow down the movement of structural elements in the crystal [159, 160].

However the diffusion rate depends not only on the energy barier, to be overcome by an atom changing its position in the lattice, but also on the concentration of defects. This factor is also a function of the external pressure, since the formation of vacancies or interstitial atoms is accompanied by an increase of the volume of crystal. Thus the formation of a defect requires work to be done against the external pressure; this work is equal to the product of the pressure and the increase of volume accompanying the formation of the defect.

It was shown in a previous chapter that the mole fraction of defects in a crystal is related to the free energy of their formation according to the following equation:

$$N_d = \exp\left(-\frac{\Delta F_f}{RT}\right) \tag{2.5.1}$$

where ΔF_f is the free energy of formation of one mole of defects.

In order to take into account the effect of the external pressure, equation (2.5.1) must be modified by adding the term corresponding to the work against the external pressure to the free energy of formation of defects:

$$N_d^p = \exp\left(-\frac{\Delta F_f + p\Delta V_f}{RT}\right) \tag{2.5.2}$$

where N_d^p is the mole fraction of defects at an external pressure p and ΔV_f is the increase of the volume of crystal accompaing the formation of one mole of defects. Dividing equations (2.5.1) and (2.5.2) we obtain:

$$N_d^p = N_d \exp\left(-\frac{p\Delta V_f}{RT}\right) \tag{2.5.3}$$

It follows from equation (2.5.3) that the equilibrium concentration of point defects in the crystal decreases with increasing external pressure.

An analogous equation describes the effect of the external pressure on the frequency of jumps of atoms in a crystal:

$$\omega^p = \omega \exp\left(-\frac{p\Delta V_m}{RT}\right) \tag{2.5.4}$$

where ΔV_m is the increase of the volume of the crystal accompanying the displacement of one mole of defects and ω and ω^p are the frequencies of jumps neglecting and taking into account the hydrostatic pressure.

The increase in volume accompanying the migration process is due to the elastic deformation of the crystal during the passage of an atom from one stable position to another. Irrespective of the mechanism of jumps (vacancy or interstitial), an atom situated in the activated position approaches its nearest neighbours and causes them to displaced from their central positions. This deviation is transmitted elastically to other atoms in an attenuated form and as a result the volume of the crystal increases.

Introducing equations (2.5.3) and (2.5.4) in the general formula (2.1.7) we obtain the following relation between the diffusion coefficient and the external pressure:

$$D^p = D \exp\left(-\frac{p\Delta V_f + \Delta V_m}{RT}\right) \tag{2.5.5}$$

where D^p and D are the diffusion coefficients in the presence and in the absence of hydrostatic pressure.

Equation (2.5.5) can be transformed by replacing the sum of the two molar volumes with one term called the activation volume:

$$D^p = D \exp\left(-\frac{p\Delta V_a}{RT}\right) \tag{2.5.6}$$

It follows from equations (2.5.5) and (2.5.6) that the diffusion coefficient and consequently the diffusion rate decrease as the external pressure acting on the crystal increases.

Figure 120 shows an example of the results of studies on the effect of external pressure on the self-diffusion coefficient of lead in polycrystalline lead at several temperatures [161]. The results of these studies confirm the correctness of the above theoretical conclusions. It should be stressed that equation (2.5.6) has the general character, independent of the mechanism of diffusion, although the physical meaning of ΔV_a depends on this mech-

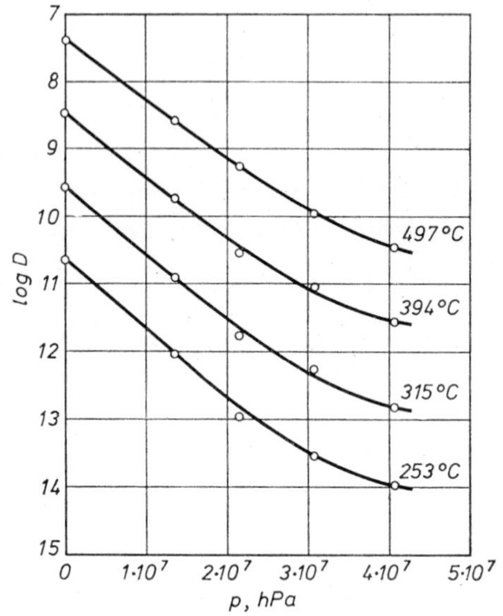

Fig. 120. The effect of pressure on the self-diffusion rate of lead in polycrystalline lead at several temperatures (after [161]).

anism. In the case of the vacancy mechanism ΔV_a represents the increase of the volume of crystal due to the formation of vacancies and their migration in the crystal structure, whereas in the case of the interstitial mechanism it represents the increase of the volume caused by the appearance of atoms (or ions) in interstitial positions and their migration in the crystal.

Since the activation volume depends on the mechanism of diffusion, it becomes possible to investigate the mechanism of diffusion in solids by measuring the dependence of the coefficient D on pressure and by comparing the values of activation volume thus obtained with the values calculated on theoretical grounds. Such calculations can be based on the theory of elasticity.

2.6 The chemical diffusion coefficient and the effective diffusion coefficient

So far our considerations have been limited to the description of diffusion in solids in terms of three diffusion coefficients, i.e. the self-diffusion coefficient of the atoms or ions, the coefficient of diffusion of defects and the tracer diffusion coefficient. All these coefficients characterise the mobility of the corresponding structural elements in crystal lattices of substances in a state of thermodynamic equilibrium with the surroundings. Consequently each can be expressed by the following equation [162]:

$$D_i = \lim_{\frac{dc}{dx} \to 0} \left(\frac{J_i}{\frac{dc_i}{dx}} \right) c_i \tag{2.6.1}$$

where D_i is the self-diffusion coefficient of the *i*-th element of the lattice (an atom, an ion or a defect), J_i is the diffusional flux, c_i is the concentration of the diffusing species and x is the position coordinate.

In many cases it is necessary to determine the mobility of the structural elements of the crystal lattice in conditions far removed from the thermodynamic equilibrium. This situation is frequently encountered during sintering processes, oxidation of metals, preparation of ferrites etc. During such processes a concentration gradient of lattice defects exists in the crystals of compound constituting the reacting system. Under the influence of this gradient, diffusion processes take place. Similar processes take place when crystals are heated to certain temperatures under constant pressure in order to obtain a required concentration of point defects corresponding to the conditions of the heat treatment.

The time required for the establishment of thermodynamic equilibrium in the system, i.e. the time necessary for reaching the defect equilibrium under the above conditions, depends on the rate of diffusion of defects. Since the diffusion takes place as a result of the concentration gradient of defects, its rate cannot always be determined from the values of self-diffusion coefficients.

On the other hand a knowledge of the appropriate value of the diffusion coefficient for the calculation of the homogenisation time of a substance under given conditions or of the rate of a solid state reaction controlled by diffusion, is required in fundamental studies and in many technological processes. Such a diffusion coefficient is called the "chemical diffusion coefficient" [162–164] in the case of the movement of defects and the "effective diffusion coefficient" in the case of the movement of atoms or ions.

The problem of chemical diffusion has already been mentioned in a previous chapter during the discussion of the Kirkendall–Frenkel effect.

In the case of compounds of the MeX type, which are of interest to us in the present work, the physical sense of the chemical diffusion coefficient can be explained as follows: if we place a crystal of a substance of this type at a definite temperature and we maintain it at a constant oxidant pressure, defects will start to diffuse in the crystal in order to establish the concentration of defects corresponding to the thermodynamic equilibrium with the surroundings.

In the case of a compound with deviations from stoichiometry the equilibrium concentration of defects is reached as a result of binding a certain amount of oxidant on the surface of the crystal or as a result of release of this component to the surroundings. In the former case we are dealing with oxidation of the compound and in the latter with its reduction. Oxidation of a compound of the $Me_{1-y}X$ type causes an increase of the concentration of cation vacancies [a shift of the defect equilibrium (1.3.85) to the right], whereas oxidation of a compound of the $Me_{1+y}X$ type is equivalent to a decrease of the concentration of interstitial cations [a shift of equilibrium (1.3.79] to the left). Obviously in both cases the reduction process causes changes in the concentration of defects in the opposite direction. In the case of compounds with a stoichiometric composition, the equilibrium concentration of defects depends practically exclusively on temperature and, for this reason, the effect of the partial pressure of the oxidant on this concentration can be neglected to a first approximation (see Chapter 1.3).

If it is assumed that both sublattices of crystals of the MeX type contain defects, the total diffusional flux of defects is the algebraic sum of the diffusional fluxes of the cationic and anionic defects:

$$J = J_k + J_a \tag{2.6.2}$$

where J_k and J_a are the diffusional fluxes of the cationic and anionic defects, respectively.

The diffusional flux defined in this way is related to the concentration gradient of the diffusing components by Fick's first law:

$$J = \tilde{D}\frac{\partial c}{\partial x} \tag{2.6.3}$$

where $\frac{\partial c}{\partial x}$ is the concentration gradient of defects and $\tilde{D}$ is the chemical diffusion coefficient.

It follows from the above considerations that the chemical diffusion coefficient is the mean diffusion coefficient of the two kinds of defects in crystal MeX, when this crystal is not in the state of thermodynamic equilibrium with its surroundings. It has been shown in previous chapters that, in compounds of the MeX type and in particular in oxides and sulphides of metals, as a rule only one sublattice (cationic or anionic) contains defects. Hence, in such cases, the chemical diffusion coefficient is the diffusion coefficient of the cationic or anionic defects and it characterises the mobility of these defects, when a concentration gradient exists in the crystal.

The relation between the chemical diffusion coefficient and the diffusion coefficient of defects is given by the following equation:

$$\tilde{D} = (1+|p|)D_d \tag{2.6.4}$$

where p is the mean effective charge of the defects with respect to the lattice. In the case of compounds of the $Me_{1-y}X$ type, p is the degree of ionisation of cation vacancies and, in the case of compounds of the $Me_{1+y}X$ type, it is the degree of ionisation of interstitial cations. In the case of compounds of the MeX_{1-y} or MeX_{1+y} type, p is the degree of ionisation of anion vacancies or of interstitial anions respectively.

It follows from equation (2.6.4) that when the predominant defects have no effective charge ($p = 0$), the chemical diffusion coefficient is equal to the diffusion coefficient of defects. When the cationic or anionic defects have a definite effective charge with respect to the lattice ($|p| > 0$), the chemical diffusion coefficient is greater than the diffusion coefficient of defects.

This is understandable in view of the fact that the chemical diffusion coefficient characterises the mobility of defects, where there is a gradient in their concentration in the crystal, whereas the self-diffusion coefficient of defects refers to conditions of thermodynamic equilibrium between the crystal and its surroundings. When the defects have no charge with respect to the lattice, the forces that could affect the diffusion coefficient of defects are absent, irrespective of whether the diffusion takes place in equilibrium conditions or in the presence of a concentration gradient of defects.

Obviously this is only true if the diffusion coefficient of the defects is independent of their concentration. In the case of the movement of defects in ionic crystals of the MeX type, this condition is usually fulfilled, since the concentration of defects is usually so small in comparison with the number of occupied lattice sites, that the crystal can be regarded as a very dilute solution of defects in the lattice. It is known that in this type of solutions the diffusion coefficient is practically independent of the concentration of the diffusing substance.

When the defects in the crystal structure have a definite charge with respect to the lattice, the situation is quite different. In the case of compounds of the $Me_{1-y}X$ or MeX_{1+y} type, this means that in addition to completely or partly ionised cation vacancies or interstitial anions, the compound contains an equivalent number of electron holes, whereas, in the case of crystals of the $Me_{1+y}X$ or MeX_{1-y} type, this means that an equivalent number of quasi-free electrons are present in the crystal. In these cases, the more mobile—electron defects migrate faster than the ionic defects as a result of the presence of a chemical potential gradient in the crystal. Consequently, an electrostatic field appears in a system with diffusing ionic and electronic defects. This field accelerates the diffusion of ionic defects and slows down the diffusion of electronic defects. This is called as ambipolar diffusion. The acceleration of the diffusion of ions increases with increasing degree of ionisation of ionic defects [higher absolute value of p in equation (2.6.4)]. On the other hand, when the crystal is in a thermodynamic equilibrium with the surroundings, the self-diffusion coefficient of defects, D_d, does not depend on the degree of ionisation of the defects, since, under these conditions, the movement of defects has a random walk character, in which there are no favoured directions. Obviously, in this case there is no reason, why the migration of electronic defects should dominate that ionic defects and, as a result, the mobility of the defects does not depend on their degree of ionisation.

It was shown in a previous chapter that the self-diffusion coefficient of defects is related to the self-diffusion coefficient of atoms or ions by equation (2.1.8). Introducing equation (2.6.4) in equation (2.1.8), we obtain the relation between the chemical diffusion coefficient and the self-diffusion coefficient of atoms:

$$\tilde{D} = (1+|z|)D\frac{1}{N_d} \qquad (2.6.5)$$

Since the self-diffusion coefficient of atoms or ions is related to the tracer diffusion coefficient by equation (2.2.1), by taking into account the correlation effect, we obtain the following relation between the coefficient D_t and the chemical diffusion coefficient:

$$\tilde{D} = (1+|z|)\frac{1}{f}D_t\frac{1}{N_d} \tag{2.6.6}$$

These considerations lead to the conclusion that the chemical diffusion coefficient can be calculated when the self-diffusion coefficient of atoms or ions is known and when the concentration of defects, as well as their degree of ionisation, is also known.

The chemical diffusion coefficient can also be determined experimentally by measuring the changes of electrical conductivity or changes of mass taking place during the period in which thermodynamic equilibrium between the crystal and its surroundings is being established. These methods are described in the third chapter of the present work.

It has already been mentioned that the movement of atoms or ions taking place in the presence of a concentration gradient in an ionic crystal is characterised by the effective diffusion coefficient D_{ef}. The numerical value of this coefficient is equal to the product of the chemical diffusion coefficient and the concentration of defects N_d:

$$D_{ef} = \tilde{D}N_d \tag{2.6.7}$$

Using equation (2.6.4) we can express equation (2.6.7) in the following form:

$$D_{ef} = (1+|p|)D_dN_d \tag{2.6.8}$$

By comparing this equation with equation (2.1.8), we observe that the effective diffusion coefficient can be equal to or greater than the self-diffusion coefficient and that the difference can be due only to the ambipolar character of diffusion ($|p| > 0$). This is understandable in view of the fact that the difference between D_{ef} and D is caused solely by the difference between $\tilde{D}$ and D_d.

Before concluding the consideration of the kinetics of diffusion in ionic crystals we should mention the general form of the equations derived for five types of diffusion coefficients. Three of them characterise the mobility of defects and atoms in the absence of a concentration gradient, i.e. in conditions corresponding to a state of equilibrium between the crystal and its surroundings, and the other two the mobility of defects and atoms in conditions far removed from the equilibrium state.

(a) *In a state of thermodynamic equilibrium*:

1. The diffusion coefficient of defects:

$$D_d = \alpha a_0^2 \chi \nu \exp\left(\frac{\Delta S_m}{R}\right) \exp\left(-\frac{\Delta H_m}{RT}\right) \tag{2.6.9}$$

2. The diffusion coefficient of atoms (the self-diffusion coefficient):

$$D = \alpha a_0^2 \chi \nu p_{X_2}^{\pm 1/n} \exp\left(\frac{\frac{2}{n}\Delta S_f + \Delta S_m}{R}\right) \exp\left(-\frac{\frac{2}{n}\Delta H_f + \Delta H_m}{RT}\right) \tag{2.6.10}$$

In the case of compounds of the $Me_{1-y}X$ and MeX_{1+y} type, the exponent $1/n$ is positive and, in the case of the compounds $Me_{1+y}X$ and MeX_{1-y}, it is negative. In the case of stoichiometric crystals, the coefficient, D, is independent of the equilibrium pressure of the oxidant and hence $1/n = 0$. In the case of non-stoichiometric compounds the numerical value of this coefficient depends on the degree of ionisation of defects p according to the following equation:

$$\frac{1}{n} = \frac{1}{2(1+|p|)} \tag{2.6.11}$$

3. The diffusion coefficient of tracers:

$$D_t = \alpha a_0^2 \chi \nu p_{X_2}^{+1/n} \cdot f \cdot \exp\left(\frac{\frac{2}{n}\Delta S_f + \Delta S_m}{R}\right) \exp\left(\frac{\frac{2}{n}\Delta H_f + \Delta H_m}{RT}\right) \tag{2.6.12}$$

(b) *In non-equilibrium conditions*:

1. The chemical diffusion coefficient:

$$\tilde{D} = \alpha a_0^2 \chi \nu (1+|p|) \exp\left(\frac{\Delta S_m}{R}\right) \exp\left(-\frac{\Delta H_m}{RT}\right) \tag{2.6.13}$$

2. The effective diffusion coefficient:

$$D_{ef} = \alpha a_0^2 \chi \nu (1+|p|) p_{X_2}^{\pm 1/n} \exp\left(\frac{\frac{2}{n}\Delta S_f + \Delta S_m}{R}\right) \exp\left(-\frac{\frac{2}{n}\Delta H_f + \Delta H_m}{RT}\right) \tag{2.6.14}$$

It follows from the above equations, that there is a definite mathematical relation between the five diffusion coefficients. Hence when one of them is known, the others can be calculated on condition that the necessary information regarding the crystal structure and, in particular, the concentration of defects, their degree of ionisation and the correlation coefficient is available.

2.7 Empirical relationships

It follows from the discussion in Chapter 2.1, that it is possible to calculate the activation energy of diffusion E_D and the pre-exponential factor D_0 in the Arrhenius equation (2.1.11) from theory only for special cases involving the interstitial mechanism of diffusion in metals. For this reason a search for empirical relations between E_D and D_0, as well as other measurable physical properties of solids has continued for over 50 years [173]. Already in 1922 Dushman and Langmuir [165] proposed the following empirical relationship for metallic systems:

$$D = \frac{E_D a_0^2}{Nh} \exp\left(-\frac{E_D}{RT}\right) \tag{2.7.1}$$

where a_0 is the lattice constant, N is the Avogadro number and h is Planck's constant.

It follows from this equation that the pre-exponential factor in the Arrhenius equation is given by the following relationship:

$$D_0 = \frac{E_D a_0^2}{Nh} = 1.04 \times 10^{-3} E_D a_0^2 \tag{2.7.2}$$

Hence the general Arrhenius equation for diffusion processes (equation 2.7.1) can be expressed in the following detailed form:

$$D = 1.04 \times 10^{-3} E_D a_0^2 \exp\left(-\frac{E_D}{RT}\right) \tag{2.7.3}$$

It follows from this relationship that the activation energy and the pre-exponential factor, D_0, can be determined on the basis of one diffusion coefficient measurement carried out at any temperature. It has been shown that the empirical Dushman–Langmuir equation gives fairly good results for pure metals (the error does not exceed 30%). The pre-exponential factor, D_0, estimated in this way varies from 0.2 to 0.9 cm^2/s.

By using the average value of this factor, equal to 0.5 cm^2/s, we can estimate the activation energy for diffusion, E_D, by means of the following simple empirical relationship:

$$E_D = -RT \ln 2D \tag{2.7.4}$$

The activation energy for diffusion can also be estimated on the basis of other simple empirical equations in which the activation energy is related to such physical properties of solids as melting point, heat of fusion, heat of evaporation and elastic modulus, since all these properties are connected with the bond energies of the atoms forming the crystal lattices of the solid. Using the symbol P as the general symbol for all these properties of solids we can express the relation between them and the activation energy of self-diffusion by the following equation:

$$E_D = AP \tag{2.7.5}$$

or

$$E_D = \frac{B}{P} \tag{2.7.6}$$

where A and B are proportionality constants. Certain of the above detailed relationships make it possible to estimate the energy of activation of diffusion quite accurately, but they do not possess any fundamental physical meaning.

As has already been mentioned, one of these detailed relationships relates the melting points of solids to their self-diffusion coefficients. The melting points are fairly closely connected with the transport properties of solids. In this case the general equation (2.7.5) becomes:

$$E_D = AT_m \tag{2.7.7}$$

where T_m is the melting point on the Kelvin scale. Figure 121 shows a plot of this relationship for several metals. It can be seen that this relationship

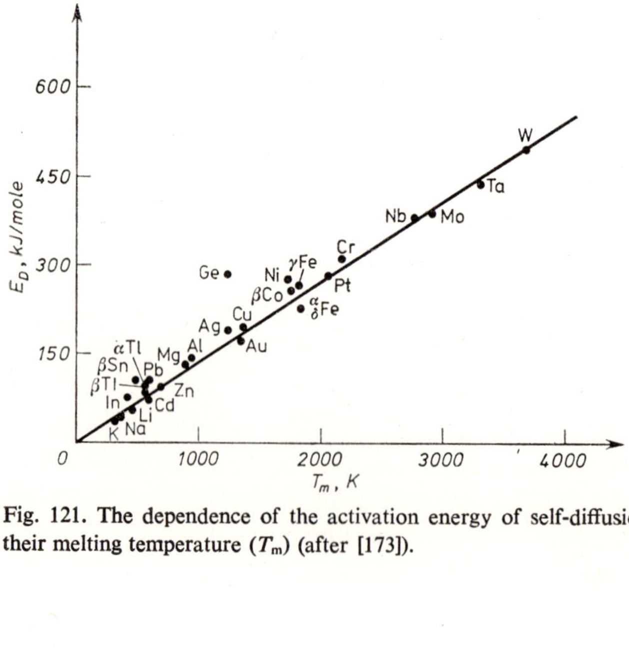

Fig. 121. The dependence of the activation energy of self-diffusion (E_D) in metals on their melting temperature (T_m) (after [173]).

is linear over a very wide temperature range, which makes it possible to estimate the numerical value of the proportionality constant, A, in equation (2.7.7) as equal to 141.1 kJ/(mole · K) (33.7 kcal/mole · K). Hence equation (2.7.7) can be expressed in the following form:

$$E_D = 33.7T_m \tag{2.7.8}$$

However it has been found that, for metals having cubic face centred structures, better results are obtained when it is assumed that A is equal to 159.1 kJ/(mole · K) (38.0 kcal/mole · K). In the case of cubic body centred structures, it is best to assume that $A = 136.1$ kJ/(mole · K) (32.5 kcal/mole · K). Equation (2.7.8) is not valid for the following elements: Sn, Ge, γ-U, β-Ti and β-Zr.

Oshyarin [166] has derived a more detailed relationship between E_D and T_m, taking into account atomic radii, force constant and packing effects. He obtained the following equation:

$$E_D = \frac{\text{const}}{f^{2/3}} T_m \tag{2.7.9}$$

where f is the atomic packing fraction, which for cubic face centred structures and cubic body centred structures is equal to 0.78 and 0.34, respectively.

It can be seen by comparing equations (2.7.7) and (2.7.9) that the constant, A, in the general equation (2.7.7) is given by:

$$A = \frac{\text{const}}{f^{2/3}} \tag{2.7.10}$$

Calculation shows that the Oshyarin equation gives better results than those obtained by means of equation (2.7.8).

By introducing equation (2.7.8) in equation (2.7.1), it is possible to derive an important empirical equation relating the self-diffusion coefficient of a metal to the temperature:

$$D = 3.4 \times 10^{-5} T_m a_0^2 \exp\left(-\frac{17.0T_m}{T}\right) \tag{2.7.11}$$

This equation makes it possible to calculate approximate values of self-diffusion coefficients of a metal over a wide temperature range.

Attempts have been made to calculate the activation energy of diffusion on the basis of the correlation between the heat of fusion L_f and the heat of sublimation L_s of solids. The relations between these physical properties can be expressed approximately as follows:

$$E_D = 16.5L_f \tag{2.7.12}$$

$$E_D = 0.65L_s \tag{2.7.13}$$

It should be stressed that the errors in the results of calculations based on the above relationships are considerably greater than those resulting from the use of equation (2.7.4) and (2.7.9).

Another physical property of solids related to the activation energy of diffusion is the linear expansion coefficient α. Askill [167] has shown that at room temperature the relation between this coefficient and the activation energy of diffusion is as follows:

$$E_D = \frac{700}{\alpha}\ \text{kJ/mole} \tag{2.7.14}$$

where α is expressed in ppm/°C.

Calculations based on this equation have shown that the errors in the determination of the activation energy of self-diffusion in metals do not exceed $\pm 10\%$.

Another empirical relationship exists between the elastic modulus of a solid and the activation energy and consequently, the rate of diffusion in its crystal lattice. Buffington has shown that the activation energy can be expressed by the following equation:

$$E_D = KE_0 a_0^3 \tag{2.7.15}$$

where K is a proportionality constant, which depends only on the structure of the lattice, E_0 is the elasticity constant and a_0 is the lattice constant.

On the basis of the theory of elasticity Zener [169] has derived the following relationship between the pre-exponential factor of the Arrhenius equation and the activation energy of diffusion:

$$D_0 = a_0^2 \nu \exp\left(\frac{\lambda \beta E_D}{RT_m}\right) \tag{2.7.16}$$

or

$$\ln \frac{D_0}{a_0 \nu} = \frac{E_D \lambda \beta}{T_m R} \tag{2.7.17}$$

where λ is the fraction of energy used for the deformation of the lattice in the elementary act of diffusion and the dimensionless quantity β is defined as follows:

$$\beta = -\frac{d\left(\frac{u}{u_0}\right)}{d\left(\frac{T}{T_m}\right)} \tag{2.7.18}$$

where u is the elastic modulus of the lattice. At absolute zero $u = u_0 \cdot \nu$ is the frequency of vibrations of the atoms in the lattice.

An analogous relationship was given by Diennes for $K = 10^{-6}$:

$$D_0 = a_0^2 \nu K \exp \frac{E_D}{RT_m} \tag{2.7.19}$$

More general relationships, making it possible to obtain more accurate results, have been derived by Sherby and Simnad [171] and by Le Clair [170]. These authors took into account the effect of the structure of crystal and the mean valence of the metal and found that the activation energy of self-diffusion in metals depends not only on the melting point but also on the structure of the lattice, characterised by the structure coefficient K_0 and the valence Z. When these parameters are taken into account it appears that three groups of metals can be distinguished with the following structures:

(a) regular body centred, (b) regular face centred, (c) diamond structure, for which the values of the coefficient K_0 are 14, 17 and 21, respectively. The effective valence, Z, of transition metals varies within relatively narrow limits and has the following values: for the metals of group IV*B* (Ti, Zr, Hf) 1.5, for metals of group V*B* (V, Nb, Ta) 3.0, for metals of group IV*B* (Cr, Mo, W) 2.8, for metals of group VII*B* (Mn, Re) 2.6. and for other transition elements 2.5.

Using these data, Sherby and Simnad [171] obtained the following empirical equation for the activation energy of self-diffusion in metals:

$$E_D = RT_m K_0 + Z \tag{2.7.20}$$

According to Askill [173] this equation is the best empirical approximation and has the most general significance. A similar equation was derived by Le Clair [172]:

$$E_D = RT_m K + 1.5Z \tag{2.7.21}$$

The significance of the above empirical relationships is varies from case to case. It should be stressed that so far the range and limitations of their application have been checked only in the case of metals and elemental semiconductors. In the case of compounds of the MeX type the situation is, unfortunately, much more complex. According to Askill the empirical equations discussed above can be arranged in the following order according to their real value:

Activation energy of self-diffusion:

$$E_D = (K_0+Z)RT_m \tag{2.7.22}$$

$$E_D = 33.7T_m \tag{2.7.23}$$

$$E_D = -RT\ln 2D \tag{2.7.24}$$

$$E_D = \frac{700}{\alpha} \tag{2.7.25}$$

$$E_D = 16.5L_f \tag{2.7.26}$$

$$E_D = 0.65L_s \tag{2.7.27}$$

The self-diffusion coefficient:

$$D = 3.4\times 10^{-5}T_m a_0^2 \exp\left(-\frac{17T_m}{T}\right) \tag{2.7.28}$$

Pre-exponential factor D_0 in equation $D = D_0\exp\left(-\frac{E_D}{RT}\right)$

$$D_0 = a_0^2\nu\exp\left(\frac{\lambda\beta E_D}{RT_m}\right) \tag{2.7.29}$$

$$D_0 = 1.04\times 10^{-3}E_D a_0^2 \tag{2.7.30}$$

where:

D = diffusion coefficient (cm^2/s),
D_0 = frequency factor (cm^2/s),
a_0 = lattice constant expressed in nm (0.1×10^{-9} m),
T = temperature in the Kelvin scale,
T_m = melting point in the Kelvin scale,
ν = frequency of vibrations of atoms in the lattice (s^{-1}),
R = gas constant (8.319 J/(mole · K))
L_f = heat of fusion in kJ/mole,
L_s = heat of sublimation in kJ/mole,
λ = constant (0.6 for metals having a regular body centred structure and 0.8 for metals having a regular face centred structure),

β = constant ($\approx 1/2$),
α = linear expansion coefficient in ppm/°C,
K_0 = constant (14 for metals having a regular body centred structure, 17 for metals having a regular face centred structure and 21 for elements having a diamond structure),
z = valency.

2.8 Diffusion in crystals containing extended defects

In our considerations of the mechanism of diffusion in solids so far we have only been taking into account the point defects that give rise to the movement of atoms in crystal lattices. However it has been shown in Chapter 1.4 that in practice in many cases the structure of defects is much more complex and that defects may appear in the form of complexes and cluster or special planar defects called Wadsley defects. Thus the question arises: what is the mechanism of migration of atoms or ions in crystal structures containing such defects? Unfortunately we must admit that our knowledge of mass transport processes in crystals containing extended defects is limited to hypotheses of a qualitative character. Since this problem is of fundamental importance from both theoretical and practical points of view, an extensive programmes of work aiming at the elucidation of of the mechanism of diffusion in materials containing extended defects are being carried out in many research establishments. Although the results of this work are still limited to hypotheses, their brief discussion seems to be indicated.

2.8.1 Diffusion in crystals containing complexes and clusters of defects

The classical example of a well known compound containing complexes and clusters of defects is wustite, $Fe_{1-y}O$. It has already been shown in Chapter 1.4 that Roth complexes, which can form larger assemblies, called Koch-Cohen clusters, are present in the cationic sublattice of this oxide. The Roth complex consists of two cation vacancies and one trivalent cation situated in an interstitial position. These complexes can increase, associating simultaneously additional vacancies, leading to the formation of Koch-Čohen clusters containing 4 trivalent interstitial cations and 13 cations vacancies.

Obviously, the mobility of cation vacancies and consequently that of iron cations in the oxide where the vacancies are combined in complexes or clusters of defects will be different from the mobility of vacancies independent of the extended defects, i.e. constituting real point defects, since the probability of the whole system of defects jumping is incomparable smaller than that of a jump of a single vacancy or a single interstitial cation. Hence the transport properties of wustite mainly depend on the mobility and concentration of quasi-free cation vacancies not combined in complexes and clusters of defects. The concentration of quasi-free vacancies in $Fe_{1-y}O$ is much smaller than their total content and is not connected with the deviation from stoichiometry (y). On the other hand the relative concentration of free vacancies depends on the total content and the temperature, since the formation of complexes of defects is a reversible process and the equilibrium between the combined and free cation vacancies shifts in the direction of the latter when the total concentration of defects decreases and when the temperature increases. For this reason, in disagreement theory (see Chapter 2.1), the diffusion coefficient of the defects and the chemical diffusion coefficient, which are the measure of the mobility of defects, are not expected to be independent of their concentration. Consequently the self-diffusion coefficient of iron in $Fe_{1-y}O$ would not be expected to be a linear function of the concentration of defects. Finally the activation energy for diffusion should depend on this concentration (i.e. on deviation from stoichiometry), since the ratio of free to combined cation vacancies is a function of their total concentration.

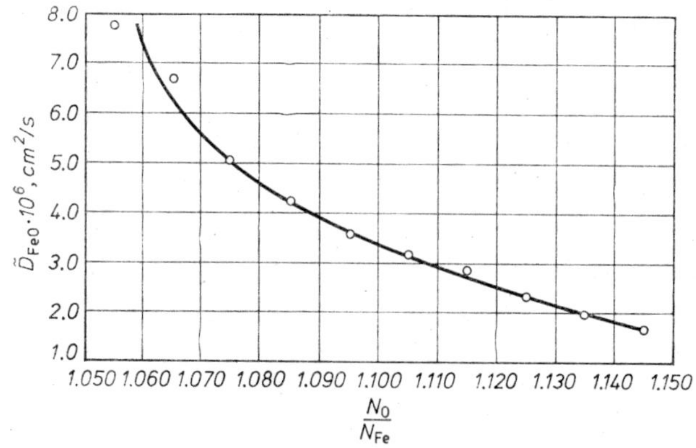

Fig. 122. The dependence of the chemical diffusion coefficient in $Fe_{1-y}O$ on the composition of the wustite phase at 1100°C (after [162]).

These conclusions have been fully confirmed by the results of experimental studies. It has been found that the chemical diffusion coefficient clearly depends on the composition of the wustite phase [162–164], the self-diffusion coefficient of iron is not a linear function of the deviation from stoichiometry [174–175] and the activation energy for this process depends on the value of y [174, 175]. Figure 122 shows, as an illustration, the results of thermogravimetric studies carried out by Hambree and Wagner (see Chapter 3.4). It can be seen from the plot that the chemical diffusion coefficient decreases with increasing deviation from stoichiometry, i.e. with increasing concentration of cation vacancies. This cannot be explained on the basis of the theory of point defects, but it follows directly from the model of extended defects, since the number of Roth complexes and consequently the concentration of Koch-Cohen clusters increase with increasing concentration of cation vacancies. These two large assemblies of defects, together with the accompanying electron hole "atmosphere" (see Chapter 1.4) considerably hinder the movement of free cation vacancies

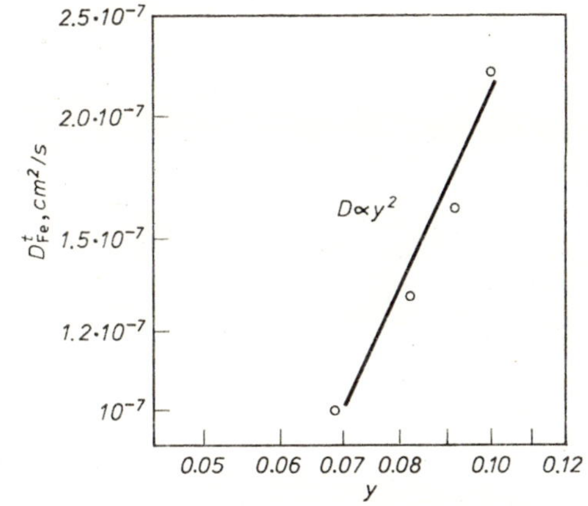

Fig. 123. The dependence of the self-diffusion coefficient of iron in $Fe_{1-y}O$ on deviation from stoichiometry y (in the semilogarithmic plot) (after [174]).

in the crystal. Hence the chemical diffusion coefficient should decrease with increasing deviation from stoichiometry, which is in fact observed.

Figures 123 and 124 show the dependence of the self-diffusion coefficient of iron on the composition of $Fe_{1-y}O$ and on the temperature [174]. It can be seen from the plot shown in Fig. 123 that D_{Fe} is not a linear function of the deviation from stoichiometry y but parabolic to a first approximation. It follows from Fig. 124 that the activation energy of self-diffusion of iron depends on the composition of the wustite phase within wide limits. In order to illustrate this effect, the dependence of E_D

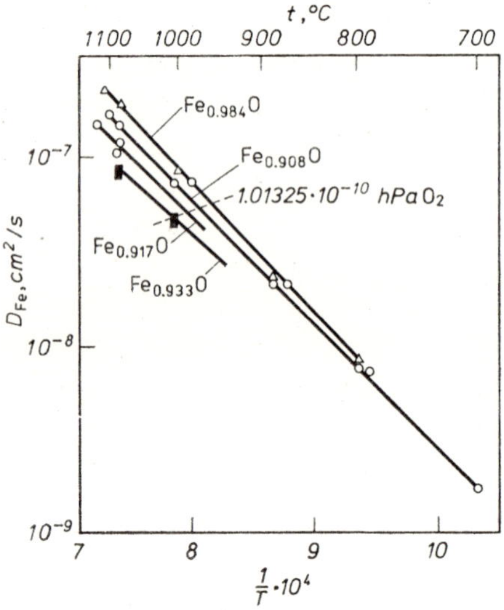

Fig. 124. The temperature dependence of the self-diffusion coefficient of iron in $Fe_{1-y}O$ for various values of y (after [174]).

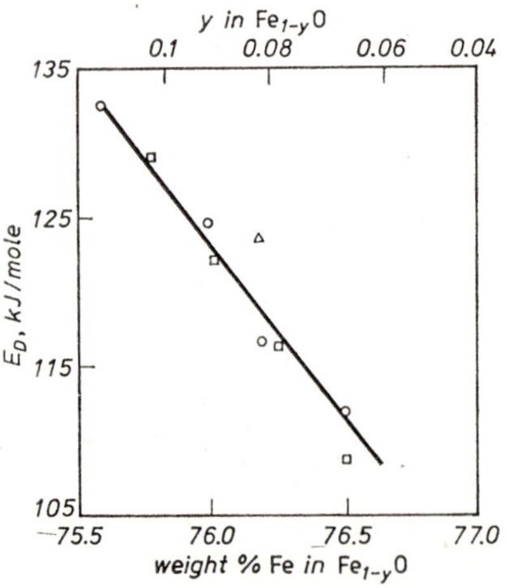

Fig. 125. The dependence of activation energy of self-diffusion of iron in $Fe_{1-y}O$ on the composition of the wustite phase (after [174]).

on the composition of $Fe_{1-y}O$ has been shown in Fig. 125 [174]. It can be seen from the plot that the activation energy of diffusion increases from 100.86 to 138.16kJ/mole with increasing deviation from stoichiometry, i.e. with increasing concentration of cation vacancies. This empirical fact is also in agreement with the model for extended defects, since, as it has already been mentioned, the increase of the concentration of Roth complexes and Koch-Cohen clusters hinders the movement of defects in the crystal, which is equivalent to raising the energy barrier for the jumps, i.e. to the increase of the activation energy for diffusion.

The results of these studies can be regarded as an indirect proof of the complexity of the structure of defects in ferrous oxide. Unfortunately it has, so far not been possible to develop, a quantitative model relating the results of studies on self-diffusion to the structure and concentration of the extended defects in this oxide. It should be mentioned that manganous oxide $Mn_{1-y}O$ behaves similarly, but the third oxide of the iron triade, $Co_{1-y}O$, has a more simple structure of defects.

An equally complex and so far unsolved problem is presented by diffusion in refractory metal oxides with the fluorite structure, in which, as in $Fe_{1-y}O$ and $Mn_{1-y}O$, complexes and clusters of defects are formed. Typical substances are uranium and cerium oxides, in which the structure of the defects has been studied more thoroughly than in other oxides of this type. In these oxides the defects are found in the anionic sublattice. Uranium oxide exemplifies substances containing an excess of oxygen (UO_{2+y}) and cerium oxide those showing an oxygen deficit (CeO_{2-y}). As was shown in Chapter 1.4, uranium oxide contains Willis complexes consisting of two anion vacancies and 3 or 4 interstitial oxygen anions (complexes of the 2:1:2 or 2:2:2 type). At sufficiently high temperatures and at relatively low concentrations these complexes can exist in the crystal lattices in a disordered state. When the temperature is decreased, the complexes show a tendency to form clusters, leading to the formation of a superstructure and intermediate phases (e.g. U_4O_9).

The mechanism of diffusion in this oxide has not yet been completely elucidated, even in the case of diffusion in the high temperature range. However there is no doubt that oxygen transport through the crystal lattice of UO_{2+y} involves not only interstitial oxygen ions but also anion vacancies. So far it is not known whether this process consists in successive jumps of whole assemblies of atoms, i.e. in the migration of Willis complexes, or in non-simultaneous jumps of interstitial ions and anion vacancies, i.e. a gradual disappearance and reformation of these complexes at various points of the crystal lattice. It is certain that the mechanism of diffusion is very complex in this case and involves at least two mechanisms, namely, the vacancy mechanism and the interstitial mechanism. For this reason the self-diffusion coefficient of oxygen in this oxide cannot be definitely related to the mobility of the excess oxygen, since the self-diffusion coefficient is a macroscopic measure of the lattice mobility (see Chapter 2.1) and reflects the average mobility of oxygen ions, which in this case consists of jumps of interstitial ions of excess oxygen and non-excess oxygen, as well as diffusion of anion vacancies, the concentration of which is also not related to the deviation from stoichiometry.

Cerium oxide shows a deficit of oxygen (CeO_{2-y}) resulting from the presence of defects in the anionic sublattice. The anion vacancies are not statistically distributed in the crystal structure, but are paired along the diagonal of MeO_8 cube. Together with six other elements of the lattice, they form a system called the Bevan complex (see Chapter 1.4). The anion vacancies in the $(MeO_6V_2)(MeO_7V)_6$ complex are completely immobile and are integral elements of the crystal structure. Therefore the diffusion of oxygen atoms involving these vacancies requires dissociation of the Bevan complexes. This process can be expressed schematically by the

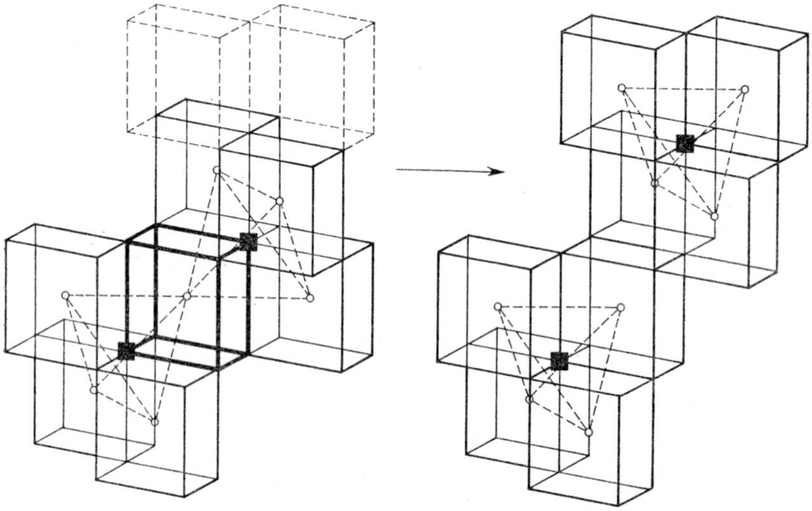

Fig. 126. The scheme of dissociation of Bevan cluster leading to the formation of mobile anion vacancies (after [48]).

following equation:

$$(MeO_6V_2)(MeO_7V)_6 + MeO_8 \rightleftharpoons 2(MeO_7V)_4 \tag{2.8.1}$$

The decomposition mechanism for Bevan complexes is shown in Fig. 126.

Obviously, the equilibrium constant of reaction (2.8.1) depends on temperature. The concentration of quasi-free anion vacancies (MeO_7V) increases with temperature and as a result the self-diffusion coefficient of oxygen in the oxide also increases. However, as in the case of uranium oxide, the self-diffusion coefficient cannot be related to the deviation from stoichiometry, y, since the concentration of quasi-free vacancies is only a fraction (so far impossible to determine) of all the vacancies, present in the crystal lattice of the oxide. Thus the self-diffusion ceofficient of vacancies, D_V, (see Chapter 2.1) calculated from D_0 and y, which is a microscopic measure of the mobility of these defects, is not a real quantity, and is undoubtedly smaller than the real mobility of quasi-free anion vacancies.

Finally it should be mentioned that the above statement regarding the fictitious values for the mobility of defects determined from diffusion measurements applies not only to oxides of the MeO_{2+y} and MeO_{2-y} types but also to oxides of the $Me_{1-y}O$ type, in which defect complexes are formed (e.g. $Fe_{1-y}O$, $Mn_{1-y}O$).

2.8.2 Diffusion in crystals containing Wadsley defects

In many refractory metal oxides, point defects are eliminated by crystallographic shear and, as a result, the so-called Wadsley defects are formed (see Chapter 1.4). When the concentration of these defects increases, they become ordered along definite crystallographic directions, which leads to the formation of intermediate phases. These phenomena are typical of oxides of molybdenum, titanium, niobium and many other metals. As a result of the formation of shear structures, the concentration of point defects (anion vacancies) in these oxides is extremely low (less than 10^{-2} at.%). In spite of this, the rate of formation of Wadsley defects, and consequently of new intermediate phases, is very fast even at temperatures below 1000°C. In this process oxygen is released by the crystal and, for this reason, it is regarded as a reduction, since the formation of a shear plane (Wadsley defect) corresponds to the removal of a whole layer of oxygen anions from the crystal lattice. This process leads to an apparent deviation from stoichiometry in the direction of an oxygen deficit (TiO_{2-y}, MeO_{3-y}, Nb_2O_{5-y}), but this deviation is not connected with the presence of anion vacancies.

What is the mechanism of oxygen transport in crystals of this type, which makes possible the rapid reconstruction of the lattice leading to the formation of a new crystallographic structure? For several years this problem fascinated many investigators for both scientific and practical reasons, since the oxides mentioned above are of interest not only to crystallochemists but also to physicists, chemists and technonologists, who have practical uses for them (electronics, heterogeneous catalysis, production of refractory materials).

At the present time, the results of theoretical considerations partly completed by the results of modern experimental studies, make it possible to draw certain conclusions about the mechanism of mass transport accompanying the formation and migration of Wadsley defects. However it should be stressed that this problem is still far from a satisfactory solution.

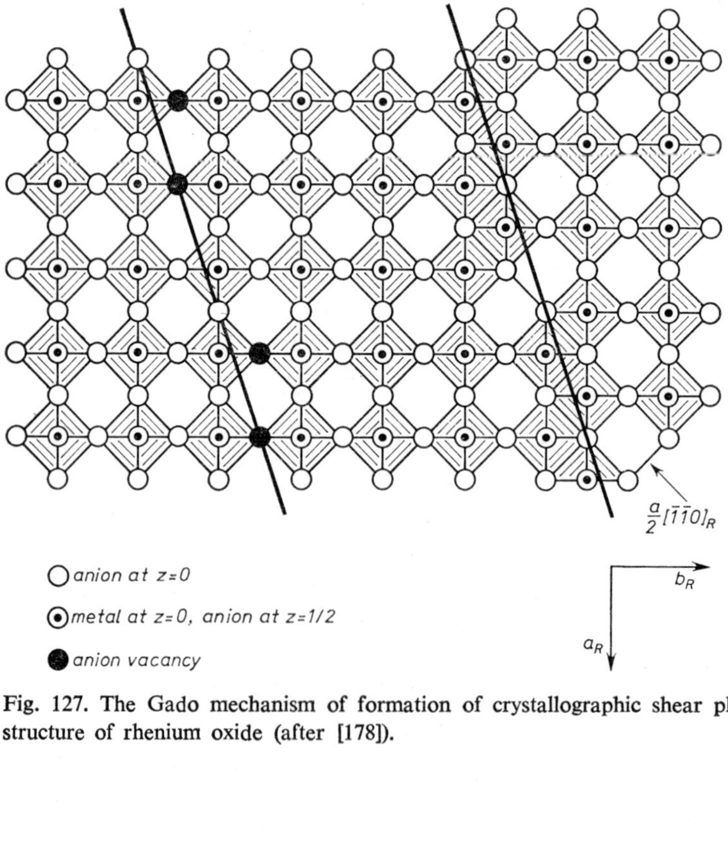

Fig. 127. The Gado mechanism of formation of crystallographic shear plane in the structure of rhenium oxide (after [178]).

There are three models for the mechanism of this process and each of them has certain advantages and disadvantages [177]. All of them concern the structure of the rhenium oxide type (ReO_3, WO_3), which is the most susceptible to crystallographic shear, but they can be also applied to other oxides, e.g. TiO_2, Nb_2O_5 and Cr_2O_3, in which the fundamental structural unit is oxygen octahedron, as in ReO_3.

The model developed by Gado [178, 179] is based on the assumption that the formation of the shear plane is preceded by diffusion of oxygen from the crystal, leading to the formation of anion vacancies. These vacancies become ordered in the bulk of the crystal in such a way that they form "walls", across which the breaking of the lattice and the formation of Wadsley defects take place. This process is shown schematically in Fig. 127. Repetition of this process leads to the formation of parallel shear planes, which divide the crystal into layers. The thickness of these layers decreases with increasing concentration of Wadsley defects.

The principal disadvantage of Gado's model results from the fact that the formation of a shear plane requires a relatively high concentration of anion vacancies, e.g. in the case of reaction:

$$WO_3 \rightarrow W_{20}O_{58} \quad (2.8.2)$$

for which this model was developed the concentration of vacancies necessary for breaking the lattice is 3.3%, whereas experimental studies indicate that the crystallographic shear process already takes place at concentrations of anion vacancies at least one order lower. Hyde and Bursill [177] have stated that intuition suggests that such a high concentration of anion vacancies in the loose lattice of ReO_3 is improbable. Another disadvantage of Gado's mechanism is that it is not possible to explain the process of migration and ordering of Wadsley defects in the crystal.

Another model for the mechanism of formation of Wadsley defects was proposed by Anderson and Hyde [180, 181]. As in Gado's model, these authors assume that the crystallographic shear is preceded by the formation of disordered anion vacancies (point defects). When a certain critical concentration of these defects, much lower than that required by Gado's model, is exceeded, the anion vacancies become ordered in the form of a flat disc at which the breaking of the lattice and the formation of Wadsley defects take place. This is illustrated in Fig. 128. The resulting shear plane fragment is surrounded by a dislocation loop. The stress in the structure along this loop, i.e. at the perimeter of the Wadsley defect, is a sink for vacancies, which thus migrate from the bulk of the lattice to this region and, as a result, the shear plane expands to the surface of the crystal.

An advantage of this model is the explanation of the possibility of formation of Wadsley defects at very low concentrations of anion vacancies.

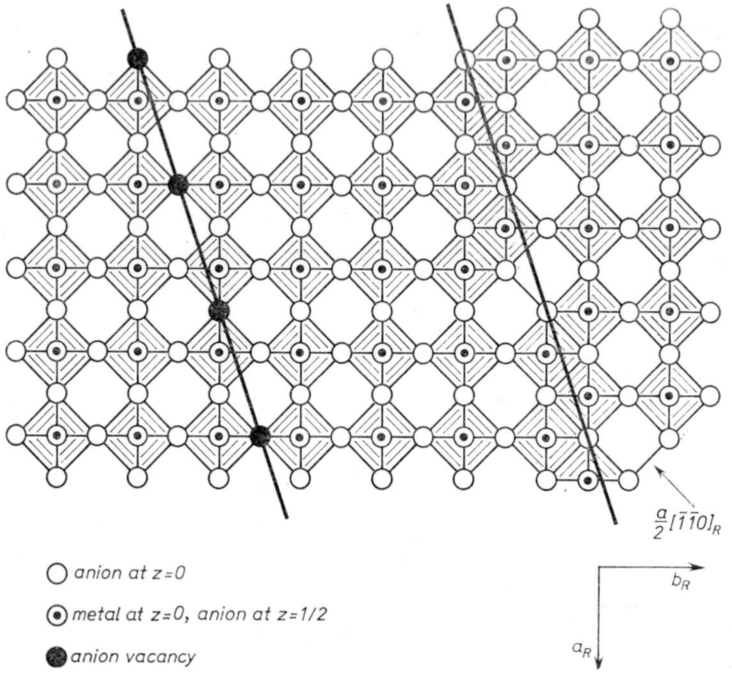

Fig. 128. The Anderson–Hyde mechanism of formation of crystallographic shear plane in the structure of ReO_3 (after [180]).

This model also explains the expansion of shear planes by postulating that this is due to the localisation of dislocations in loops around external defects. On the other hand this model does not explain the migration of Wadsley defects leading to the formation of intermediate phases by ordering of the defects.

In a third model, proposed by Anderson and Wadsley [182], it is assumed that the shear plane is formed not in the bulk but on the surface of crystal without the participation of point defects. During the first stage of reduction of a MeO_3 type compound oxygen ions present in the surface layer of the crystal leave their lattice sites and pass to the gas phase (Fig. 129*a*). Then metal and oxygen ions present in the resulting incomplete octahedra of the surface layer along the line of the defect migrate to the bulk of the lattice by leaving the "normal" lattice sites and occupying interstitial positions (Fig. 129*b*)). The resulting Wadsley defect can then migrate to

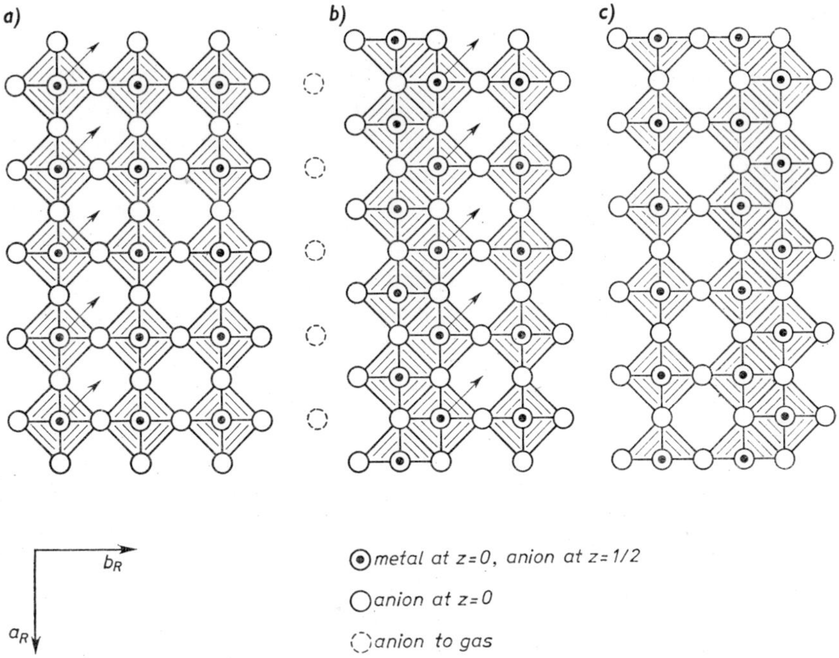

Fig. 129. The Anderson–Wadsley mechanism of formation of crystallographic shear plane in the structure of ReO_3 (after [182]).

the bulk of the crystal as a result of a concerted jump of succesive rows of cations and anions as shown schematically in Fig. 129*c*). This is equivalent to the reconstruction of the normal structure of the surface layer of the crystal (see Fig. 129*c*)) and as a result the process of formation of the next Wadsley defect can take place.

The Anderson–Wadsley model explains both the formation of shear planes and their migration and the mechanism of ordering of these defects in the crystal that follows. On the other hand this model excludes the possibility of formation of Wadsley defects in the bulk of the crystal, which is contrary to results of experimental studies. However, O'Keefe [183] has pointed out that there are no contradictions between the Anderson–Wadsley and Anderson–Hyde models. The former explains the surface "nucleation" of Wadsley defects and their migration and the latter the "nucleation" and growth of the defects in the bulk of the crystal. Therefore the two models supplement each other and, taken together, they explain the processes of formation and migration of Wadsley defects under different thermodynamic conditions.

In his discussion of the various ways in which matter can be transported in the above oxides, O'Keefe [184] states that there are no analogies between the diffusion of oxygen and the migration of Wadsley defects. The lattice diffusion process takes place under the influence of a concentration gradient and is a result of an enormous number of random jumps of oxygen ions in all possible diretions, whereas the migration of Wadsley defects takes place under the influence of a stress gradient and is a result of successive concerted jumps of metal and oxygen ions in one direction. Hence it is not possible to draw conclusions about the reactivity of an oxide containing Wadsley defects on the basis of the self-diffusion coefficient of oxygen in this oxide, since, when the value of this coefficient is low, the mobility of Wadsley defects may be high and vice versa. O'Keefe [184] also points out that the low mobility of anion vacancies, resulting from the low values of the self-diffusion coefficient, need not necessarily mean that the rate of growth of shear planes in the crystal is slow. The necessary condition for a fast growth of a Wadsley defect is its direct contact with the free surface of the crystal. In this case the dislocation loop surrounding the defect

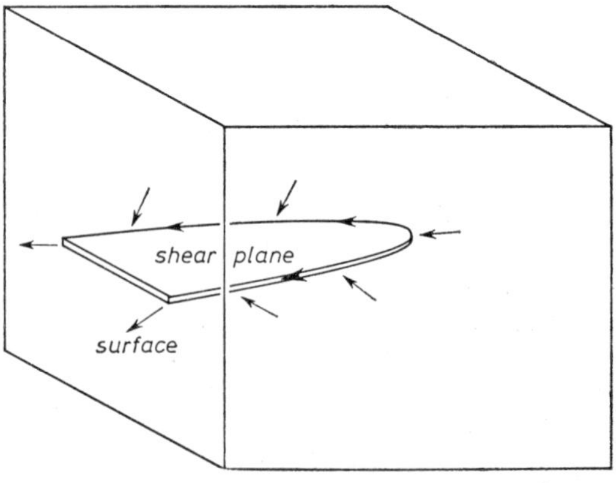

Fig. 130. The scheme of high diffusivity paths of oxygen transport to the surface of a crystal containing Wadsley defect (after [184]).

is in contact with the surface and as a result the crystal contains favoured paths for the diffusion of oxygen ions to the surface. This is shown schematically in Fig. 130.

Summing up it can be stated that the formation and migration of Wadsley defects in certain transition metal oxides is a complex process, which shows no analogies to lattice diffusion although this type of diffusion participates in the process. The complete elucidation of the mechanism of these processes requires further extensive studies.

2.9 Results of some experimental studies

The equations derived in previous chapters make it possible to interpret the experimental data and to draw some conclusions regarding not only the mechanism of diffusion in solids, but also the degree of ionisation of point defects, their mobility, electrical conductivity, etc.

In view of the considerable experimental difficulties connected with studies on self-diffusion and defect equilibria the experimental data regarding these problems are still relatively scarce. However, since these problems are of fundamental importance for the theory of solids and in particular for the theory of oxide and sulphide semiconductors as well as for many solid state reactions, which are important from the practical point of view, a brief discusion of a few selected examples seems to be indicated.

Studies on self-diffusion in compounds of the MeX type must be carried out either at a constant partial pressure of component X_2 at variable temperature or at a constant temperature at variable pressure of the oxidant if they are to be interpreted in the light of the theory of diffusion. However a complete interpretation of the data is possible provided that the self-diffusion coefficient is determined as a function of both the above parameters. Unfortunately necessity for this was not realised in the early studies of these problems and, consequently, only some of the results of measurements of self-diffusion are suitable for rational interpretation. Among these results should be mentioned first of all the results of extensive and already classic studies on self-diffusion of iron in ferrous oxide ($Fe_{1-y}O$) carried out by Birchenall and his collaborators [185] and the data obtained by Carter and Richardson [186] for cobaltous oxide ($Co_{1-y}O$).

The results of these studies, obtained in both cases by the tracer method, are shown graphically in Figs. 131 and 132.

Both these oxides are *p* type semiconductors with a metal deficit ($Me_{1-y}O$). Ferrous oxide (wustite) shows particularly large deviations from stoichiometry, reaching 0.12 mole of iron per mole of FeO ($Fe_{0.88}O$). The deviations of cobaltous oxide from stoichiometric composition are many times smaller.

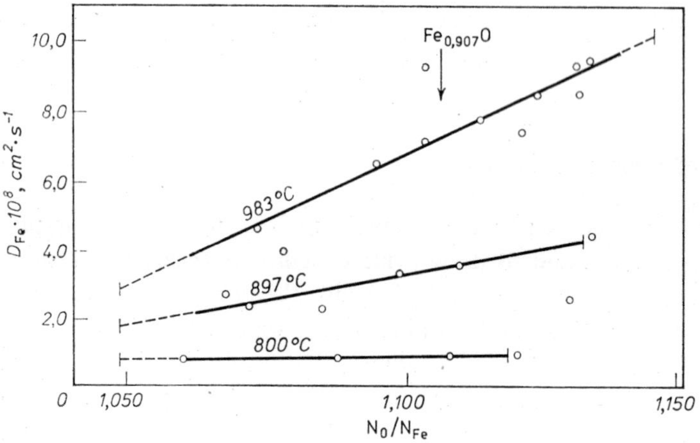

Fig. 131. The dependence of self-diffusion coefficient of iron in $Fe_{1-y}O$ on deviation from stoichiometry for several temperatures (after [185]). N_O and N_{Fe} stand for the mole fractions of oxygen and iron in the wustite phase respectively.

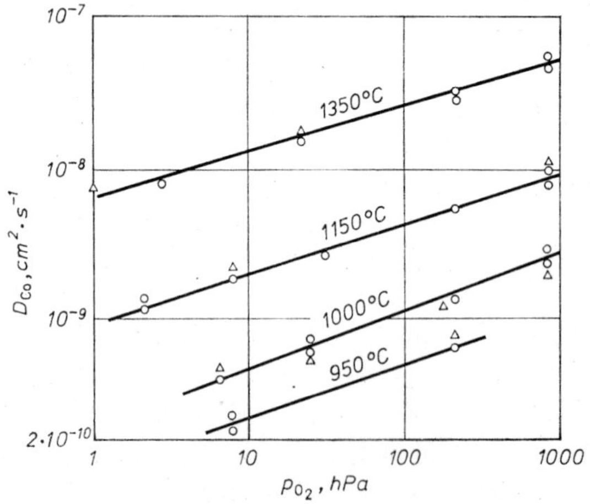

Fig. 132. The dependence of self-diffusion coefficient of cobalt in cobaltous oxide on equilibrium oxygen pressure for several temperatures (after [186]).

It was shown in Chapter 2.1, that the activation energy for diffusion, determined experimentally, is the sum of two quantities: the activation energy (enthalpy) of formation of defects, the concentration of which increases exponentially with temperature, and the activation energy for the migration of these defects in the lattice. Hence the determination of the temperature dependence of the diffusion coefficient at constant pressure gives the following empirical relationship:

$$D = D_0 \exp\left(-\frac{E_D}{RT}\right) \qquad (2.9.1)$$

where E_D is the sum of the above two activation energies.

The separation of these two terms requires either knowledge of the enthalpy of formation of the given type of defect (ΔH_f) or measurements of diffusion at a constant concentration of defects in the given temperature range in order to determine the second term (ΔH_m). This is possible provided that a suitable amount of a dopant controlling the concentration of defects is introduced to the crystal lattice.

However the results of such studies are not fully representative, since the activation energy for diffusion may depend on the concentration of defects. When the diffusion coefficient is determined as a function of temperature and also as a function of pressure of the component X_2 in the atmosphere surrounding the crystal, there is a possibility of selecting data from the results corresponding to the same stoichiometric composition and hence the same concentration of defects at different temperatures.

The arrow in Fig. 131 shows the composition of wustite ($Fe_{0.907}O$) for which the values of D_{Fe}, measured at various temperatures, were selected. The temperature dependence of these values can be represented by the following equation:

$$D_{Fe} = 0.118 \exp\left(-\frac{29\,700}{RT}\right) \qquad (2.9.2)$$

It follows from this equation that the activation energy of diffusion of iron ions through cation vacancies is: $\Delta H_m = 124.35$ kJ/mole (29.7 kcal/mole).

On the other hand, the data corresponding to a constant oxygen pressure, i.e. different concentrations of defects at various temperatures, lead to the following relationship:

$$D_{Fe} = \text{const} \exp\left(-\frac{40\,500}{RT}\right) \qquad (2.9.3)$$

If we assume that the cation vacancies are singly ionised, we obtain the following equation:

$$E_D = \left(\frac{\Delta H_{V'_{Fe}}}{2} + \Delta H_m\right)$$
$$= 169\,565.4 \text{ kJ/mole } (40500 \text{ kcal/mole}) \qquad (2.9.4)$$

It follows from equation (2.9.2) and (2.9.3) that the activation energy of formation of cation vacancies in the crystal lattice of wustite is :

$$\Delta H_{V'_{Fe}} = 90.44 \text{ kJ/mole } (21.6 \text{ kcal/mole})$$

It should be stressed that the mechanism of diffusion of iron in wustite is probably much more complex than appears from the above considerations, since complexes and clusters of defects are formed (see Chapters 1.4 and 2.8).

The data shown in Fig. 132 indicate that the self-diffusion coefficient of cobalt in cobaltous oxide, like that of iron in $Fe_{1-y}O$, increases with increasing equilibrium oxygen pressure and that this relationship is temperature dependent. According to the general equation (2.1.26) this dependence is linear on a log–log plot and, in this case, it can be represented by the following equation:

$$D_{Co} = \text{const}\, p_{O_2}^{1/n} \qquad (2.9.5)$$

It can be seen from the graph that the exponential coefficient in equation (2.9.5) decreases with increasing temperature. At 1000, 1150 and 1350°C its values are 1/2.2, 1/3.3 and 1/3.6, respectively. This is understandable in view of the fact that the degree of ionisation of defects increases with increasing temperature [see equations (1.3.112)–(1.3.115)].

The results of the above studies on self-diffusion are in agreement with more recent data, obtained by Tannhauser and Fisher [38, 39], on the dependence of the concentration of cation vacancies in cobaltous oxide on oxygen pressure and temperature. These data indicate that the exponential coeficient $1/n$ in equation (2.9.5) increases with increasing oxygen pressure. In the range of moderate pressures of the order of 101.325 Pa (10^{-3} atm) this exponent is almost equal to 1/4. Carter and Richardson carried out their studies in a relatively narrow range of pressures close to 101.325 kPa (1 atm). The results obtained by Tannhauser and Fisher indicated that in this range of oxygen pressures the values of the exponent are higher than 1/4. Therefore, assuming, like Tannhauser and Fisher, that the exponent $1/n$ in equation (2.9.5) is equal to 1/4 for pressures of the order of 101.325 Pa

(10^{-3} atm) (the lower limit in Carter and Richardson work), i.e. that the cation vacancies are singly ionised under these conditions, we can express the diffusion coefficient of cobalt in cobaltous oxide as a function of the temperature and the oxygen pressure by the following equation [cf. equation (2.1.28)]:

$$D_{Co} = \text{const}\, p_{O_2}^{1/4} \exp\left(-\frac{\frac{1}{2}\Delta H_{V'_{Co}} + \Delta H_m}{RT}\right) \tag{2.9.6}$$

At the above oxygen pressure, the temperature dependence of D_{Co} can be expressed by the following empirical equation:

$$D_{Co} = \text{const} \exp\left(-\frac{38\,000}{RT}\right) \tag{2.9.7}$$

In their studies on the temperature dependence of the concentration of cation vacancies in $Co_{1-y}O$, Fisher and Tannhauser have shown that the heat of formation of singly ionised vacancies is: $\Delta H_{V'_{Co}} = 62.8$ kJ/mole (15 kcal/mole) [see equation (1.3.122)]. It follows from comparison of equations (2.9.6) and (2.9.7) that:

$$\tfrac{1}{2}\Delta H_{V'_{Co}} + \Delta H_m = 159\,098.4 \text{ kJ/mole (38000 kcal/mole)} \tag{2.9.8}$$

Substituting the value of the enthalpy of formation of singly ionised cation vacancies determined by Tannhauser and Fisher for $\Delta H_{V'_{Co}}$ in this equation, we can calculate the activation energy for diffusion of these vacancies in the crystal lattice of $Co_{1-y}O$. The results of such calculations show that:

$$\Delta H_m = 110.95 \text{ kJ/mole (26.5 kcal/mole)} \tag{2.9.9}$$

Recently Fryt, Mrowec and Walec [187, 188] used the Rosenburg's kinetic method (see Chapter 3 of the present work) for the determination

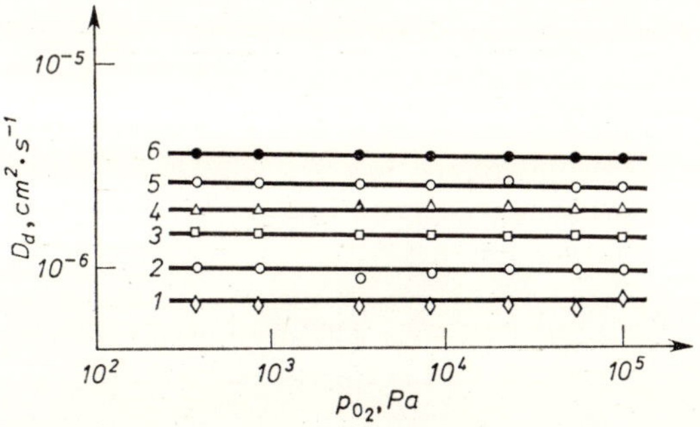

Fig. 133. The dependence of self-diffusion coefficient of cation vacancies in cobaltous oxide on oxygen pressure (after [187]). Curves *1–6* correspond to 1000, 1050, 1100, 1150, 1200 and 1250°C.

of the chemical diffusion coefficient in cobaltous oxide as a function of temperature and equilibrium oxygen pressure. The results of this work are shown in Figs. 133 and 134.

It follows from Fig. 133 that the diffusion coefficient of defects is, in accordance with expectation, independent of the equilibrium oxygen pressure

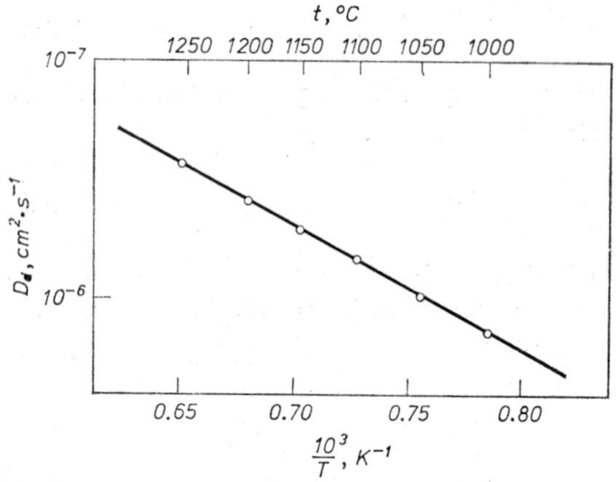

Fig. 134. The temperature dependence of self-diffusion coefficient of cation vacancies in cobaltous oxide (after [187]).

within wide limits of temperature and pressure, i.e. it is independent of the concentration of defects. The activation energy of diffusion determined from the slope of the plot of $\log \tilde{D}$ against $1/T$ (Fig. 134) is $\Delta H_m = 102.58$ kJ/mole (24.5 kcal/mole). Comparison of this value with that calculated by means of equation (2.9.6), on the basis of data obtained by Tannhauser and Fisher as well as by Carter and Richardson, shows that the agreement between them is satisfactory. This indicates that the model of defect structure in cobaltous oxide is correct.

According to this model the predominant defects in cobaltous oxide over a wide range of temperature and oxygen pressure are singly ionised cation vacancies formed in a quasi-chemical surface process:

$$\tfrac{1}{2}O_2 \rightleftharpoons V'_{Co} + h^{\bullet} + O_O \tag{2.9.10}$$

This means that the deviation from stoichiometric composition is practically equal to the total concentration of defects in the cationic sublattice:

$$y \cong [\text{def.}] \tag{2.9.11}$$

where [def.] is the total concentration of defects in the cationic sublattice expressed in mole fractions.

However, recently Przybylski and Mrowec [189, 190] showed that, at temperatures above 1000°C, this model must be modified. These authors calculated the total concentration of defects in cobaltous oxide [def.] as a function of temperature and oxygen pressure and showed that above 1000°C it is higher than that determined from deviations from stoichiometry [191]. They also showed that this difference increases with temperature. This is illustrated in Fig. 135, which also shows the results of measurements

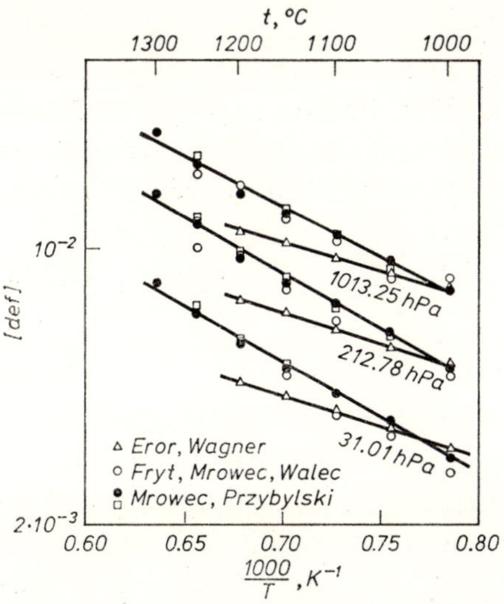

Fig. 135. The temperature dependence of the concentration of point defects in cobaltous oxide for several oxygen pressures (after [189]).

of the total concentration of defects [def.] recently obtained by the Rosenburg's kinetic method [187, 188] (see Chapter 3.2.5). It follows from the plot that the calculated values of [def.] are in excellent agreement with the values determined experimentally.

The total concentration of defects in $Co_{1-y}O$ has been calculated by means of the following equation:

$$[\text{def.}] = \frac{D_{Co}(1+|p|)}{\tilde{D}_{CoO}} \tag{2.9.12}$$

This equation follows directly from the relationship between the self-diffusion and chemical diffusion coefficients and the concentration of defects [see equation (2.6.5)]. The concentration of defects calculated in this way is the total defect concentration, since all the defects present in the cationic sublattice and not only the cation vacancies resulting from the deviation from stoichiometry, y, participate in the diffusion process. Up to the present, it has been assumed that the singly ionised cation vacancies formed in the quasi-chemical surface reaction:

$$\tfrac{1}{2}O_2 \rightleftharpoons V'_{Co} + h^{\bullet} + O_O \qquad (2.9.13)$$

are the predominant defects in the oxide within wide limits of temperature and oxygen pressure, i.e. that:

$$[\text{def.}] = y \qquad (2.9.14)$$

Results of calculations carried out by Mrowec and Przybylski, as well as the experimental data obtained by Fryt, Mrowec and Walc by Rosenburg's method, indicate that this assumption is correct only for temperatures not exceeding 1000°C. Above this temperature (see Fig. 135) the intrinsic ionic disorder of the Frenkel type:

$$Co_{Co} \rightleftharpoons Co_i^{\bullet\bullet} + V''_{Co} \qquad (2.9.15)$$

or of the Schottky type:

$$\text{zero} \rightleftharpoons V''_{Co} + V_O^{\bullet\bullet} \qquad (2.9.16)$$

plays a significant part.

Studies on the mechanism of oxidation of cobalt [190, 192] carried out by the tracer method (see Chapter 3.1) have shown that the compact layer of cobaltous oxide is formed as a result of outward diffusion of the metal irrespectively of temperature (900–1350°C) and oxygen pressure (0.101325–101.325 kPa, i.e. 10^{-3}–1 atm). This indicates that the anionic sublattice is practically free from defects compared with the cationic sublattice. Consequently it is possible to accept one of the alternatives given by equations (2.9.15) and (2.9.16) by rejecting the Schottky type disorder. Thus in cobaltous oxide, at temperatures above 1000°C, the disorder in the crystal lattice consists in the occurrence of cation vacancies, electron holes and interstitial cations. The concentration of cation vacancies is greater than that resulting from the deviation from stoichiometry, since these vacancies are formed as a result of both the quasi-chemical surface reaction [equation (2.9.13)] and the intrinsic Frenkel disorder [equation (2.9.15)].

It also follows from Fig. 135 that the total enthalpy of formation of defects, ΔH_f^s, is greater than the enthalpy of formation of singly ionised cation vacancies in the surface process [ΔH_f^s = 117.23 kJ/mole (28 kcal/mole) and ΔH_f = 60.71 kJ/mole (14.5 kcal/mole)]. The fact that the total enthalpy of formation of defects in the oxide is almost twice as high as that determined on the basis of the deviation from stoichiometry indicates that the heat of formation of a pair of Frenkel defects is considerably higher than that of a singly ionised cation vacancy. This is understandable in view of the fact that in the Frenkel process an interstitial defect is formed in addition to the cation vacancy. The incorporation of this defect in an interstitial position requires a relatively large amount of energy.

Complications of another kind are encountered in the case of nickel oxide which, like cobaltous oxide, is a typical metal deficit *p*-type semiconductor ($Ni_{1-y}O$). Studies of its electrical conductivity [193, 195] have shown that the following equation is valid even a relatively wide range of temperature and oxygen pressure:

$$\sigma = \text{const}\, p_{O_2}^{1/5} \exp\left(-\frac{E_\sigma}{RT}\right) \tag{2.9.17}$$

where E_σ is the activation energy for the electrical conductivity.

The results of these studies are shown in Fig. 136. The value of the exponent in equation (2.9.17) indicates that in nickel oxide singly and

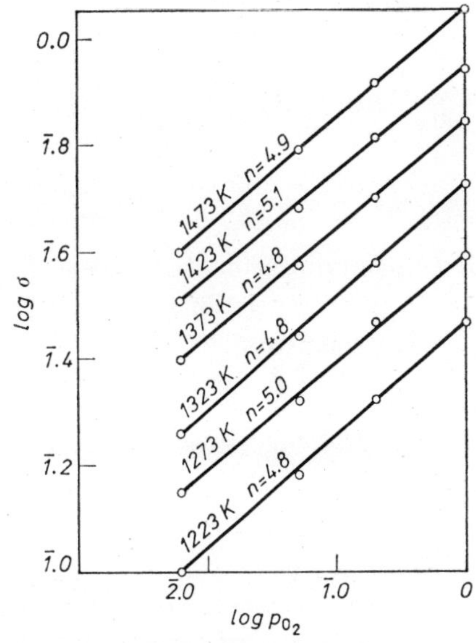

Fig. 136. The dependence of the electrical conductivity of nickel oxide on oxygen pressure (after [193]); [doubly logarithmic plot; n is the denominator of the exponent $\frac{1}{n}$ in general equation (1.5.7)].

doubly ionised cation vacancies occur simultaneously in comparable amounts. This conclusion is based on the following considerations:

If it is assumed that, under the above conditions, practically all the cation vacancies are ionised and that one half of them are doubly ionised, the defect equilibria can be expressed by means of the following partial equations:

$$\tfrac{1}{2}O_2 \rightleftharpoons V'_{Ni} + h^{\bullet} + O_O \tag{2.9.18}$$

and

$$\tfrac{1}{2}O_2 \rightleftharpoons V''_{Ni} + 2h^{\bullet} + O_O \tag{2.9.19}$$

Adding these equations together and dividing by two in order to obtain the expression for one mole of defects we obtain the total reaction for the formation of singly and doubly ionised vacancies and the equivalent number of electron holes:

$$\tfrac{1}{2}O_2 \rightleftharpoons \tfrac{1}{2}V'_{Ni} + \tfrac{1}{2}V''_{Ni} + \tfrac{3}{2}h^{\bullet} + O_O \tag{2.9.20}$$

According to the above assumption, it follows from equation (2.9.20) that:

$$[V'_{Ni}] = [V''_{Ni}] \tag{2.9.21}$$

Hence the electroneutrality condition of the lattice is in this case as follows:

$$[V'_{Ni}] + [V''_{Ni}] = \tfrac{2}{3}[h^{\bullet}] \tag{2.9.22}$$

Applying the law of mass action to equation (2.9.20) we obtain:

$$K = \frac{[V'_{Ni}]^{1/2}[V''_{Ni}]^{1/2}[h^{\bullet}]^{3/2}}{p_{O_2}^{1/2}} \tag{2.9.23}$$

Using uquation (2.9.21) and the electroneutrality condition given by equation (2.9.22) we finally obtain the following relation between the concentration of electron holes and the oxygen pressure:

$$[h^{\bullet}] = 6K^{2/5}p_{O_2}^{1/5} \tag{2.9.24}$$

where K is the equilibrium constant of reaction (2.9.20), given by equation (2.9.23). Using equation (1.3.61) we can express equation (2.9.24) in the following form:

$$[h^{\bullet}] = \mathrm{const}\, p_{O_2}^{1/5} \exp\left(-\frac{\tfrac{2}{5}\Delta H_f}{RT}\right) \tag{2.9.25}$$

where ΔH_f is the total enthalpy of formation of one mole of singly and doubly ionised cation vacancies according to equation (2.7.20), "const" contains the entropy term for this process and the number 6 resulting from mathematical transformations.

In order to relate the above relationship to the electrical conductivity of the oxide, it must be introduced to the general equation (1.5.1) and the mobility of the current carriers in this equation must be replaced with equation (1.5.9), since it has been shown that in nickel oxide the transport of the current carriers (electron holes) is an activated mechanism (see Chapter 1.5). After these transformations we obtain the following relation between the electrical conductivity, the equilibrium oxygen pressure and the temperature:

$$\sigma = \text{const} p_{O_2}^{1/5} \exp\left(-\frac{\frac{2}{5}\Delta H_f + \Delta H_m^{h^\bullet}}{RT}\right) \tag{2.9.26}$$

Comparison of this equation with the empirical formula (2.9.17) confirms the correctness of the assumption regarding the defect structure of this oxide. On the basis of the above model of disorder, it is possible to calculate the total enthalpy of formation of singly and doubly ionised cation vacancies ΔH_f, using the experimentally determined value of the activation energy for the electrical conductivity of $Ni_{1-y}O$, E_σ, and the value of the activation energy of the movement of electron holes $\Delta H_m^{h^\bullet}$, determined in a different way, since it follows from the comparison of equation (2.9.26) and (2.9.17) that:

$$\Delta H_f = 2.5(E_\sigma - \Delta H_m^{h^\bullet}) \tag{2.9.27}$$

Experimental studies have shown that $E_\sigma = 82.9$ kJ/mole (19.8 kcal/mole) [194] and $\Delta H_m^{h^\bullet} = 23.03$ kJ/mole (5.5 kcal/mole) [196] and hence $\Delta H_f = 149.47$ kJ/mole (35.7 kcal/mole).

In order to check the correctness of the defect model above described for the crystal lattice of nickel oxide, the results of studies on the self-diffusion of nickel and on chemical diffusion in this material can be used. If the model is correct, the self-diffusion coefficient of nickel in nickel oxide should be a function of temperature and oxygen pressure analogous to that for the electrical conductivity:

$$D_{Ni} = \text{const}\, p_{O_2}^{1/5} \exp\left(-\frac{\frac{2}{5}\Delta H_f + \Delta H_m^{d}}{RT}\right) \tag{2.9.28}$$

where ΔH_m^d is the activation energy of diffusion of cation vacancies according to equation (2.1.6).

Equation (2.9.28) results directly from the fact that the macroscopic diffusion coefficient D_{Ni} is, according to equation (2.1.8), the product of the miscroscopic diffusion coefficient—in this case the diffusion coefficient of cation vacancies $D_{V_{Ni}}$—and the concentration of these vacancies expressed in mole fractions. Applying the law of mass action to equilibrium (2.9.20) and taking into account equation (2.9.21) as well as the electroneutrality condition (2.9.22) we come to the conclusion that the concentration of cationic vacancies is, like that of electron holes, a function of oxygen pressure and temperature as follows:

$$[V_{Ni}]_c = \mathrm{const}\, p_{O_2}^{1/5} \exp\left(-\frac{\frac{2}{5}\Delta H_f}{RT}\right) \tag{2.9.29}$$

where $[V_{Ni}]_c = [V'_{Ni}] + [V''_{Ni}]$ is the total concentration of cation vacancies.

Introducing equation (2.9.29) in equation (2.1.8) and using equation (2.1.6), we obtain equation (2.9.28).

It follows from this equation that the empirical activation energy of diffusion, E_D, [equation (2.1.22)] should be related to the enthalpy of formation of cation vacancies and the activation energy of their migration by the following equation:

$$E_D = \tfrac{2}{3}\Delta H_f + \Delta H_m^d \tag{2.9.30}$$

The value of E_D has been determined by various authors [197–201] from the temperature dependence of the self-diffusion coefficients of nickel in the oxide. This value varies from 185.06–254.56 kJ/mole (44.2 to 60.8 kcal/mole), but the results of the majority of recent studies indicate that it is close to 209.34 kJ/mole (50 kcal/mole). Putting this value in equation (2.9.30) and using the calculated value of ΔH_f equal to 149.47 kJ/mole (35.7 kcal/mole), we can calculate the activation energy for diffusion of cation vacancies in nickel oxide. According to such calculations ΔH_m^d = 149.47 kJ/mole (35.7 kcal/mole).

This activation energy has also been determined experimentally by means of measurements of the temperature dependence of the chemical diffusion coefficient $\tilde{D}$ [193]. It follows from equations (2.6.4) and (2.1.6) that the activation energy of diffusion of defects is equal to the activation energy for chemical diffusion:

$$\tilde{D} = (1+|z|)D_d = \mathrm{const}\exp\left(-\frac{\Delta H_m}{RT}\right) \tag{2.9.31}$$

The value of ΔH_m^d determined in this way is equal to 141.1 kJ/mole (33.7 kcal/mole). The difference between this value and the value of ΔH_m^d calculated on the basis of the assumed defect model does not exceed 6%. This agreement can be regarded as completely satisfactory. It confirms the correctness of the defect model for the crystal lattice of $Ni_{1-y}O$.

This example is a good illustration of the method of investigating the defect structure of compounds of the MeX type and of the utilisation of the results of studies on diffusion for checking the hypotheses regarding the defect structure model in the material under consideration.

It is noteworthy that equations (2.9.25) and (2.9.29), as well as relationships (2.9.26) and (2.9.28) resulting from these equations, are particular forms of general solutions (1.3.117) and (1.5.12) as well as (2.1.28). It follows from the general solutions that the exponent of the oxidant pressure is related in a definite way to the term involving ΔH_f; $(/1n)$ has the value equal to one half of the factor $2/n$, appearing in the term involving ΔH_f.

In the next example, we shall discuss the defect structure and the semiconducting properties of cuprous oxide against a background of results of studies on the deviation from stoichiometry as well as measurements of self-diffusion and electrical conductivity as a function of temperature and equilibrium oxygen pressure. This oxide, like NiO, is a metal deficit p-type semiconductor ($Cu_{2-y}O$). Hence the predominant defects are cation vacancies and electron holes. The results of studies on the deviations from stoichiometry and on self-diffusion of copper in $Cu_{2-y}O$, obtained by Mrowec, Godlewski and Stokłosa [202], can be represented by the following empirical equations:

$$y = 0.886\, p_{O_2}^{1/4.2} \exp\left(-\frac{12\,000}{RT}\right) \tag{2.9.32}$$

and

$$D_{Cu} = 2.15 \times 10^{-3} p_{O_2}^{1/3.9} \exp\left(-\frac{24\,000}{RT}\right) \tag{2.9.33}$$

The interpretation of the results of these studies first requires the consideration of defect equilibria in the crystal lattice of cuprous oxide. If it is assumed that all the cation vacancies are associated with electron holes, i.e. that the disorder in the crystal lattice of $Cu_{2-y}O$ consists in the occurrence of non-ionised cation vacancies, the equilibrium between the defects and the oxygen pressure can be expressed by the following equation:

$$\tfrac{1}{2}O_2 = O_O + 2V_{Cu}^{x} \tag{2.9.34}$$

where V_{Cu}^{x} is a non-ionised cation vacancy and O_O is an oxygen atom in a normal lattice site.

Applying the law of mass action to this equation we obtain the following relation between the concentration of non-ionised cation vacancies and oxygen pressure:

$$[V_{Cu}^{x}] = p_{O_2}^{1/4} \exp\left(\frac{\Delta S_f}{R}\right) \exp\left(-\frac{\Delta H_f}{RT}\right) \tag{2.9.35}$$

where ΔS_f and ΔH_f are the entropy and the enthalpy of formation of one mole of non-ionised cation vacancies. $[V_{Cu}^{x}]$ is the concentration of non--ionised cation vacancies expressed in mole fractions.

Since in this case the concentration of non-ionised cation vacancies resulting from reaction (2.9.34) corresponds to the predominant disorder in the lattice, it is practically equal to the deviation from stoichiometry $y \cong [V_{Cu}^{x}]$.

If it is assumed that there is practically no interaction between the ionic and electronic defects, i.e. that cation vacancies are completely ionised, the defect equilibrium can be expressed by the following scheme:

$$\tfrac{1}{2}O_2 = O_O^{x} + 2V'_{Cu} + 2h^{\bullet} \tag{2.9.36}$$

where V'_{Cu} and $h^{\bullet}$ are an ionised cation vacancy and an electron hole respectively.

In this case the electroneutrality condition for the lattice has the following simplified form:

$$[V'_{Cu}] = [h^{\bullet}] \tag{2.9.37}$$

Applying the law of mass action and the above electroneutrality condition to equation (2.9.36) we obtain the following expression for the dependence of the concentration of ionised cation vacancies in cuprous oxide on temperature and oxygen pressure:

$$y = [V'_{Cu}] = p_{O_2}^{1/8} \exp\left(\frac{\frac{1}{2}\Delta S_f + \Delta S'_f}{R}\right) \exp\left(\frac{\frac{1}{2}\Delta H_f + \Delta H'_f}{RT}\right) \tag{2.9.38}$$

where $\Delta S'_f$ and $\Delta H'_f$ are the entropy and the enthalpy of ionisation of one mole of cation vacancies respectively.

Comparison of the theoretical relationships (2.9.35) and (2.9.38) with the empirical equation (2.9.32) indicates that in the range of temperature and oxygen pressure examined, cuprous oxide mainly contains non-ionised cation vacancies, since the exponent of the oxygen pressure in the empirical equation (2.9.32) is close to 1/4. On the other hand the results of high temperature studies on the electrical conductivity of cuprous oxide indicate that in these conditions the current carriers are electron holes resulting from the defect equilibrium represented by equation (2.9.36), since it has been found that the electrical conductivity increases with increasing equilibrium oxygen pressure according to the following equation:

$$\sigma = \text{const} p_{O_2}^{1/7} \qquad (2.9.39)$$

The electrical conductivity of semiconductors of this type is related to the concentration of electron holes by the following relationship:

$$\sigma = [h^{\cdot}]\mu_{h_x^{\cdot}} q^{+} \qquad (2.9.40)$$

If it is assumed that the mobility of electron holes, $\mu_{h_x^{\cdot}}$, is practically independent of their concentration, the increase of electrical conductivity, accompanying the increase of oxygen pressure, is caused by the increase of the concentration of electron holes.

It follows from the above considerations that, in cuprous oxide, a certain fraction of the cation vacancies is ionised. However, since the deviation from stoichiometry increases with increasing oxygen pressure according to equation (2.9.32), it could be assumed that only an insignificant fraction of cation vacancies is ionised (the value of the exponent of the oxygen pressure is only slightly lower than 1/4). In order to check this conclusion it was necessary to estimate the ratio of the concentration of ionised to non-ionised cation vacancies and to determine the extent of the effect of the real value of the concentration ratio of the two kinds of vacancies on the exponent at the equilibrium oxygen pressure.

The data obtained by Trivich and his collaborators [203] indicate that the concentration of electron holes in cuprous oxide at 1000°C and oxygen pressure equal to 66.47 hPa (6.56×10^{-2} atm) is equal to 1.2×10^{18} cm^{-3}. This concentration expressed in mole fractions is equal to $[h^{\cdot}] = 1.5 \times 10^{-4}$. Since the concentration of electron holes is equal to that of ionised cation vacancies [see equation (2.9.37)], the concentration of these defects under the above conditions is: $V'_{Cu} = 1.5 \times 10^{-4}$. The total concentration of cation vacancies in $Cu_{2-y}O$ can be calculated on the basis of the empirical equation (2.9.32). At 1000°C and 66.47 hPa (6.56×10^{-2} atm) oxygen pressure it is equal to $y = [V_{Cu}]_{total} = 2.17 \times 10^{-3}$. Comparison of this value with that calculated on the basis of other data [203] indicates that, as was expected, the non-ionised cation vacancies are predominant defects in the range of temperature and oxygen pressure examined.

By using the ratio of concentrations of non-ionised to ionised cation vacancies estimated by the above method, we can calculate the predicted dependence of the total concentration of vacancies on oxygen pressure.

By assuming that the total concentration of cation vacancies in cuprous oxide is equal to the following sum:

$$[V'_{Cu}]_{total} = [V^{x}_{Cu}] + [V'_{Cu}] \tag{2.9.41}$$

we can express the dependence of the concentration of vacancies on oxygen pressure by the following equation:

$$[V_{Cu}]_{total} = Ap_{O_2}^{1/n} = Bp_{O_2}^{1/4} + Cp_{O_2}^{1/8} \tag{2.9.42}$$

where constant A can be calculated for given temperature from the empirical equation (2.9.32) and the constant C can be evaluated on the basis of knowledge of the dependence of the concentration of electron holes on oxygen pressure, which, as has already been mentioned, was determined by Trivich *et al.* The constant B is eliminated when the ratio of the concentrations of defects is introduced in equation (2.9.42):

$$a = \frac{[V^{x}_{Cu}]}{[V'_{Cu}]} = \frac{Bp_{O_2}^{1/4}}{Cp_{O_2}^{1/8}} = \frac{B}{C}p_{O_2}^{1/8} \tag{2.9.43}$$

Using this relationship we can express equation (2.9.42) in the following form:

$$Ap_{O_2}^{1/n} = aCp_{O_2}^{1/8} + Cp_{O_2}^{1/8} = Cp_{O_2}^{1/8}(a+1) \tag{2.9.44}$$

By rearranging this equation and taking logarithms we obtain the following relationship between the exponent $1/n$, the ratio a and the oxygen pressure:

$$\frac{1}{n} = \frac{\log\left[\frac{C}{A}p_{O_2}^{1/8}(a+1)\right]}{\log p_{O_2}} \tag{2.9.45}$$

In order to calculate the value of the exponent, $1/n$, it is necessary to put suitable values of A and C in equation (2.9.45) and to replace a with the ratio of the concentrations of the vacancies for a particular oxygen pressure. It follows from equation (2.9.38) that the concentration of ionised cation vacancies at constant temperature increases with increasing oxygen pressure according to the following equation:

$$[V'_{Cu}] = Cp_{O_2}^{1/8} \tag{2.9.46}$$

The results of studies [203] of the electrical conductivity confirm the character of the above relationship for a temperature of 1000°C.
Hence on introducing the calculated concentration of ionised cation vacancies and the corresponding value of oxygen pressure 66.47 hPa (6.56 ×

$\times 10^{-2}$ atm) in equation (2.9.46), we can calculate constant C. It is equal to 1.65×10^{-1}, Then, by putting the value of the temperature, equal to 1000°C, in the empirical relationship (2.9.32), we can calculate constant A. This is equal to 7.8×10^{-3}. Using the results of the above calculations we can express equation (2.9.45) in the following detailed form:

$$\frac{1}{n} = \frac{\log[2.12\times10^{-2}p_{O_2}^{1/8}(a+1)]}{\log p_{O_2}} \tag{2.9.47}$$

By replacing the symbol a in this equation with the value of the concentration ratio of defects $[V_{Cu}^{x}]/[V_{Cu}']$ calculated for several values of oxygen pressure, it is possible to calculate the dependence of exponent $1/n$ on oxygen pressure. This dependence is shown graphically in Fig. 137. It

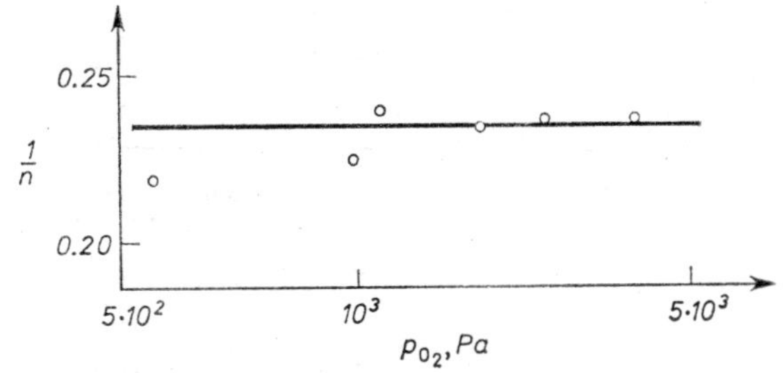

Fig. 137. The dependence of the exponent $\frac{1}{n}$ in equation (2.9.47) on oxygen pressure (after [202]); the continuous line shows experimental results.

follows from the plot that the calculated value of the exponent is practically independent of oxygen pressure in the pressure range examined and that it is very close to the experimentally determined value.

The results of the above discussion confirm the correctness of the conclusion that in the range of temperature and oxygen pressure examined, non-ionised cation vacancies are the predominant defects in the crystal lattice of cuprous oxide.

On the basis of this defect model for the crystal lattice of $Cu_{2-y}O$, it is possible to calculate the entropy and the enthalpy of formation of non-ionised cationic vacancies. Such calculations have been carried out with the aid of the empirical equation (2.9.32) and the theoretical relationship (2.9.35). The results are shown in Table 12.

The results of the studies mentioned above together with the data obtained by Trivich *et. al.* [203] make it possible to draw certain conclusions

TABLE 12

The results of high temperature studies on cuprous oxide

Kind of study	Temperature range °C	E kJ/mol	E kcal/mol	ΔH_f kJ/mol	ΔH_f kcal/mol
Self-diffusion of	800–1050	151.1	36.1		
copper	900–1050	100.5	24.0		
Self-diffusion of oxygen	1030–1120				
Electrical conductivity	500–1100	62.8	15.0		
	750–1000	61.1	14.6		
	850–1100	92.1	22.0		
	650–1100	62.8	15.0		
Deviation from	900–1000			66.9–50.2	16.0–12.0
stoichiometry	850–1100			90.9	21.7
	950–1100			50.2	12.0
Ionisation energy of				62.8	15.0
acceptor levels				48.1	11.5

about the mechanism of electrical conductivity in cuprous oxide. It is still not known at present, whether this oxide is a broad band semiconductor, i.e. in which the transport of the current carriers is a non-activated process, or whether it is an activated process, in which the movement of electron holes requires an activation energy, i.e. takes place according to the small polarons or the hopping model. In the former case the temperature dependence of electrical conductivity is practically exclusively caused by changes in the concentration of the charge carriers, since the mobility of these carriers is either only slightly temperature dependent (it is a power function) or is independent on temperature. Therefore, if cuprous oxide were a broad band semiconductor, the dependence of its electrical conductivity on temperature and oxygen pressure would be able to be expressed by the following simplified equation:

$$\sigma = \text{const}\, p_{O_2}^{1/8} \exp\left(-\frac{\frac{1}{2}\Delta H_f + \Delta H_f'}{RT}\right) \tag{2.9.48}$$

This equation was obtained by transforming the general expression (2.9.40) by means of equations (2.9.37) and (2.9.38) and assuming that the mobility of electron holes does not depend on temperature.

On the other hand, if the transport of charge carriers was activated as in the small polarons theory, the mobility of the carriers, like their con-

ΔH_m kJ/mol	ΔH_m kcal/mol	ΔS_f J/(mol·K)	ΔS_f cal/(mol·K)	ΔS_m J/(mol·K)	ΔS_m cal/(mol · K)	$1/n$
50.2	12.0					1/3.9
164.5	39.3	−63.2	−15.1	62.0	14.8	1/2
						1/7
						1/7
						1/8
						1/8
						1/3.7–1/5
50.6	12.1	20.5	4.9	−6.3	−1.50	1/3.7
50.2	12.0	−0.96	−0.23	5.4	1.3	

centration, would be an exponential function of temperature and, in the simplest case, could be expressed by equation (1.5.9).

By introducing this equation in equation (2.9.40) and using equations (2.9.37) and (2.9.38), we obtain the following relationship:

$$\sigma = \text{const}\, p_{O_2}^{1/8} \exp\left(-\frac{\frac{1}{2}\Delta H_f + \Delta H_f' + \Delta H_m^{h\cdot}}{RT}\right) \tag{2.9.49}$$

It follows from equations (2.9.48) and (2.9.49) that, in the case of non-activated transport of electron holes, the activation energy of the electrical conductivity of cuprous oxide would be related to the heat of formation of cation vacancies and the energy of their ionisation according to the following equation:

$$E_\sigma = \tfrac{1}{2}(\Delta H_f + \Delta H_f') \tag{2.9.50}$$

On the other hand, in the case of activated transport of current carriers, taking place according to the hopping mechanism, E_σ could be expressed by the following equation:

$$E_\sigma = \tfrac{1}{2}(\Delta H_f + \Delta H_f') + \Delta H_m^{h\cdot} \tag{2.9.51}$$

The results of studies carried out by Trivich *et al.* [203] as well as by other authors [204–210] indicate that, in the range of temperature and oxygen pressure considered the activation energy for electrical conduction in $Cu_{2-y}O$ is equal to 62.8 kJ/mole (15 kcal/mole). The value of $\Delta H'_f$, determined by many authors from measurements of optical and electrical properties of this oxide, is equal to 62.8 kJ/mole (15 kcal/mole). Finally the value of ΔH_f determined by Godlewski, Mrowec and Stokłosa [202] is equal to 50.24 kJ/mole (12 kcal/mole) [see the empirical relationship (2.9.32)]. By putting these values in equation (2.9.51) it is possible to calculate the eventual activation energy of the movement of electron holes in the crystal lattice of $Cu_{2-y}O$. It is equal to about 6.28 kJ/mole (1.5 kcal/mole). This value is within the limits of the experimental error in the determinations of both the electrical conductivity and the heat of formation and ionisation of cation vacancies.

Thus this discussion leads to the conclusion that at high temperatures cuprous oxide is a broad band semiconductor. In order to confirm this conclusion it is necessary to carry out further studies on the electrical conductivity and particularly on the energy of ionisation of cation vacancies in this oxide.

Further conclusions regarding the thermodynamic and kinetic properties of point defects in cuprous oxide can be drawn from self-diffusion measurements of copper in $Cu_{2-y}O$. It was shown in Chapter 2.1 that self-diffusion coefficient is related to the concentration of defects and their diffusion coefficient according as follows:

$$D = D_d N_d \qquad (2.9.52)$$

In the case of $Cu_{2-y}O$, this expression has the following form:

$$D_{Cu} = D_{V_{Cu}}[V_{Cu}] \qquad (2.9.53)$$

It follows from this equation that the diffusion coefficient of cation vacancies, $D_{V_{Cu}}$, which is the measure of their mobility in the crystal lattice of $Cu_{2-y}O$, can be calculated as a function of temperature and concentration of defects on the basis of measurements of deviations from stoichiometry, y, and the self-diffusion coefficients of copper in this material [empirical relationships (2.9.32) and (2.9.33)].

Comparison of these data leads to the conclusion that the dependence of the concentration of cation vacancies on oxygen pressure hardly differs from the analogous dependence of the self-diffusion coefficient of copper in cuprous oxide. Consequently the diffusion coefficient of cation vacancies in this oxide is practically independent of their concentration. This conclusion is of particular importance, since it implies that there are no interactions between ionic defects in the crystal lattice of cuprous oxide (the situation is analogous to that in the case of $Co_{1-y}O$). For this reason the possibility of formation of complexes and clusters of defects can be excluded.

By assuming that the dependence of D_{Cu} on oxygen pressure is the same as that of V_{Cu}, we can calculate the diffusion coefficient of defects by dividing the empirical equation (2.9.33) by (2.9.32). This leads to the following empirical temperature dependence of the diffusion coefficient of cation vacancies:

$$D_{V_{Cu}} = 2.43 \times 10^{-3} \exp\left(-\frac{12\,000}{RT}\right) \tag{2.9.54}$$

It follows from the theory of diffusion in solids that the diffusion coefficient of cation vacancies in cuprous oxide is the following function of temperature [see equation (2.1.6)]:

$$D_{V_{Cu}} = \alpha a_0^2 \chi \nu \exp\left(\frac{\Delta S_m}{R}\right) \exp\left(-\frac{\Delta H_m}{RT}\right) \tag{2.9.55}$$

where ΔS_m and ΔH_m are the entropy and the enthalpy of jumps of copper ions from lattice sites to neighbouring vacancies.

By introducing equations (2.9.55) and (2.9.35) in equation (2.9.53) the self-diffusion coefficient of copper in $Cu_{2-y}O$ can be expressed by the following equation:

$$D_{Cu} = \alpha a^2 \chi \nu p_{O_2}^{1/4} \exp\left(\frac{\Delta S_f + \Delta S_m}{R}\right) \exp\left(-\frac{\Delta H_f + \Delta H_m}{RT}\right) \tag{2.9.56}$$

It follows from this equation that the experimentally determined activation energy of diffusion is the sum of the heat of formation of the cation vacancies and their migration in the periodic field of the lattice.

Comparison of equation (2.9.55) with equation (2.9.54) shows that the activation enthalpy of diffusion ΔH_m is equal to 50.24 kJ/mole (12 kcal/mole). In order to estimate the entropy of diffusion, ΔS_m, it is assumed, following Moore and O'Keefe [212], that coefficients α and χ are equal to unity and the vibration frequency ν is calculated from the formula:

$$\nu = (2/\pi a_0)(\Delta H_m/M)^{1/2} \tag{2.9.57}$$

where a_0 and M are the lattice constant of cuprous oxide and the atomic weight of copper respectively. The vibration frequency estimated in this way is equal to $1.33 \times 10^{12}\ s^{-1}$ and the lattice constant is equal to 0.426 nm. By putting these values in equation (2.9.55) and comparing it with the empirical equation (2.9.54) it is possible to calculate the activation entropy of diffusion. It is equal to 5.443 J(mole · K) (1.3 cal/moleK).

The above example indicates that the combined measurements of deviations from stoichiometry, self-diffusion and electrical conductivity offer wide possibilities for the investigation of the defect structure and semiconducting properties of oxides.

Before ending this chapter we shall discuss the effect of chromium as a trivalent dopant on the defect structure and semiconducting properties of nickel oxide. The solubility of chromium oxide in nickel oxide is very limited. At about 1200°C it does not exceed 2 at.% Cr [213]. For this

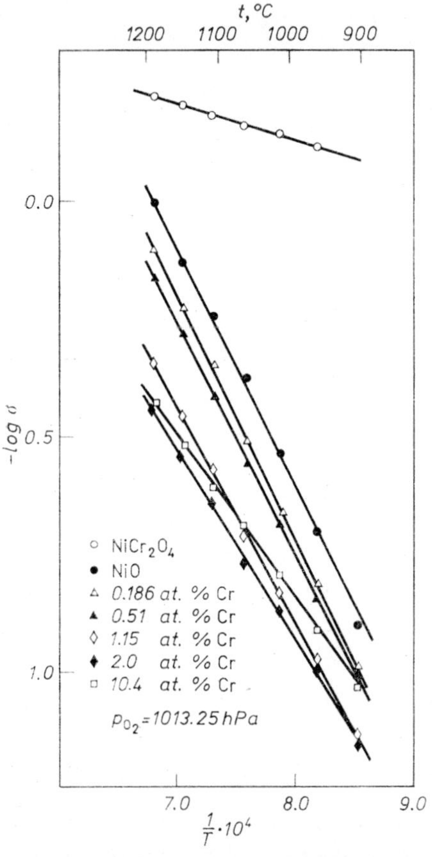

Fig. 138. The temperature dependence of the electrical conductivity of nickel oxide, chromium–nickel spinel and solid solutions $NiO–Cr_2O_3$ for oxygen pressure equal to 101.325 kPa (1 atm) (after [215]).

reason the measurements were carried out on solid solutions containing from 0.8 to 2.0 at.% Cr and additionally on a heterogeneous mixture containing 10.4 at.% Cr as well as on pure spinel $NiCr_2O_4$.

These studies, carried out by Mrowec, Wójcikiewicz and Walec consisted of measurements of the dependence of electrical conductivity on the temperature and equilibrium oxygen pressure. The measurements were carried out by the four probe method at temperatures from 950 to 1200°C and partial oxygen pressures from 101.325–0.101325 kPa (1 to 10^{-3} atm) in a flowing gas apparatus. The required partial pressures of oxygen were obtained by maintaining a constant flow of suitable mixtures of argon and oxygen through the reaction chamber. The total pressure of gas in

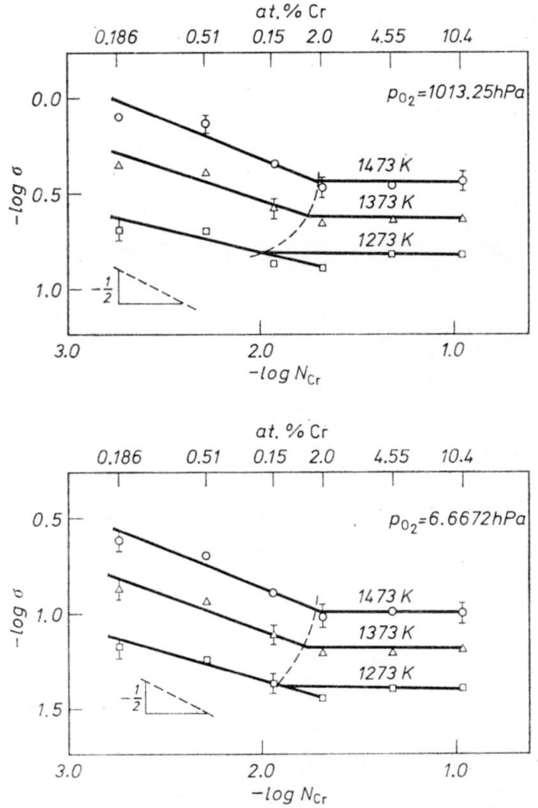

Fig. 139. The dependence of the electrical conductivity of solid solutions $Ni–Cr_2O_3$ on their composition for several temperatures and two oxygen pressures (after [215]).

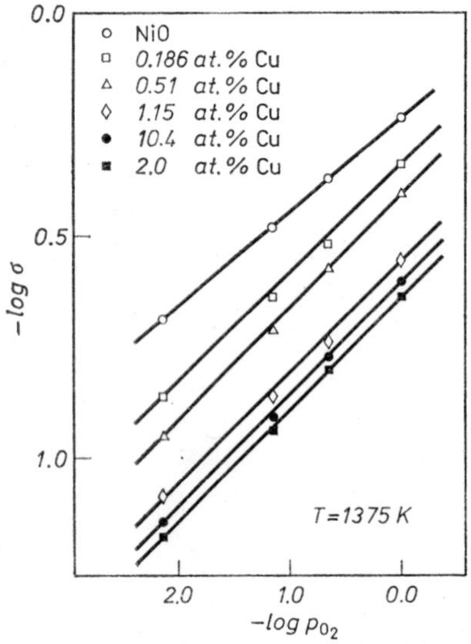

Fig. 140. The dependence of the electrical conductivity of pure and chromium doped nickel oxide on oxygen pressure at 1373 K (after [215]).

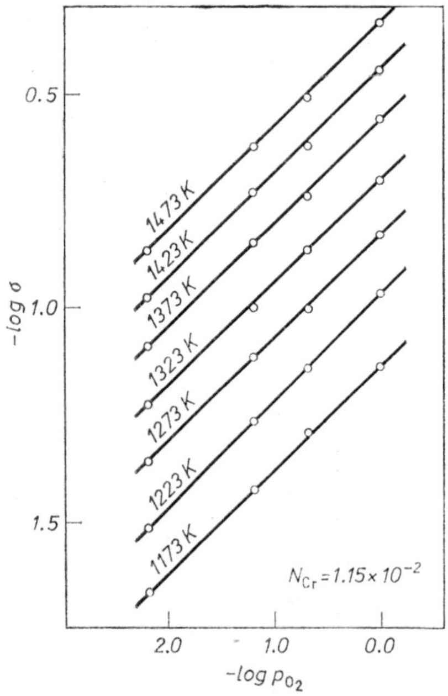

Fig. 141. The dependence of the electrical conductivity of solid solution NiO–Cr_2O_3 containing 1.15 at.% Cr on oxygen pressure for several temperatures (after [215]).

the apparatus was always atmospheric. The results of these measurements are shown graphically in Figs. 138–141.

It follows from Fig. 138 that the electrical conductivities of pure nickel oxide, NiO–Cr_2O_3 solid solutions, heterogeneous mixtures and pure spinel were linear functions of temperature of the Arrhenius type. The activation energy for the electrical conductivity of solutions containing up to 1.15 at.% Cr hardly differed from that of pure nickel oxide and was equal to about 92.11 kJ/mole (22 kcal/mole).

It also follows from Fig. 138 that the electrical conductivity decreases with increasing amount of the dopant up to about 2 at.% Cr and then increases reaching, in the case of pure spinel, a value considerably higher than that of pure nickel oxide. This indicates that in the temperature range examined the solubility of chromium oxide in nickel oxide does not exceed about 2 at.% Cr. This solubility can be estimated from log–log plot in Fig. 139, which shows the electrical conductivity of the materials examined as a function of chromium content. The broken line shows the temperature dependence of the limit of the solubility of Cr_2O_3 in NiO. It follows from the graph that the solubility of chromium oxide in nickel oxide slightly increases with increasing temperature, but is practically independent of oxygen pressure.

It follows from Figs. 140–141 that the electrical conductivity of the solid solutions and of pure nickel oxide increases with increasing oxygen pressure and that the character of this function is independent of the amount of dopant and temperature. For this reason the dependence of electrical conductivity on these two parameters can be expressed by the following empirical equations (for pure NiO and for the solid solutions respectively):

$$\sigma_{NiO} = \mathrm{const}\, p_{O_2}^{1/5} \exp\left(-\frac{E_{\sigma NiO}}{RT}\right) \tag{2.9.58}$$

and

$$\sigma_{NiO-Cr_2O_3} = \mathrm{const}\, p_{O_2}^{1/4} \exp\left(-\frac{E_{\sigma NiO-Cr_2O_3}}{RT}\right) \tag{2.9.59}$$

where $E_{\sigma NiO}$ and $E_{\sigma NiO-Cr_2O_3}$ are the activation energies of the electrical conductivity of pure nickel oxide and the solid solution respectively.

As it was mentioned above, the electrical conductivity of the solid solutions decreases with increasing amount of dopant. It follows from Fig. 139 it is a power function. Hence equation (2.9.59) can be generalized by introducing the third parameter, i.e. the amount of dopant in the solutions:

$$\sigma_{NiO-Cr_2O_3} = \text{const}\frac{1}{[\text{Cr}]^{1/n}} p_{O_2}^{1/4} \exp\left(-\frac{E_{\sigma NiO-Cr_2O_3}}{RT}\right) \tag{2.9.60}$$

where [Cr] is the concentration of chromium in the solid solution. The exponent $1/n$ slightly increases with temperature. Its value changes from 1/3.0 at 1000°C to 1/2.5 at 1200°C.

The interpretation of the above results requires the consideration of defect equilibria in pure and chromium doped nickel oxide. In the last example it was shown that in nickel oxide the predominant point defects in a certain range of temperature and oxygen pressure are singly and doubly ionised cation vacancies occurring in comparable amounts. For this reason the electrical conductivity of this oxide, according to the empirical equation (2.9.17), is a function of temperature and partial oxygen pressure as follows:

$$\sigma = \text{const}\, p_{O_2}^{1/5} \exp\left(-\frac{\frac{2}{3}\Delta H_f + \Delta H_m^{h^\bullet}}{RT}\right) \tag{2.9.61}$$

In order to elucidate the defect structure in $NiO-Cr_2O_3$ solid solutions, it is necessary to consider the effect of trivalent chromium ions on defect equilibria in nickel oxide for three hypothetical limiting cases and to compare the function $\sigma = f(p_{O_2}, T, [\text{Cr}])$ with the empirical equation (2.9.60).

Case 1. *Doubly ionised cation vacancies predominate*

$$[V''_{Ni}] \gg [V'_{Ni}]$$

$$2Cr_2O_3 + 2h^\bullet \rightleftharpoons 4Cr^\bullet_{Ni} + 5O_O + \tfrac{1}{2}O_2 + V''_{Ni} \tag{2.9.62}$$

Case 2. *Singly ionised cation vacancies predominate*

$$[V'_{Ni}] \gg [V''_{Ni}]$$

$$Cr_2O_3 + h^\bullet \rightleftharpoons 2Cr^\bullet_{Ni} + V'_{Ni} + 3O_O \tag{2.9.63}$$

Case 3. *The amounts of singly and doubly ionised cation vacancies are comparable*

$$[V'_{Ni}] \cong [V''_{Ni}]$$

$$3Cr_2O_3 + 3h^\bullet \rightleftharpoons 6Cr^\bullet_{Ni} + V'_{Ni} + V''_{Ni} + 8O_O + \tfrac{1}{2}O_2 \tag{2.9.64}$$

The electroneutrality condition for the lattice in the above three limiting cases has the following forms:

1. $[Cr^{\bullet}_{Ni}]+[h^{\bullet}] = 2[V''_{Ni}]$ (2.9.65)

2. $[Cr^{\bullet}_{Ni}]+[h^{\bullet}] = [V'_{Ni}]$ (2.9.66)

3. $[Cr^{\bullet}_{Ni}]+[h^{\bullet}] = 2/3[V_{Ni}]_c$ (2.9.67)

where $[V_{Ni}]_c = [V'_{Ni}]+[V''_{Ni}]$

It follows from the above equations that, irrespective of the degree of ionisation of cation vacancies, the concentration of electron holes in $NiO–Cr_2O_3$ solid solutions is lower than in pure nickel oxide, whereas the concentration of cation vacancies is higher.

The concentration of electron holes in pure $Ni_{1-y}O$ is, in the range of oxygen pressure and temperature considered, of the order of 10^{-2} at.% ($[Ni_{V_{Ni}}] \cong 10^{-4}$). This means that the concentration of electron defects in $NiO–Cr_2O_3$ solid solutions is considerably lower than the dopant concentration, i.e. the following condition is fulfilled:

$$[h^{\bullet}] \ll [Cr^{\bullet}_{Ni}] \tag{2.9.68}$$

Hence electroneutrality condition for the lattice, given by equations (2.9.65), (2.9.66) and (2.9.67), is simplified to the following forms:

$$[Cr^{\bullet}_{Ni}] = 2[V''_{Ni}] \tag{2.9.69}$$

$$[Cr^{\bullet}_{Ni}] = [V'_{Ni}] \tag{2.9.70}$$

$$[Cr^{\bullet}_{Ni}] = \tfrac{3}{2}[V_{Ni}]_c \tag{2.9.71}$$

By using these relationships, we obtain the following expression for the dependence of the concentration of electron holes in the solid solution on temperature, equilibrium oxygen pressure and dopant concentration:

For case 1: $[V''_{Ni}] \gg [V'_{Ni}]$:

$$[h^{\bullet}] = \text{const}\left(\frac{1}{[Cr^{\bullet}_{Ni}]}\right)^{1/2} p_{O_2}^{1/4}\exp\left(-\frac{\frac{1}{2}\Delta H_{f(1)}}{RT}\right) \tag{2.9.72}$$

For case 2: $[V''_{Ni}] \ll [V'_{Ni}]$:

$$[h^{\bullet}] = \text{const}\left(\frac{1}{[Cr^{\bullet}_{Ni}]}\right) p_{O_2}^{1/2}\exp\left(-\frac{\Delta H_{f(2)}}{RT}\right) \tag{2.9.73}$$

For case 3: $[V''_{Ni}] \cong [V'_{Ni}]$:

$$[h^{\bullet}] = \text{const}\left(\frac{1}{[Cr^{\bullet}_{Ni}]}\right)^{2/3} p_{O_2}^{1/3}\exp\left(-\frac{\frac{2}{3}\Delta H_{f(3)}}{RT}\right) \tag{2.9.74}$$

where $\Delta H_{f(1)}$, $\Delta H_{f(2)}$ and $\Delta H_{f(3)}$ are respectively the enthalpies of formation of one mole of doubly ionised cation vacancies, singly ionised cation vacancies and singly and doubly ionised cation vacancies present in comparable amounts in the $NiO–Cr_2O_3$ solid solution.

By introducing the above equations in the general relationship (1.5.1) and using equation (1.5.8) we finally obtain the following relationships between the electrical conductivity of the solid solutions, the dopant content, the temperature and the equilibrium oxygen pressure for the three limiting cases:

$$\sigma_1 = \text{const}\left(\frac{1}{[\text{Cr}_{\text{Ni}}^{\bullet}]}\right)^{1/2} p_{O_2}^{1/4} \exp\left(-\frac{\frac{1}{2}\Delta H_{f(1)}+\Delta H_m}{RT}\right) \tag{2.9.75}$$

$$\sigma_2 = \text{const}\left(\frac{1}{[\text{Cr}_{\text{Ni}}^{\bullet}]}\right) p_{O_2}^{1/2} \exp\left(-\frac{\Delta H_{f(2)}+\Delta H_m}{RT}\right) \tag{2.9.76}$$

$$\sigma_3 = \text{const}\left(\frac{1}{[\text{Cr}_{\text{Ni}}^{\bullet}]}\right)^{2/3} p_{O_2}^{1/3} \exp\left(-\frac{\frac{2}{3}\Delta H_{f(3)}+\Delta H_m}{RT}\right) \tag{2.9.77}$$

It follows from the above considerations that, depending on the degree of ionisation of cation vacancies, the electrical conductivity of $NiO–Cr_2O_3$ solid solutions depends in different ways on both the oxygen pressure and the dopant content. Comparison of the above three theoretical relationships with the empirical equation (2.9.60) indicates that the experimental data are in best agreement with equation (2.9.75). This leads to an important conclusion about the degree of ionisation of cation vacancies in $NiO–Cr_2O_3$ solid solutions and the structure of these solutions. In the derivation of equations (2.9.75)–(2.9.77) it was assumed that chromium ions are incorporated at the cation sublattice sites of the crystal lattice of NiO and not in interstitial positions. The agreement between the theoretical predictions and the experimental data confirms this assumption and makes it possible to state that chromium doped nickel oxide is a substitutional and not an interstitial solid solution. Apart from this, the value of the exponent at the oxygen pressure in equation (2.9.60) indicates that the cation vacancies in the solid solution are doubly ionised. The last conclusion is apparently in disagreement with the results of studies carried out on pure nickel oxide, since it follows from the dependence of the electrical conductivity on oxygen pressure that in pure nickel oxide, in the same range of temperature and oxygen pressure, the singly and doubly ionised cation vacancies are present in comparable amounts. Since the addition of chromium causes an increase of the concentration of cation vacancies, it appears that the degree of association of ionic and electronic defects should increase, i.e. that the singly ionised cation vacancies may be predominant in the solid solution. However it should be borne in mind that the addition of chromium simultaneously causes a decrease of the concentration of electron holes and that, in the range of dopant concentrations discussed here, the concentration of cation vacancies is several orders of magnitude higher than that of electron holes. Hence, in these conditions, only a small fraction of cation vacancies can be associated with electron holes and consequently the completely ionised cation vacancies must be predominant.

It follows from the above discussion that the activation energy of the electrical conductivity of $NiO–Cr_2O_3$ solid solutions is practically independent of the amount of dopant and is approximately equal to 92.11 kJ/mole (22 kcal/mole). At the same time the electrical conductivity increases with increasing oxygen pressure and temperature according to equation (2.9.60). Thus we can express this dependence by the following detailed equation:

$$\sigma_{NiO-Cr_2O_3} = \text{const}\frac{1}{[Cr^{\bullet}_{Ni}]^{1/2}}p_{O_2}^{1/4}\exp\left(-\frac{22\,000\pm1000}{RT}\right) \tag{2.9.78}$$

Comparison of this equation with the theoretical equation (2.9.75) gives the following expression:

$$E_{\sigma NiO-Cr_2O_3} = \tfrac{1}{2}\Delta H_{f(1)}+\Delta H_m^{h^\bullet} \tag{2.9.79}$$

It follows from this relationship that, in order to calculate the enthalpy of formation of cation vacancies in the solid solution, it is necessary to know not only the value of $E_{\sigma NiO-Cr_2O_3}$, but also the activation energy for the movement of electron holes $\Delta H_m^{h^\bullet}$. The value of this activation energy was determined by Mitoff [196], Bransky and Tallan [216] and by Sockel and by Ilschner [217] ,who obtained the values equal to 23.03. (5.5 kcal/mole), 35.59 (8.5 kcal/mole) and 24.28 kJ/mole (5.8 kcal/mole) respectively. On substituting for $\Delta H_m^{h^\bullet}$ in equation (2.9.79) with the mean value of 27.63 kJ/mole (6.6 kcal/mole) and for $E_{\sigma NiO-Cr_2O_3}$ with the value 92.11 kJ/mole (22 000 cal/mole), determined in the work discussed, it is possible to calculate the enthalpy of formation of doubly ionised cation vacancies in the solid solution. It is equal to 128.95 kJ/mole (30.8 kcal/mole). The enthalpy of formation of cation vacancies in pure nickel oxide, calcullated analogously, is equal to 149.47 kJ/mole (35.7 kcal/mole).

It follows from the above considerations that the enthalpy of formation of cation vacancies in NiO–Cr_2O_3 solid solutions does not differ appreciably from the enthalpy of formation of these defects in pure nickel oxide. However it might be expected that it would be higher in the case of the solid solution than in the case of the pure oxide, whereas the results of calculations show that the reverse is true. This is due to the fact that in the solid solution completely ionised vacancies are formed, whereas in the pure oxide half the vacancies are signly ionised. It should be stressed that the results of the above calculations are not sufficient for drawing quantitative conclusions about the thermodynamic properties of the system examined, since in these calculations it was assumed that the activation energy for the movement of electron holes in the pure nickel oxide is the same as that in the chromium doped oxide and that this movement is an activated process, which so far has not been checked experimentally. Apart from this it should be remembered that, in the solid solutions, the formation of defect complexes is possible, since the concentration of cation vacancies in the solutions is much higher than in the pure oxide. It has been shown in Chapter 1.4.2 that when the concentration of point defects exceeds about 0.1 at.% they begin to interact, which leads to the formation of more or less complex defect clusters having thermodynamic properties considerably different from those of simple point defects which do not interact with each other. Nevertheless the above example illustrates the importance of electrical conductivity measurements in studies on the defect structure in solid solutions of metal oxides.

The results of experimental studies quoted in this chapter are limited to oxides in which the concentration of defects does not exceed 1 at.% and in the majority of cases is lower than 0.1 at.% ($Ni_{1-y}O$, $Cu_{2-y}O$). For this reason it is possible to consider these systems as strongly dilute solid solutions and to apply the point defect theory to them. Application of this theory has also been attempted to many oxides and sulphides of metals in which the concentration of defects is considerably higher. This may lead to erroneous conclusions. A classical example of this situation is wustite ($Fe_{1-y}O$) which, as has been mentioned above, shows very large deviations from stoichiometry ($Fe_{0.88}O$), equivalent to a concentration of cation vacancies exceeding 10 at.%. This oxide has been extensively studied for many years. Figure 142 shows the dependence of the deviation from stoichiometry, y, on the equilibrium oxygen pressure at several temperatures on a double logarithmic plot. It can be seen from the plot that, in contrast to the majority of oxides, the deviation of the wustite phase from stoichiometric composition at a constant oxygen pressure decreases with increasing temperature. In addition, y is not a linear function of pressure and the form of this dependence changes with temperature.

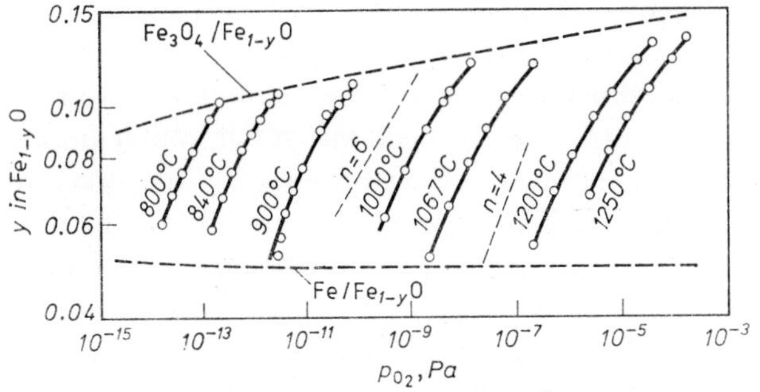

Fig. 142. The dependence of the deviation from stoichiometry y of the $Fe_{1-y}O$ phase on oxygen pressure for several temperatures (after [174]).

In spite of these anomalies, wustite was regarded as a solid solution of cation vacancies and electron holes in the $Fe_{1-y}O$ lattice and the theory of point defects was applied to this system for many years. According to this defect model $Fe_{1-y}O$ lattice the process of formation of cation vacancies can be expressed by the following equation:

$$\tfrac{1}{2}O_2 \rightleftharpoons V^p_{Fe} + ph^{\cdot} + O_O \qquad (2.9.80)$$

where V^p_{Fe} is a cation vacancy having effective charge p within the limits of zero and -2. On applying the law of mass action and a suitable electroneutrality condition for the lattice to the above relationship, we can obtain the general relation between the concentration of defects and oxygen pressure:

$$y = [V^p_{Fe}] = \mathrm{const}\, p_{O_2}^{1/(n+1)2} \qquad (2.9.81)$$

where the exponent $1/n$ can assume values from $1/2$ to $1/6$ corresponding to non-ionised and doubly ionised cation vacancies respectively. The non-linear character of the isotherms shown in Fig. 142 was explained as being due to the change of the degree of ionisation of defects accompanying the increase of oxygen pressure. Although the theoretical predictions were in formal agreement with experimental data, the fact that the point defect theory cannot be applied without limitations to a system having so high degree of disorder has been long realised, since it was difficult to suppose that cation vacancies having effective electrical charge do not interact with each other and that they are statistically distributed in the crystal at concentrations reaching 12 at.%. These doubts were confirmed during the last decade. Using neutron and X-ray diffration methods Roth and also Koch and Cohen and other authors have shown that the defect structure in the wustite phase is much more complex. This problem has been discussed in greater detail in Chapter 1.4.2. Here we should only mention that this oxide contains complexes of defects consisting of two doubly ionised octahedral cation vacancies and one trivalent interstitial cation in the tetrahedral position. Thus the extrinsic chemical disorder of the structure resulting from the deviation from stoichiometry overlaps with the Frenkel type intrinsic disorder, which leads to the formation of extended defects called Roth complexes. It should be mentioned that for this reason the concentration of cation vacancies is twice as high as the deviation from stoichiometry, since the formation of each cation vacancy in the quasi-chemical reaction (2.9.80) is accompanied by the formation of another cation vacancy in the Frenkel process:

$$Fe_{Fe}^{\bullet} \rightleftharpoons Fe_i^{\bullet\bullet\bullet} + V_{Fe}'' \qquad (2.9.82)$$

where $Fe_{Fe}^{\bullet} = h^{\bullet}$ is a trivalent iron cation in the normal lattice site.

Taking into account the existence of the above complexes Kofstad and Hed [218, 219] have made an attempt to apply the classical point defect theory to wustite. They represented the formation of the defect complex by the following quasi-chemical equation:

$$Fe_{Fe} + V_i + \tfrac{1}{2}O_2 \rightleftharpoons Fe_c^{x} + O_O \qquad (2.9.83)$$

where $Fe_c^{x} = (V_{Fe}'' Fe_i^{\bullet\bullet\bullet} V_{Fe}')$ is a defect complex, electrically neutral with respect to the lattice, and V_i is a tetrahedral interstitial vacancy. This complex can be ionised according to the following equation:

$$Fe_c^{x} + Fe_{Fe} \rightleftharpoons Fe_c' + Fe_{Fe}^{\bullet} \qquad (2.9.84)$$

where $Fe_c' = (V_{Fe}''\ Fe_i^{\bullet\bullet\bullet}\ V_{Fe}'')$.

The total concentration of complexes expressed in mole fractions is equal to the deviation from stoichiometry y, which can be expressed by the following equation:

$$y = [Fe_c^{x}] + [Fe_c'] = \alpha + \beta \qquad (2.9.85)$$

In this case the condition for electroneutrality of the lattice has the following form:

$$[h^{\bullet}] = [Fe_{Fe}^{\bullet}] = [Fe_c'] \qquad (2.9.86)$$

In view of the considerable degree of disorder in the crystal lattice of wustite, the concentration of divalent iron cations in the normal lattice sites must be allowed for, since a considerable part of the cation sites are empty or are occupied by trivalent cations (electron holes). Hence:

$$[Fe_{Fe}] = 1-\beta-2y \tag{2.9.87}$$

The concentration of interstitial vacancies $[V_i]$ can also be expressed by y assuming that some of the interstitial positions are occupied by complexes and that the six interstitial vacancies surrounding the complex cannot be occupied by other complexes for geometrical reasons. Hence:

$$[V_i] = 2-7y \tag{2.9.88}$$

By applying the low of mass action to reactions (2.9.82) and (2.9.83) and using equations (2.9.86), (2.9.87) and (2.9.88), Kofstad and Hed have derived the following relationships:

$$p_{O_2} = \left[\frac{K_c(1-2y)-\beta(2-7y)}{y-\beta}\right]^{-2} \tag{2.9.89}$$

and

$$\beta = \frac{K_c(1-y)\pm[K_c'^2(1-3y)^2+4K_{cy}'(1-2y)]^{1/2}}{2(1-K_c')} \tag{2.9.90}$$

where K_c and K_c' are the equilibrium constants for reactions (2.9.83) and (2.9.84) respectively. The numerical solutions of these equations are in excellent agreement with the experimetal data.

Kofstad and Hed's model was fully confirmed by extensive studies carried out by Janowski *et al.*, who determined thermogravimetrically the deviation from stoichiometry in wustite phase over a wide ranges of temperature and oxygen pressure and used the same samples for the determination of the total concentration of cation vacancies by the X-ray method. The results of these studies are shown in Fig. 143. It follows from the graph that, in agreement with the results of neutronographic studies carried out by Roth, the concentration of cation vacancies in wustite is about twice as high as the deviation from stoichiometry. Assuming, in agreement with Kofstad and Hed, that at high temperatures the Roth complexes are singly ionised, Janowski *et al.* suggested that they are formed according to the following reaction:

$$\tfrac{1}{2}O_2+Fe_{Fe}+V_i \rightleftharpoons Fe_k^{\cdot}+h^{\cdot}+O_O \tag{2.9.91}$$

By applying the law of mass action to this reaction we obtain:

$$K = \frac{[Fe_K^{\bullet}][h^{\bullet}][O_O]}{[Fe_{Fe}][V_i]\,p_{O_2}^{1/2}} \tag{2.9.92}$$

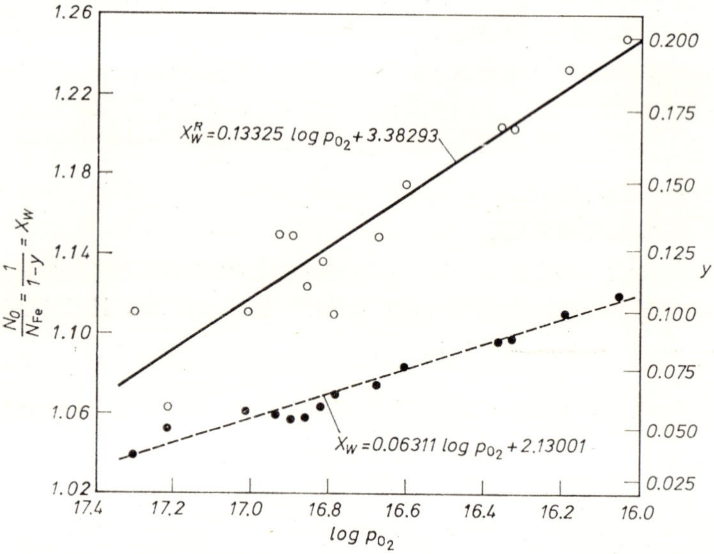

Fig. 143. The dependence of the concentration of cation vacancies in ferrous oxide ($Fe_{1-y}O$) on the equilibrium oxygen pressure (after [76]). Curve *1* shows the concentration of cation vacancies resulting from the deviation from stoichiometry [equation (2.9.80)]. Curve *2* characterises the total concentration of cation vacancies resulting from both the quasi-chemical surface reaction (2.9.80) and the intrinsic disorder of the Frenkel type [equation (2.9.82)].

Assuming that $[Fe_{Fe}] = 1-3y$, $[O_O] = 1$ and $[h^{\bullet}] = y$ and using equation (2.9.88), we obtain:

$$K = \frac{y^2}{(1-3y)(2-7y)p_{O_2}^{1/2}} \tag{2.9.93}$$

It follows from this equation that, at a constant temperature, $y^2/(1-3y)(2-7y)$ should be a linear function of the square root of oxygen pressure. This conclusion is fully confirmed by the data shown in Fig. 144.

The temperature dependence of the equilibrium constant, K, of reaction (2.9.91), calculated by Janowski *et al.*, can be expressed as follows:

$$K = 1.01 \times 10^{-6} \exp\left(\frac{59.3}{RT}\right) \tag{2.9.94}$$

This relationship is in very good agreement with an analogous expression given by Kofstad and Hed:

$$K = 1.45\times 10^{-6}\exp\left(\frac{64.0}{RT}\right) \tag{2.9.95}$$

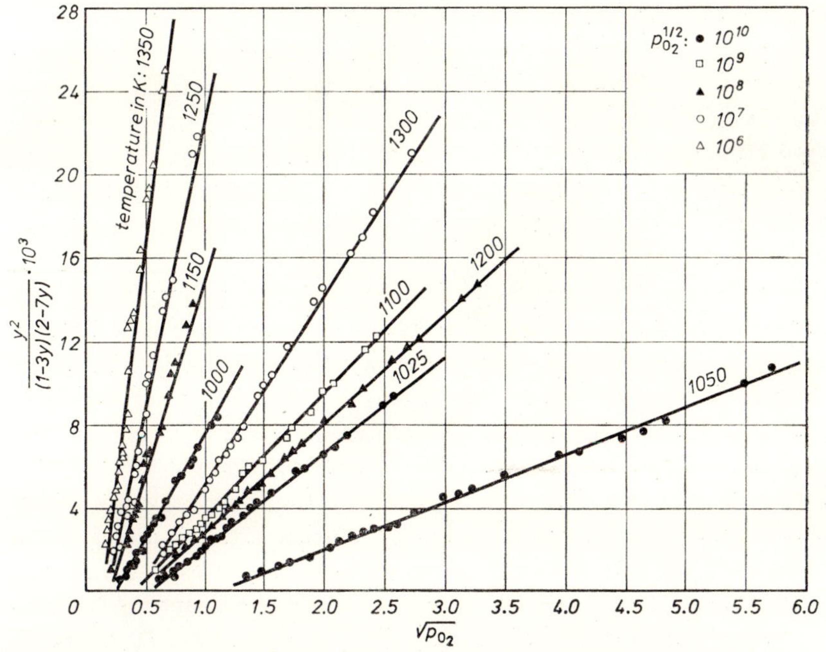

Fig. 144. The plot of the function expressed by equation (2.9.93) (after [76]).

The above results can be regarded as another convincing proof for the existence of defect complexes in wustite. However in Chapter 1.4.2 it was shown that this oxide may contain not only Roth complexes but also more complicated and larger agglomerations of defects called Koch-Cohen clusters. This means that the problem is very complex and hence it requires further experimental and theoretical studies.

2.10 Diffusion along grain boundaries

So far the discussion of diffusion in solids has been limited to bulk crystals in which point defects are the only irregularities in the periodic structure of the lattice. This is true in the case of monocrystals where the density of dislocations is negligibly small. However in practice we usually have to deal with polycrystalline substances, in which, in addition to point defects, other kinds of structural disorder occur, in particular, grain boundaries, dislocations as well as external and internal surfaces. For this reason, it is necessary to discuss the effect of these linear and planar defects on the diffusion process.

The primary difference between the diffusional properties of mono- and polycrystalline substances results from the fact that the latter consist of aggregates of crystals oriented completely at random (this does not apply to polycrystalline systems with a growth texture). For this reason substances of this type never show the anisotropy of diffusion, characteristic of monocrystals with non-cubic structures. However these effects are not of great importance.

The principal difference results from the fact that polycrystalline substances contain additional linear and planar defects. along which diffusion can take place much faster than through point defects, For this reason defects of this type are called high-diffusivity paths. The most important of them are grain boundaries. Theoretical considerations indicate that the density of packing of atoms along grain boundaries is smaller than in the bulk of crystal. This suggests that the activation energy of diffusion along grain boundaries should be lower than that of lattice diffusion and consequently the rate of inter-granular diffusion should be faster. Results of experimental studies confirm these conclusions (see Chapter 2.1). However measurements of grain boundary diffusion coefficients D_g are much more difficult to make than those of lattice diffusion coefficients and so far they have only been successfully carried out for a few metals. In the case of compounds of the MeX type, there is only indirect information obtained from differences between diffusion coefficients in polycrystalline substances and single crystals. Information about the structure of grain boundaries and the thickness of the zone defined by this term is also very scarce. However there are a number theories that help explain the fact that the rate of diffusion along grain boundaries is much faster than that of diffusion in the bulk of crystal. Three models of inter-granular diffusion deserve attention. One of them, based on the concept of dislocation structure at grain boundaries, seems to provide the best explanation.

In all the theories, it is assumed that the thickness of the region constituting the grain boundary does not exceed a few atomic diameters and is equal to 0.3–0.5 nm. In one of the models it is assumed that the atoms in this region form a substance having properties similar to those of liquids. Consequently the mechanism of diffusion of atoms or ions along grain boundaries is quite different from the mechanism of diffusion within the grains.

The second model is a modification of the first and is called the "island–basin" model. According to this theory the grain boundary consist of regions (basins) in which neighbouring grains are not in contact with each other and as a result the configuration of the atoms in these regions resembles the structure of liquids. The basins are separated from each other by the so-called "good fitting islands" in which the neighbouring grains are in direct contact with each other. In these places, the configuration of the atoms hardly differs from their distribution within grains. For this reason diffusion takes place mainly in the liquid like phase.

Both these models are very speculative, since, apart from the fact that the diffusion rates predicted on the basis of these models are in qualitative agreement with experimental data, there are very few experimental proofs confirming the structure assumed for the grain boundaries.

The third theory of intergranular diffusion is based on the dislocation model of grain boundaries; the experimental evidence for this model is the most extensive. In this case the grain boundary is regarded as a number of edge dislocations. Their density (number per unit length) increasses with increasing disorientation angle θ between two grains in contact with one another. This model can be represented as two-dimensional lattices meeting at a small angle θ (Fig. 145).

Geometrical considerations lead to the following relation between the disorientation angle θ, the distance, d, between neighbouring dislocations and the lattice constant, b,:

$$\sin\left(\frac{\theta}{2}\right) = \frac{1}{2}\frac{b}{d} \qquad (2.10.1)$$

It follows from Fig. 145 that the distance between neighbouring dislocations decreases with increasing disorientation angle θ. It can be shown that for angles greater than 45° the dislocation model of grain boundaries becomes meaningless, since the distance d would have to be smaller than the lattice constant. In the case of boundaries large angles the "island–basin" model, based on a liquid like structure, is more probable.

The low angle grain boundary, consisting of many edge dislocations, can be also regarded as a row of parallel channels in which the packing

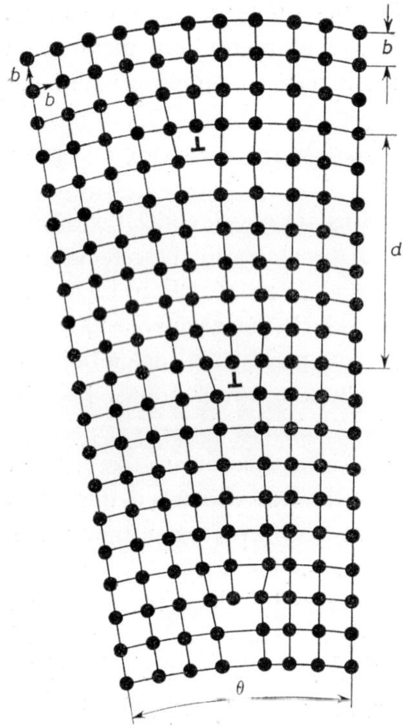

Fig. 145. The model of a low angle dislocative grain boundary (d is the distance between the neighbouring edge dislocations, θ is the disorientation angle and b is the lattice constant).

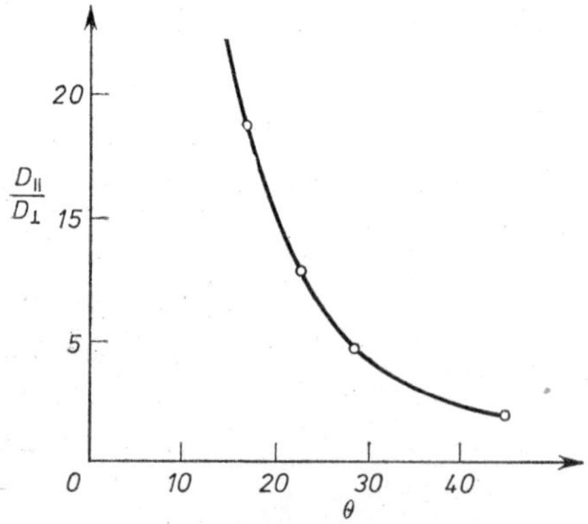

Fig. 146. The effect of the disorientation angle on the rate of diffusion along grain boundaries (after [763]) ($D_{\parallel}$ and $D_{\perp}$ are the coefficients of diffusion parallel and perpendicular to the dislocation line and θ is the disorientation angle).

of atoms is loosest. Such a channel forms the immediate surroundings of the dislocation line. Since in this region the strains are the highest and the packing is the loosest, it would be expected that the diffusion rate would be faster along the dislocation lines than in the direction parallel to their rows (the dislocation channel model).

It follows from the above model that the diffusion along grain boundaries should have an anisotropic character, in contrast to the "liquid" models. The differences should be the greatest in the case of small angles θ and should gradually decrease to zero as this angle approaches 45°.

The results obtained for silver at 450°C fully confirm the above theoretical conclusions and provide an important experimental proof for the correctness of the dislocation model for low angle grain boundaries. The results of these studies are shown in Fig. 146 in the form of a plot of the ratio of the diffusion coefficients for movement parallel and perpendicular to the dislocation lines.

The relation between dislocation channel diffusion coefficient (D_p) and the grain boundary diffusion coefficient D_g (the "liquid" model) is given by the following equation (assuming that anisotropy is absent):

$$D_g \delta = 2D_p h^2 \left[\frac{\sin\theta/2}{b} \right] \tag{2.10.2}$$

where δ is the thickness of the diffusion layer (the thickness of the grain boundary), estimated to be 5×10^{-8} cm, and h^2 is the cross sectional area of the dislocation channel.

It follows from this equation that the grain boundary diffusion coefficient should be a function of disorientation angle. The first systematic studies of the effect of angle θ on the grain boundary diffusion coefficient were carried out by Smoluchowski *et al.* [220]. In these studies the authors used so-called column samples of polycrystalline silver and measured the rate of penetration of copper along the grain boundaries. Columnar polycrystallites consist of grains so oriented that they all had in approximately the same crystallographic direction. The results of these studies are shown in Fig. 147.

In agreement with the predictions resulting from the dislocation model of grain boundaries, the data obtained by Smoluchowski *et al.* [220] indicate that the rate of diffusion along the grain boundaries increases with increasing disorientation angle θ and reaches its maximum value at θ equal to about 45°. At higher values of θ the grain boundaries dislocation structure begins to disappear.

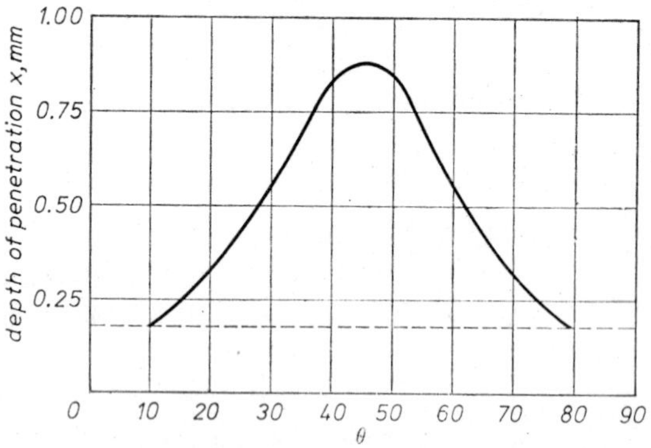

Fig. 147. The effect of the disorientation angle on the rate of diffusion of silver along grain boundaries of copper crystals (after [220]) (x is the depth of penetration after two weeks of annealing at 675°C).

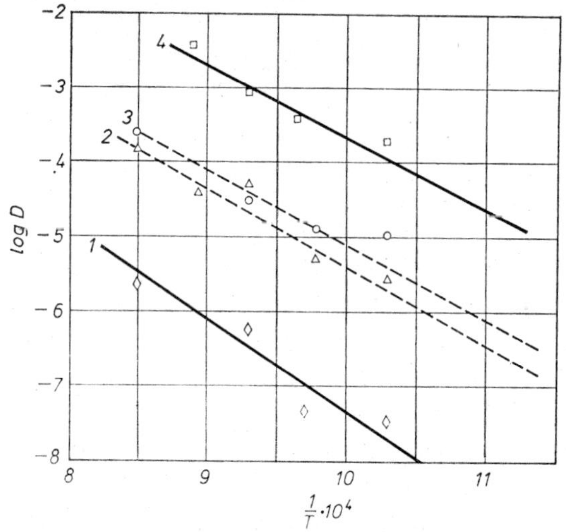

Fig. 148. The effect of the size of grains on the rate of diffusion of zinc in a Cu–Zn alloy (after [764]); curve *1* corresponds to the monocrystal and curves *2*, *3* and *4* to the polycrystalline material having the grain size 0.48, 0.37 and 0.13 mm respectively.

It is noteworthy that at very high and very low values of the disorientation angle the rate of diffusion along the grain boundaries becomes similar to that of lattice diffusion. The results obtained by Smoluchowski *et al.* were later confirmed for various other systems by other authors.

Since the rate of diffusion along grain boundaries is considerably faster than the rate in the bulk of crystal there should be a definite relationship between the mean value of the coefficient of diffusion in a polycrystalline material and the size of grains, since as the size of grains decreases, the number of grain boundaries per unit surface area increases and consequently the fraction of the surface occupied by the grain boundaries also increases. Therefore the mean value of the diffusion coefficient at constant temperature should increase with decreasing size of grains in the polycrystalline material. This conclusion was confirmed as the results in Fig. 148 show [62]. They indicate that the rate of diffusion of zinc in a polycrystalline sample of brass increases with decreasing size of grains.

As was mentioned at the beginning of this chapter, the activation energy for lattice diffusion E_v is considerably higher than that for diffusion along grain boundaries E_g. Shewmon [131] has shown that in the case of metals $E_g = \frac{1}{2}E_v$. This indicates that as the temperature increases, the grain boundary diffusion coefficient increases much more slowly than that of lattice diffusion.

Thus the diffusion coefficient in a polycrystalline material, which is the sum of the values of D_g and D_v, has a different physical meaning in different temperature ranges. At low temperatures diffusion along grain boundaries predominates in the overall process of transport of matter and the mean value of D is almost the same as the value of D_g. At sufficiently high temperatures the reverse situation exists and $D \cong D_v$ (see Fig. 2).

The passage from the low temperature to the high temperature region mainly depends on the size of grains. According to the considerations above the temperature, at which the change over occurs, increases with decreasing grain size. This is illustrated in Fig. 149.

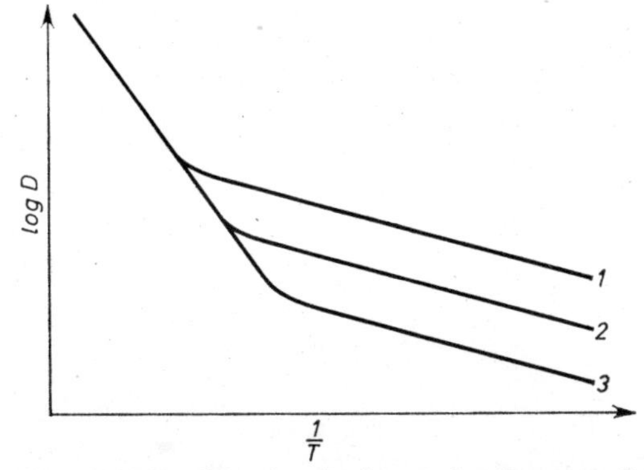

Fig. 149. The effect of grain size on the character of diffusion in polycrystalline material. Curves *1*, *2* and *3* correspond to materials of increasing grain size.

Dislocations can also play an important role in low temperature lattice diffusion. When the density of dislocations is considerable, as in the case of metals and alloys, and when they are distributed at random in the crystal, the diffusion involving these defects has a character analogous to that of real lattice diffusion taking place by means of point defects. The mathematical analysis of this problem was carried out by Hart [222], who gave the following equation for the lattice diffusion coefficient taking into account the effect of dislocations:

$$D = D_f f + D_v(1-f) \tag{2.10.3}$$

where D is the mean coefficient of diffusion in crystal, D_f and D_v are the coefficients of diffusion involving dislocations and point defects, respectively, and f is the fraction of time which the diffusing atom spends within the dislocation.

Shewmon [131] estimated the value of f as equal to 10^{-7} and showed that the mean value of the diffusion coefficient D of metals is principally determined by diffusion at dislocations at temperatures lower than one half the melting point of the material. For this reason, Shewmon did not believe that, the extrapolation of the results of lattice diffusion measurements obtained at high temperatures to low temperatures provided real picture of the mobility of structural elements in the crystal lattices of metals.

In view of the considerably smaller density of dislocations in ionic crystals of the MeX type, the above effect undoubtedly plays a much less important role in the case of these substances. However it should be taken into account at temperatures less than half the melting point.

2.11 Surface diffusion

Diffusion of atoms or ions can also take place along external and internal surfaces of solids. This process requires a lower activation energy than those for other forms of diffusion and for this reason its rate is larger. Thus:

$$\Delta H_p < \Delta H_g < \Delta H_v \quad \text{and} \quad \Delta D_p > \Delta D_g > \Delta D_v \tag{2.10.4}$$

Studies on surface diffusion are among the most difficult experimental problems in solid state physics. The difficulties are mainly connected with the effect of traces of impurities present at the surface on the diffusion process. For this reason little is yet known about the mechanism of surface diffusion. In recent years the use of modern experimental techniques and in particular low energy electron diffraction (LEED) and field ion microscopy, has yielded valuable new information that has contributed to gradual elucidation of the mechanism of this process.

It has been found that, in the case of metals, the activation energy for surface diffusion can be of the same order as the activation energy for lattice diffusion. Thus in the case of copper, at temperatures ranging from the melting point T_m to 0.87 T_m, the activation energy for lattice diffusion ΔH_v = 196.78 kJ/mole (47 kcal/mole) and the activation energy for surface diffusion ΔH_p = 209.34 kJ/mole (50 kcal/mole). Shewmon and Choi [223] proposed the following explanation for this anomaly: The atoms diffusing on the surface are in a state similar to that existing in the evaporation process. For this reason the activation energy of surface diffusion is related to the heat of evaporation ΔH_{ev} more closely than with the activation energy for lattice diffusion. Since the binding energy of an atom situated on the surface is proportional to the number of nearest neighbours, Shewmon and Choi pointed out that in the case of a cubic face centred lattice $\Delta H_p = \frac{2}{3}\Delta H_{ev}$. For copper ΔH_{ev} = 334.944 kJ/mole (80 kcal/mole) and hence ΔH_p = 221.9 kJ/mole (53 kcal/mole). This is in good agreement with the experimental datum.

It was assumed in the above model that the diffusion takes place on a smooth, i.e. not defected surface. This assumption seems to be justified only at high temperatures. For this reason other models for surface diffusion have been proposed for the low temperature region. One of them is based on the vacancy mechanism. The problem of the mechanism of surface diffusion is very complex, since in addition to point defects, the surface of a crystal can contain other imperfections, such as grooves, steps etc., which have a pronounced effect on the migration of atoms on the surface. These problems still await both a theoretical solution and experimental investigations.

2.12 Reactive diffusion

In the case discussed in the previous chapters diffusion processes did not cause any changes in the systems considered apart from changes in concentration and in certain physical and chemical properties, such as density, hardness, electrical conductivity. The character of the solid solution and its structure remained unchanged. However in many cases diffusion processes lead to allotropic transformations or to the formation of new phases with structures and physical properties different from those of the starting material. This type of diffusion, leading to the formation of new phases in the system, is called reactive diffusion [224].

The alltropic transformations caused by diffusion are due to changes in composition, leading to supersaturation of the solid solution. In the regions of supersaturation a new solid solution is formed in the system according to thermodynamic conditions defined in the phase diagram.

A typical example of an allotropic transformation caused by diffusion is the formation of the α phase during the diffusion of chromium in γ iron at 1200°C [225]. Figure 150 shows a plot of the concentration of chromium

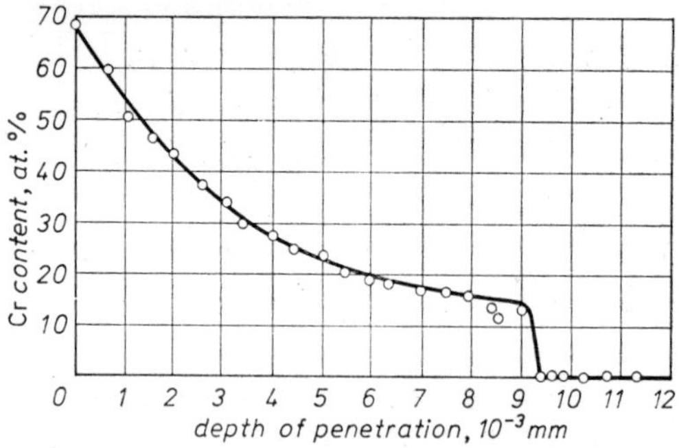

Fig. 150. The distribution of concentration of chromium in a sample of iron after 96 hours of diffusion annealing at 1200°C (after [225]).

in γ iron against the distance from the surface of the sample after 96 hours of annealing at 1200°C. It can be concluded that the α phase forms a layer at the surface of the sample and that the thickness of this layer depends on the annealing time. The concentration of chromium in the γ phase at the boundary between the two phases is about 14 at.% and decreases stepwise to about 1 at.% after entering the α phase. This kind of sudden change in the concentration of the diffusing species in the originally homogeneous system indicates that the diffusion process is accompanied by alltropic transformations.

As was mentioned above, reactive diffusion can also lead to the formation of new phases consisting of chemical or intermetallic compounds. The course of diffusion leading to the formation and development of a new phase consisting of a chemical compound can be divided into several stages. In the initial stage the reacting system consists of two one-component phases which are in contact with one another. In these phases the process of mutual disolution of components begins. As a result of this process solid solutions of component A in phase B and component B in phase A are formed. When saturation is reached, the process of nucleation of a new phase A_nB_m begins and leads to the formation of a continuous layer of this phase, which separates pure phase A from pure phase B. The new phase A_nB_m is not strictly stoichiometric, but is a solid solution having the highest concentration of component A at the A/A_nB_m interface and the highest concentration of component B at the B/A_nB_m interface. Thus in all three phases of the system there are concentration gradients causing diffusion of both components (Fig. 151) and hence the thickness of the

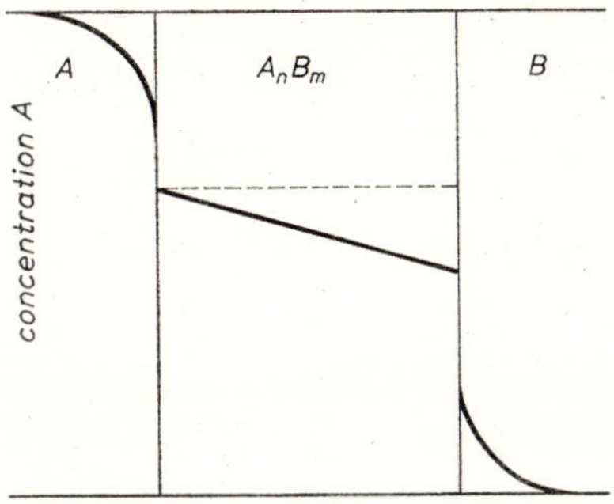

Fig. 151. The distribution of concentration of component A in a heterogeneous reacting system of the A–A_nB_m–B type.

A_nB_m phase increases. After passing through the layer of A_nB_m atoms of component A combine with atoms of component B on the B/A_nB_m phase boundary and form new lattice planes consisting of the reaction product. An analogous process takes place at the other boundary. As a result of these processes the growth of the new phase continues until one of the components becomes exhausted.

Reactive diffusion may lead to the formation of more than one phase of the reaction product. In this case the heterogeneous system is more complex. When the partial diffusion coefficients of the individual components in the reaction product are not the same, phenomena analogous to the Kirkendall and Frenkel effects take place. This is mainly observed in cases when new phases consisting of chemical or intermetallic compounds are formed as a result of lattice diffusion. The degree of disorder in the sublattices of this type of compound is as a rule different and as a result the partial diffusion coefficients of components A and B usually differ from one another by several orders of magnitude. In such conditions the formation of the new phase takes place mainly or practically exclusively at one of the phase boundaries only. This is equivalent to the formation of pores and cracks at the opposite phase boundary, where the loss of the substrate is not compensated by the formation of the product. In this situation it is unjustified to quantitative conclusions about the partial diffusion coefficients in the A_nB_m phase on the basis of the kinetics of thickening of this phase and the rate of shift of the A/A_nB_m or B/A_nB_m phase boundaries as in the case of metallic systems when the Frenkel effect is present.

It should be added that reactive diffusion takes place not only when two solid phases react with one another, but also in the case of reactions of solids with gases or liquids when the reaction product is a solid. Thus the the oxidation of metals and alloys can also be regarded as reactive diffusion phenomena.

3. Methods for Determining the Defect Equilibria and Diffusion Coefficients in Crystals

The experimental determination of the concentration of defects in crystal lattices and their mobility is very difficult. The principal difficulties are caused by the fact that from the point of view of measuring technique even in highly defected crystals the concentration of defects is extremely small. This concentration must be determined under conditions of thermodynamic equilibrium between the crystal and the surrounding atmosphere, i.e. at high temperatures at which equilibrium is established in a relatively short time. In addition, the concentration must be determined as a function of both the temperature and the oxidant pressure. Only then can the results of the determination be used to estimate the degree of ionisation of defects and the entropy and enthalpy of their formation.

Analogous difficulties are encountered in determinations of coefficients of self-diffusion, defect diffusion and of chemical diffusion in crystals. However, since knowledge of both the structure of defects and their mobility in crystals is needed for the characterisation of many important physico-chemical properties of solids, methods for studying the structures and mobility of defects have been continually extended and improved over many years. A comprehensive discussion of these methods is outside the scope of the present work. For this reason the discussion will be limited to a brief description of a few selected methods for determining the defect equilibria and rates of diffusion of atoms in crystals. This will enable the reader to appreciate the difficulties facing experimentalists dealing with these problems and will help him to make a suitable choice of method with the aid of the original publications.

3.1 Determination of the type of disorder

It follows from the discussion in the first part of the present work that, in crystal lattices of compounds of the MeX type, disorder is practically always limited to the anionic or the cationic sublattice. For this reason the investigator of the structure of defects in a given substance must first find out which sublattice is defected. There are two principal methods for solving this problem. One of them is directly related to the determination of self-diffusion coefficients of the metal and the oxidant by means of radioactive or stable isotopes. This method will be discussed in the next chapter. Here we will only state that in order to draw conclusions regarding the type of structural disorder, i.e. to find out which sublattice contains the defects, it is necessary to determine the self-diffusion coefficients of both the metal D_{Me} and the oxidant D_X under the same conditions. When $D_{Me} \gg D_X$, the disorder is practically limited to the cationic sublattice and when $D_{Me} \ll D_X$ the defects are only present in the anionic sublattice. These determinations are very tedious, not only because it is necessary to determine both D_{Me} and D_X, but mainly because the determination of self-diffusion coefficient in a lattice practically free from defects is extremely difficult, since in this case the values of D are very low. Thus whenever possible the type of disorder in crystal structures of the MeX type is determined by another method, called the marker method. It consists in oxidising the metal forming the compound MeX in X atmosphere and examining of the participation of the individual reagents in the general process of transport of matter through the layer of the oxidation product (scale).

The principles of this method were already worked out in 1930 by Pfeil [226] and were later developed by Sachs [227], Archarow [228], Birchenall [229] and Mrowec [230–232]. The application of this method by Pfeil to studies on the mechanism of formation of oxide scale on iron in the nineteen thirties caused a radical change of view about the mechanism of oxidation of metals and created the experimental basis for the modern theory of these processes. At the present time the marker method is widely used, not only in studies on the mechanism of oxidation of metals and alloys, but also in studies on the mechanism of diffusion in metal alloys (the Kirkendall effect) and in investigations of solid state reactions, particularly in studies on the mechanism of formation of spinels, ferrites, sintering processes etc.

The principle of the marker method consists in the following procedure: a substance called the tracer, which, under the conditions of oxidation, does not react with the metal or alloy to be examined or with the components of the scale and does not dissolve in them, is placed on the smooth surface of the metal or alloy in the form of a thin porous layer or a thin wire. After oxidation the position of the marker in the scale is determined and on this basis conclusions regarding the mechanism of the scale formation are drawn.

The principal condition for the successful use of this method is that the contact between the marker and the surface of the sample before the start of the oxidation process should be extremely good. When the marker is used in the form of a wire, Czerski [223] recommended attaching it to the surface of the metal by means of a thin, electrolytically deposited layer of the same metal. This technique is suitable for pure metals, but cannot be used in the case of alloys, since it is very difficult to obtain electrolytically a layer of an alloy having the required composition, whilst coating of the sample of the alloy with pure metal may disturb the oxidation process.

According to Cagnet and Moreau [234], good contact of the wire to the surface of the sample can be obtained by welding. A disadvantage of this technique is the formation of an alloy consisting of the marker and the metal of the sample arround the weld, which may disturb of the oxidation process and consequently to erroneous results. Good contact between the surface of the sample and the wire can be obtained by pressing the wire into the surface or by stretching it on the surface as described by Engell. Cagnet and Moreau [236] objected to the last technique, since in their opinion the wire stretched on the surface of the metal can be "bypassed" by the scale during its growth as a result of plastic flow of the reaction product, which obviously may lead to erroneous results. Extensive studies carried out by the author of the present work and his collaborators have shown that in the case of wire markers the best results are obtained by placing the wire on the polished surface of the sample in the form of short (about 1 mm) pieces and oxidising the sample in the horizontal position [230–232, 237]. This technique ensures perfect contact between the marker and the sample without any preliminary treatment which, as a rule, disturbs the surface of the sample. This technique is also the simplest.

When the marker consists of a thin, porous layer of metal deposited on the surface of the sample by evaporation, electrolysis or decomposition of a salt of a noble metal, the contact between such a thin (about 10^2 nm) layer and the surface is always satisfactory. On the other hand the localisation of the thin layer in the scale after the oxidation is much more difficult than in the case of wire markers about 10 μm thick.

When the marker consists of a compound of a metal with an oxidant, e.g. a metal oxide, the free energy of formation of this compound under the conditions of reaction must be considerably lower than the free energies of formation of the compounds constituting the scale, since only then is the marker chemically neutral with respect to the system examined, though obviously its solubility in the sample and in the scale must be practically zero.

Birchenall and collab. [238], Carter and Richardson [239], Löhberg [240] and other authors used radioactive markers deposited on the surface of the sample in the form of a thin layer of a noble metal. The radioactive isotope of silver used by Birchenall was deposited electrolytically on the surface of the sample. Carter and Richardson obtained radioactive layer of platinum by decomposition of ammonium chloroplatinate. This decomposition took place directly on the surface of the sample as a result of the reduction of platinum ions by the metal in the sample. In this case the localisation of the marker in the scale by means of autoradiography was very easy, but in view of the low resolving power of autoradiograms this technique seems less accurate than the classical methods.

The principal disadvantage of the use of a wire marker is the screening effect on the diffusion processes taking place in the scale, which can give inaccurate results in studies of fast processes. In principle the technique based on the use of thin layers is free from this disadvantage, provided the layer is porous and very thin (of the order of 10^3 nm). If this condition is not fulfilled, the screening effect of the marker is more pronounced than in the case of a wire marker. This effect was first observed by Birchenall [229] during his studies on the mechanism of formation of ferrous sulphide on iron.

After placing the marker on the surface of the sample (Fig. 152*a*), oxidation is carried out until the thickness of the scale becomes at least ten times greater than that of the marker. The oxidation then is terminated and the position of the marker in the scale is determined, usually on the

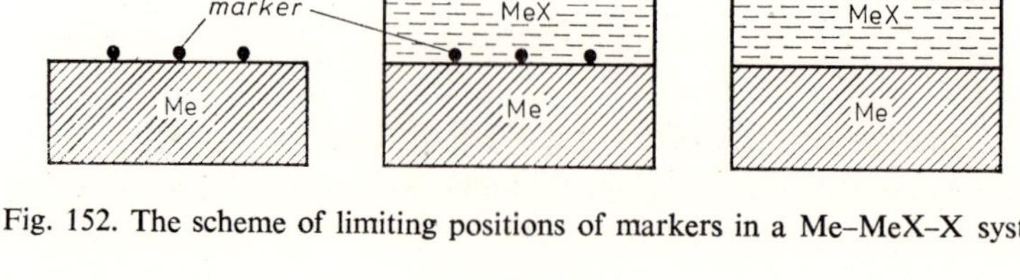

Fig. 152. The scheme of limiting positions of markers in a Me–MeX–X system.

basis of microscopic observations of the cross section of the scale perpendicular to the original surface covered with the marker. In rare cases the localisation of the marker is carried out by means of chemical or spectroscopic analyses of successive layers of the scale taken from the surface parallel to the original surface of the sample. This method was used in the classical work of Pfeil.

If, after the reaction, the marker is situated inside the scale on the phase boundary with the metallic core, it means that the reaction took place practically exclusively on the outer scale/oxidant phase boundary, i.e. as a result of outward diffusion of the metal (Fig. 152*b*). If, after the oxidation, the marker is located on the outer surface of the scale, it means that the layer of the oxidation product grew on the inner scale/metal phase boundary, i.e. as a result of inward diffusion of the oxidant (Fig. 152*c*). Since the marker does not participate in the oxidation, it must be present in the part of the scale which was formed in the initial stage of the reaction.

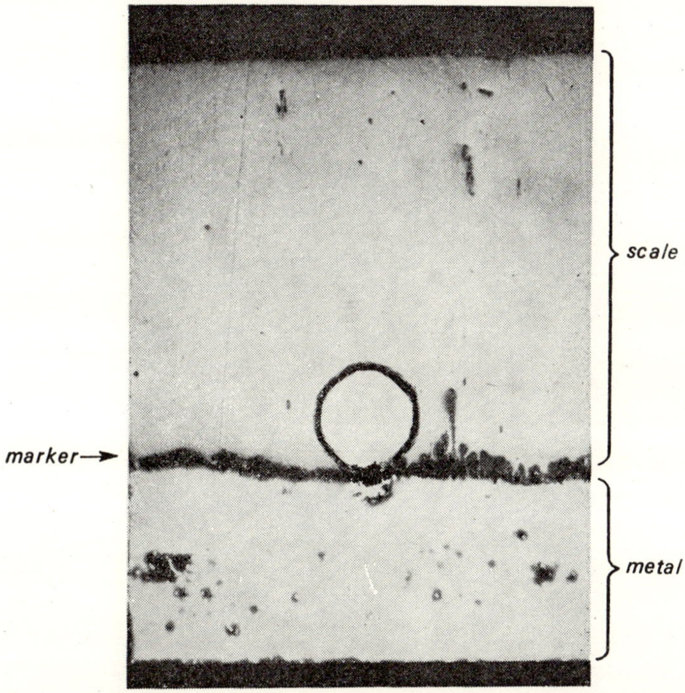

Fig. 153. The cross-section of an oxide layer formed on iron at 950°C .The cross-section of the platinum marker is visible at the metal/oxide interphase.

The location of the marker at the metal/scale interface means that compound MeX constituting the scale is defected practically exclusively in the cationic sublattice ($Me_{1-y}X$ or $Me_{1+y}X$). The presence of the marker on the outer surface of the scale indicates that the crystal lattice of MeX practically only contains defects in the anionic sublattice (MeX_{1-y} or MeX_{1+y}).

Figures 153, 154 and 155, showing microphotographs of cross sections of oxide and sulphide scales formed on several metals, illustrate the results

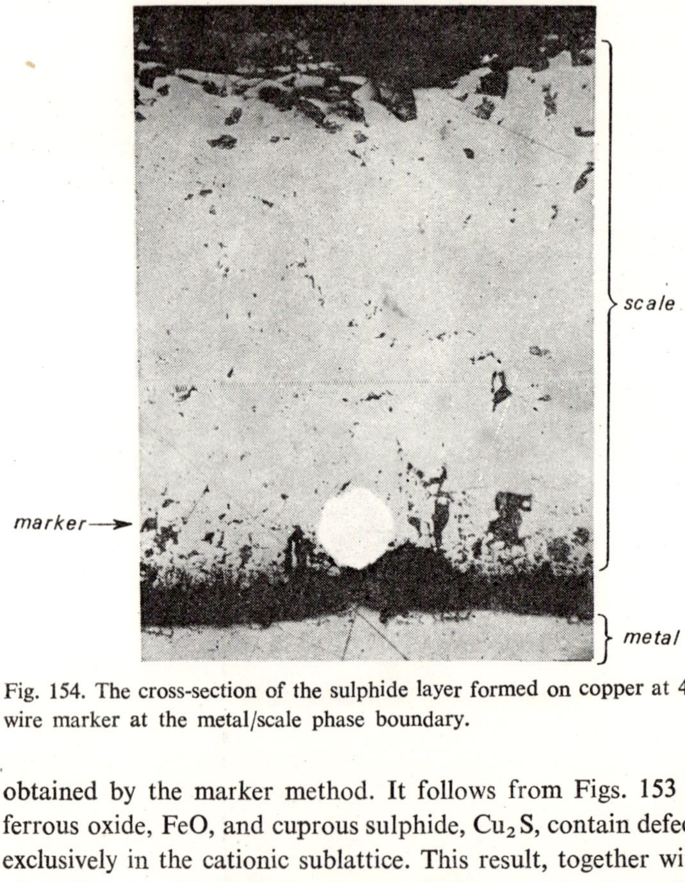

Fig. 154. The cross-section of the sulphide layer formed on copper at 450°C. Platinum wire marker at the metal/scale phase boundary.

obtained by the marker method. It follows from Figs. 153 and 154 that ferrous oxide, FeO, and cuprous sulphide, Cu_2S, contain defects practically exclusively in the cationic sublattice. This result, together with the results of studies on the deviation from stoichiometry, leads to the conclusion that both these compounds show real deficits of metal and hence their correct stoichiometric formulae are: $Fe_{1-y}O$ and $Cu_{2-y}S$. In the case of titanium the results of marking have shown (Fig. 155) that rutile (TiO_2) contains defects practically exclusively in the anionic sublattice [241]. Hence its correct stoichiometric formula is (TiO_{2-y}).

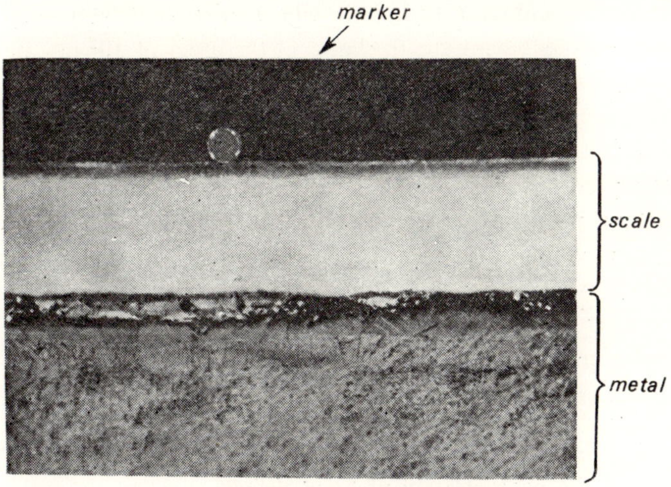

Fig. 155. The cross-section of the oxide layer formed on titanium at 900°C. Platinum wire marker at the outer surface of the scale.

It follows from the above considerations that the determination of the type of disorder (i.e. whether the cationic or the anionic sublattice is defected) is relatively easy. However for several reasons, it is not always possible to use the marker method. First of all the reaction conditions must be so chosen that a single phase compact scale forms on the smooth metal surface, since only then is the growth of the scale due to lattice diffusion of metal or of oxidant (in the form of ions together with electrons) via ionic and electronic defects according to Wagner's theory [242]. It is not always possible to obtain such a scale. Sometimes it is not possible to avoid cracking of the layer of the reaction product. In other cases microfissures are formed in the scale or it has a multilayer structure in which the layers are morphologically different, which obviously excludes the possibility of applying the marker method to the determination of the type of disorder in the crystal lattice of the oxidation product of the metal. Sometimes it is not possible to obtain the layer of scale having a thickness sufficient for the marker method. This is mainly observed with oxides and sulphides of low melting metals such as zinc, cadmium, aluminum, etc. The concentration of lattice defects in these compounds is extremely low* and since the rate of growth of a compact scale layer on the metal is determined by the diffusion of the reagents in this layer, the oxidation rate is very slow.

In the case of zinc, aluminum, cadmium, silicon, beryllium and magnesium, thin passive layers having the thickness of the order of 10^2 nm are formed on the surface of the sample. Although the oxidation rate increases exponentially with increasing temperature, even near the melting points of these metals it is too small for successfull application of the marker method. The attempted application of this method to oxidation of liquid metals failed, since a porous scale was usually formed as a result of the appreciable vapour pressure of the metals.

In spite of the above limitations the marker method is very valuable in studies on the kind of disorder in crystal lattices of many metal oxides and sulphides and of more complex compounds, such as spinels and mixed oxides and sulphides, which are formed during oxidation of binary alloys.

3.2 Determination of the concentration of point defects

When the type of defect present in the crystal structure is known, the concentration can be determined by various methods. However this is a very difficult problem and, although many methods of measuring the concentration of defects have been developed, it is not always possible to obtain satisfactory results. In the present work we shall not discuss the problem of the determination of the concentration of defects in compounds without deviations from stoichiometry, since in this case the experimental difficulties are so great, that this problem has not so far been solved. Apart from this, the group of compounds with strictly stoichiometric compositions is very small (alkali metal halides and halides of heavy metals) in comparison with the enormous number of oxides and sulphides of metals with deviations from stoichiometry in a wide range of temperature and oxidant pressures.

It has already been shown in previous chapters that the concentration of predominant defects in these substances is directly related to the deviation

* It should be mentioned that, as a general rule, the concentration of point defects in crystals decreases with increasing absolute value of the free energy of formation of compounds constituting the crystals.

from stoichiometry. In compounds showing a real deficit of metal ($Me_{1-y}X$) the deviation from stoichiometry y is equal to the mole fraction of cation vacancies, and in compounds showing a real excess of metal ($Me_{1+y}X$), y is equal to the mole fraction of interstitial cations. In compounds with a real deficit of the oxidant (MeX_{1-y}), y is equal to the mole fraction of anion vacancies. Hence by determining the dependence of the deviation from stoichiometry on temperature and equilibrium oxidant pressure it is possible to determine the absolute values of concentration of predominant defects and the enthalpy as well as the entropy of their formation and their degree of ionisation.

Among many methods for the determination of the deviation from stoichiometry of compounds of the MeX type there are three principal methods which, owing to their sensitivity and relative simplicity, give the most reliable results. They are the thermogravimetric, the volumetric and the electrochemical methods. In recent years a new method, the so-called kinetic method, has been developed. It seems to be an excellent method. The principles of the above methods will be discussed in the chapter that follow.

3.2.1 The thermogravimetric method

Although many kinds of apparatus can be used for the measurements, this method consists essentially of the determination of changes of the mass of the sample of the material examined caused by changes in external parameters. When a sample of compound MeX is heated to a suitably high temperature at a constant partial pressure of component X in the surrounding atmosphere, after a certain time, thermodynamic equilibrium for the MeX–X_2 system is established and the mass of the sample becomes constant. This is equivalent to the stabilisation of the chemical composition of the sample. Consequently the deviation from stoichiometry y also reaches the value characteristic of the experimental conditions. If one of the parameters of the system is now changed, the sample begins to take up or release the oxidant until a new equilibrium state is established. In this new equilibrium state the mass of the sample is constant, but different from that in the previous equilibrium state. For example when the pressure of component X surrounding a sample of compound $Me_{1-y}X$ is decreased, the defect equilibrium shifts to the left [see equation (1.3.85)], which is equivalent to a decrease of the concentration of cation vacancies, i.e.

a decrease of the deviation from stoichiometric composition. This is caused by the decomposition of some of the compound to free oxidant, which passes to the gas phase. The liberated ions of the metal and electrons fill the cation vacancies and electron holes. As a result of this process the mass of the sample decreases. Similarly in the case of compounds with a real excess of the oxidant (MeX_{1+y}) the decrease of the oxidant pressure causes a decrease of the deviation from stoichiometry (a decrease of the concentration of interstitial anions).

In the case of compounds of the $Me_{1+y}X$ type a decrease of pressure also causes a decrease of the mass of the sample, but this is equivalent to a shift of the defect equilibrium to the right [see equation (1.3.79)], i.e. to an increase of the concentration of interstitial cations. Obviously this means an increase of the deviation from stoichiometry. An analogous situation exists in the case of a compound with a real deficit of the oxidant (MeX_{1-y}). In this case a decrease of pressure causes an increase of the concentration of anion vacancies (an increase of the value of y).

The measurements of changes of the mass of the sample can give direct information about the concentration of defects on condition that the absolute value of y corresponding to a definite temperature and oxidant pressure is known. For this reason the first stage of the investigation is not the determination of changes of the mass of the sample of MeX, but the determination of the total mass of oxidant combined with the metal at a definite temperature and oxidant pressure. For this purpose a sample of the metal of known mass is completely oxidised and the increase of the mass is determined. The constant chemical composition of the resulting compound MeX, characteristic of the experimental conditions, is reached when the increase of the mass of the sample ceases. On the basis of the mass of the oxidant determined by this method and the known mass of the metal it is possible to calculate the mole fraction Me/X, i.e. the deviation from stoichiometry y. Then the measurements of changes of the mass discussed above are carried out.

The results of several measurements of this type give a set of values of y for various equilibrium states, which makes it possible to plot y against p_{X_2} for various temperatures. In the first part of the present work it was shown that this function has the following general form for compounds of the $Me_{1-y}X$ or MeX_{1+y} type:

$$y = \text{const}\, p_{X_2}^{1/n} \tag{3.2.1}$$

In the case of compounds of the $Me_{1+y}X$ or MeX_{1-y} type this function has the following form:

$$y = \text{const}\, p_{X_2}^{-1/n} \tag{3.2.2}$$

Double logarithmic plots of the above relationship against temperature should be a set of straight lines (isotherms) the slopes of which give the numerical values of the exponential coefficient $1/n$ for the corresponding temperatures. This enables the temperature dependence of the degree of ionisation of defects to be determined.

The results of the above studies can be also plotted in the form of isobars for various oxygen pressures in the Arrhenius system $\left[\log y = f\left(\frac{1}{T}\right)\right]$. In this case straight lines should likewise be obtained. From their slopes the enthalpy of formation of defects ΔH_f can be calcualted and from their intercepts with the axis of ordinates, the entropy of this process ΔS_f, since it follows from the first chapter that the concentration of point defects in crystals of the MeX type at a constant oxidant pressure is the following function of temperature:

$$y = N_d = \text{const}\exp\left(-\frac{E_f}{RT}\right) \tag{3.2.3}$$

where $E_f = \frac{2}{n}\Delta H_f$ and $\text{const} = \exp\left(\frac{\frac{2}{n}\Delta S_f}{R}\right)$. The numerical values of n characterising the degree of ionisation of defects can be determined from the slopes of the isotherms [equations (3.2.1) and (3.2.2)]. This simple procedure for determining the absolute values of the deviation from stoichiometry cannot be used in all cases. We shall return to this problem later, in the discussion of the results of measurements of the deviation of cuprous oxide from stoichiometry.

Thermogravimetric measurements can be used as the basis of the above calculations on condition that the changes of mass of compound MeX are determined very accurately. The measurements must be carried out only when the sample is in strictly defined thermodynamic conditions. This requires a suitable evacuated apparatus in which a balance together with the sample of the material to be examined can be placed. It is also necessary to have a device for heating the space in which the sample is placed.

The first accurate vacuum microbalance of the beam type was constructed by Gulbransen [243, 244]. A scheme of the apparatus is shown in Fig. 156. The construction of the microbalance is shown in Fig. 157.

The microbalance consists of a quartz frame to which a very thin tungsten wire is welded by means of silver chloride. The strained tungsten wire

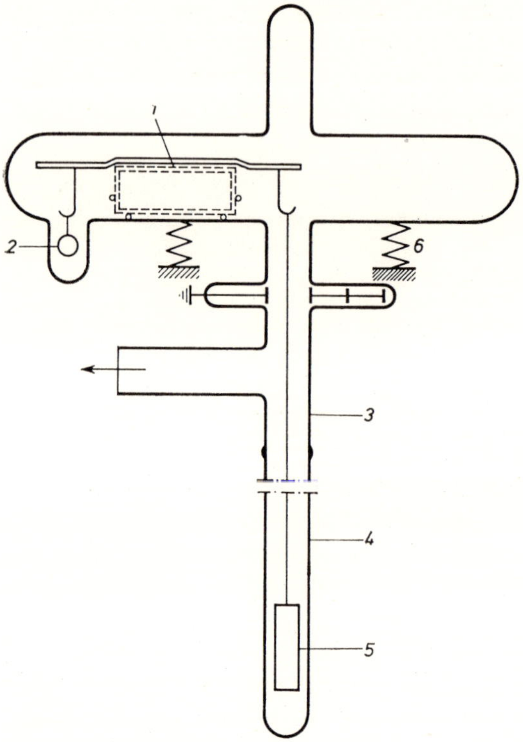

Fig. 156. The scheme of the Gulbransen apparatus (after [243]). *1*—quartz microbalance, *2*—counterweight of the sample, *3*—glass tube, *4*—quartz tube, *5*—sample, *6*—vibrations damping system.

supports a mobile thin quartz beam with forked ends to which tungsten wires are sealed. The sample and a suitable weight balancing it are suspended from the beam. On the sample side, the tungsten wire sealed to the fork is slightly longer and acts as the indicator for the swings of the balance. The reference point is a suitable indicator which is the extension of one arm of the framework of the balance. The changes of the position of the indicator with respect to a reference point, caused by changes of the mass of sample, are determined by means of a suitable microscope.

According to Gulbransen, this apparatus makes it possible to measure changes in mass with an accuracy of the order of 10^{-6} g, when the total mass of the sample is about 1 g. The balance maintains the zero point

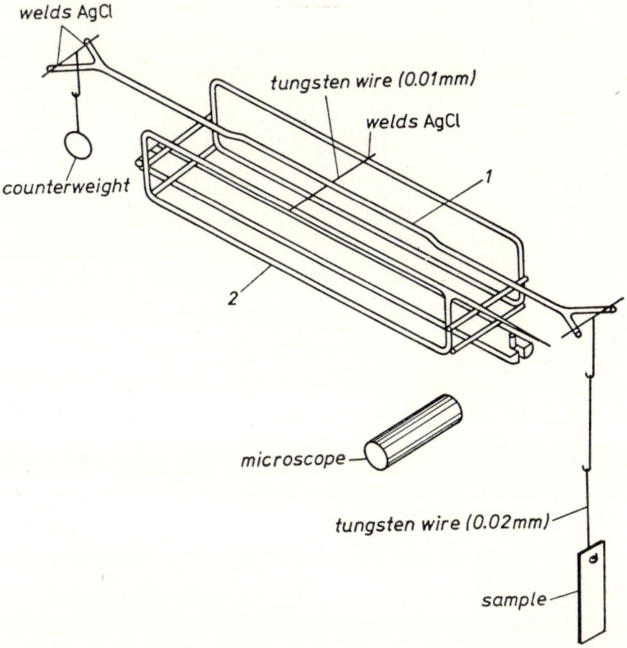

Fig. 157. The scheme of Gulbransen's quartz microbalance (after [243]). *1*—quartz beam, *2*—the framework of the balance.

perfectly; the deviations from this point, observed over many days, never exceeded 1×10^{-6} g. The pressure changes in the apparatus have only a slight effect on the behaviour of the balance. Its pressure coefficient is lower than $2.96\ 10^{-12}$ g/Pa (3×10^{-7} g/atm). On the other hand the balance is fairly sensitive to temperature changes; its temperature coefficient is equal to about 8×10^{-7} g/K. For this reason Gulbransen's apparatus requires careful thermostating.

The apparatus described above was improved by Rhodin [245] who was able to increase its sensitivity more than six times. At the present time changes in the mass of a sample can be measured with an accuracy of the order of 10^{-7} g.

A smooth surface of a metal contains 10^{15} atoms per 1 cm^2. When each atom of the metal binds one oxygen atom (NiO, CoO), the increase of the mass of the sample per unit surface area is 3×10^{-8} g/cm^2. When the whole surface of the sample is about 10 cm^2 (usually it is larger), the increase of the mass of metal after oxidation, due to the formation of the oxide monolayer, is 3×10^{-7} g. Thus by means of the apparatus described above it is possible to measure changes of the mass of sample caused by the formation or decomposition of individual lattice planes of the oxide. However this is of the same order as sensitivity of the balance and in practice considerable errors are to be expected.

The latest thermogravimetric devices make it possible to determine the mass changes with an accuracy of the order of 10^{-10} g/cm^2, but in this case the weight of the sample must be very small and the errors of the results obtained by means of such devices, caused by many factors (shocks buoyancy etc.), exclude the possibility of application of these devices to studies on defect equilibria.

Commercial electromagnetic vacuum microbalances with automatic control of the zero point are based on a different principle. In contrast to the Gulbransen system, the changes of the mass of sample are determined in this case on the basis of changes in the intensity of electric current flowing through an electromagnet creating the field in which the counterweight of the sample is placed. The magnetic field constantly maintains the balance in the zero position. Since the intensity of the current is proportional to the mass of the sample, it is possible to use this device for continuous measurements of the mass without any optical measurements.

Vacuum microbalances of this type make it possible to determine mass changes with as high an accuracy as that obtained by means of Gulbransen's device, but the measurements are much simpler, since they consist in simply taking the readings of a meter calibrated in units of mass. An apparatus of this type can be connected to an automatic recorder which continuously records the changes of mass.

The spiral quartz balances of the McBain type are also widely used in studies on defect equilibria in solids. The sensitivity of these devices varies within wide limits depending on the diameter and the number of coils and the thickness of wire from which the spiral is made. According to Hauffe a spiral quartz balance having a coil diameter 3 to 5 cm and containing about 35 coils makes it possible to determine mass changes of samples weighing about 100 mg with an accuracy of at least $\pm 10^{-6}$ g. This accuracy is as good as that obtained by means of the beam microbalance of the Gulbransen type. Since the quartz spiral microbalances are of simple construction and are easy to use, they have considerable advantages over the beam microbalances of the Gulbransen type. They are also much cheaper.

Figure 158 shows an apparatus with a spiral microbalance as used by

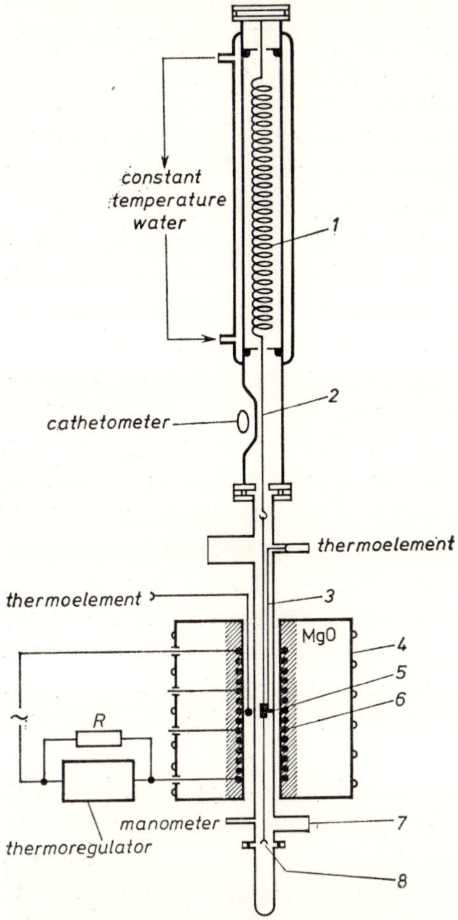

Fig. 158. The scheme of the microthermobalance with spiral quartz balance (according to Kofstad [247]); *1*—quartz spiral, *2*, *3*, *8*—platinum wire, *4*—constant temperature water, *5*—sample, *6*—alundum cement, *7*—gas inlet.

Kofstad [247] in his studies on defect equilibria in refractory metal oxides. In these studies the measurements of changes in the mass of the sample consisted of the determination of the elongation or shortening of the spiral by means of a suitable microscope or a cathetometer. The reaction space was separated from the balance space by a relatively long piece of tube in order to minimise the effect of heat from the furnace on the quartz spiral. It has been found that the readings of spiral quartz balances are sensitive to even small temperature changes and for this reason the balance compartment must be carefully thermostatted and the calibration of the spiral must be carried out in conditions corresponding to those used during the measurements.

Other errors may be caused by thermal diffusion of the gas in the lower part of the apparatus. This diffusion is particularly intense in the upper part of the furnace and in the analogous lower part where the temperature gradients are the greatest. In the case of furnaces in which the distribution of temperature is symmetrical the movement of gas caused by thermal diffusion in the two parts is equally intense, but its directions are opposite. For this reason Kofstad proposed that an additional hanger should be places under the sample. This gives a symmetrical system which is not affected by the thermal diffusion of gas in the furnace. However, measurements of deviations from stoichiometry must be often carried out in the flowing gas and not in static conditions. This is particularly necessary, when the measurements are carried out in mixtures of gases, such as C–CO_2, N_2–O_2, Ar–O_2 etc., in which a constant partial pressure of oxygen can be maintained by using a suitable rate of flow of the mixture through the reaction chamber. Under these conditions, another factor, which, in principle, cannot be eliminated, affects the performance of the balance. For this reason the calibration of the balance must be carried out under a definite pressure and gas flow rate through the reaction space, and all the elements of the suspension as well as the shape of the sample must be taken into account.

It can be seen from Fig. 158 that when Kofstad's apparatus is used in flowing gas conditions, the current of gas does not flow through the zone in which the spiral is placed. It has been found that this system is not very suitable for flowing gas conditions, since the limitation of the sensitivity of spiral quartz balances is caused not only by the factors discussed above, but also by unavoidable vibrations of the spiral, which interfere with the exact determination of its elongation and consequently with the determination of mass changes. The vibrations of the spiral can be considerably decreased by arranging a flow of gas through the space in which the spiral is placed.

An apparatus developed by Mrowec and Stokłosa, in which the gas flows through the whole device, is shown schematically in Fig. 159. It has

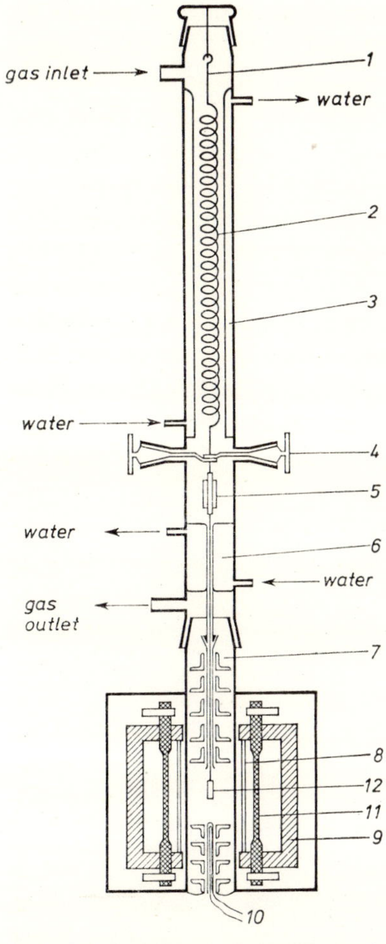

Fig. 159. The scheme of microthermobalance with spiral quartz balance (according to Mrowec and Stokłosa [248]); *1*—quartz suspension, *2*—spiral, *3*, *6*—constant temperature water, *4*—spiral damper, *5*—scale for reading the elongation of the spiral by means of a microscope with an accuracy of 2×10^{-3} mm, *7*—quartz radiation shields, *8*—ceramic tube, *9*—screen of the furnace, *10*—thermoelement, *11*—vacuum valves, *12*—sample.

been found that, at a gas flow rate of the order of 10 ml/s, the sensitivity of the balance increases about three fold as a result of the partial elimination of the vibrations of the spiral. In this apparatus a system of radiation shields has been introduced in the reaction chamber in order to decrease the fluctuation in the temperature of the sample.

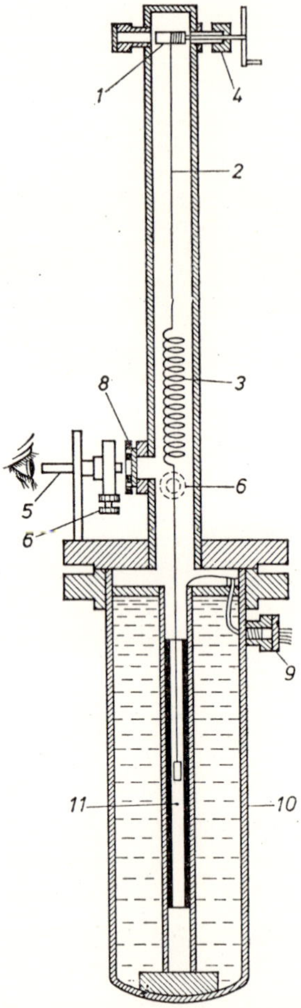

Fig. 160. The scheme of the thermogravimetric apparatus for studying the kinetics of oxidation of metals and defect equilibria at high oxygen pressures (after [249]); *1*—steel drum, *2*—gold chain, *3*—quartz spiral, *4*—valve, *5*—microscope, *6*—windows for introducing the sample, *7*—cathetometer, *8*—window made of armour glass, *9*—valve, *10*—steel pressure guard of the furnace, *11*—reaction chamber.

It should be stressed that the current of gas introduced from above, is an additional factor that improves the thermostating of the space in which the spiral is placed. Since the balance readings depend on the gas flow rate and the pressure, the calibration of the spiral must be carried out under conditions corresponding to those in which the measurements are to be made.

In many cases it is necessary to study defect equilibria at pressures up to 10.1325 MPa (100 atm). This is particularly necessary when the decomposition pressure at the given temperature of the compound examined is greater than 1013,25 hPa (1 atm) and when the dependence of the concentration of defects on the oxidant pressure is to be determined over a large range of pressure. Such measurements can be carried out by means of a special thermogravimetric apparatus developed by McKewen and Fassel [249]. This apparatus is shown schematically in Fig. 160.

The heated space in the apparatus consists od a tubular electric furnace covered with a thick steel jacket resistant to high pressures. The cables supplying the electric current to the furnace winding and the thermocouple leads are connected through a choke. In the upper part of the apparatus there is a quartz spiral suspended from a steel drum by a platinum chain. The sample is suspended on a platinum wire connected to the quartz spiral. The steel drum with the platinum chain is fixed to a movable axle which is connected to the outside of the apparatus through a special coupling. This makes it possible to wind the chain onto the drum so that the position of the sample inside the apparatus can be adjusted. The surface of the drum is threaded. The thread on the drum fits the thread on the axle moving the drum. As a result, during the winding of the chain on the drum none of the suspended parts change their horizontal positions. The sample is introduced in the reaction space through an opening which can be hermetically sealed. Above the furnace chamber there is an armour glass window opposite a measuring microscope together with a cathetometer. These instruments are permanently fixed to the frame of the furnace. The apparatus is connected to an high vacuum system and to an oxygen supply.

When measurements of mass changes are carried out at temperatures up to 1000°C, the construction of thermogravimetric devices does not present very difficult problems. However studies of this type must usually be carried out at higher temperatures, particularly in the case of metal oxides with small deviations from stoichiometry. For temperatures not exceeding 1600°C the furnace can be made of mulite which is relatively readily sealed to pyrex glass. For temperatures not exceeding 1800°C alumina tubes can be used but in this case the heating and cooling of the furnace must be carried out veiy slowly since alumina is sensitive to thermal shock. For temperatures exceeding 1800°C tubes made of zirconium oxide, thorium oxide or beryllium oxide must be used. They are even more sensitive to temperature changes than alumina. It is very difficult to ensure the air tightness of the apparatus containing elements made of these materials since they cannot be sealed to glass or to quartz.

The construction of a furnace capable of giving such high temperatures is another problem. Some form of resistance heating is always used to attain these temperatures. Up 1300°C a canthal coil (an alloy consisting of iron, chromium and aluminium) in the form of a wire or a strip wound directly on the reaction tube can be used but at 1300°C it burns out after a relatively short time (about 200 h).

For temperatures up to about 1800°C coils made of strips of a Pt–Rh alloy containing up to 40% Rh can be used. These furnaces have a short life because the rate of evaporation of the element is relatively high. Better results are obtained in this temperature range by resistance heating with sintered $MoSi_2$ which is also called supercanthal. However this material requires an oxidising atmosphere to enable a protective layer of silica to be formed on the surface. Supercanthal is brittle and thus cannot be used in the form of a coil. It is usually employed in the form of U shaped rods.

At temperatures up to 1900°C very good results are obtained with coils of molybdenum or tungsten strips wound directly on the reaction tube. Tungsten strips wound on a tube made of beryllium oxide can be used for a long time to 2000°C. However these materials require a special protective atmosphere, since in oxidising conditions they are already rapidly corroded at 800°C. The protective atmosphere usually consists of pure argon or hydrogen.

Figure 161 shows schematically the thermogravimetric apparatus for studies on defect equilibria in crystals at temperatures not exceeding 2000°C [250]. The diagram illustrates the difficulties encountered in the construction of this type of apparatus.

It should be mentioned that, in studies carried out at very high temperatures, there are difficulties in the choice of suitable materials for the thermoelements and wires used for suspending the samples. The classical Pt–Rh thermocouples are unsuitable above 1400°C, since the rapid evaporation of platinum or its oxidation cause appreciable changes in the thermoelectric force, leading to false temperature readings. For this reason various combinations of refractory metals and alloys are used. The tungsten–rhenium system gives good results, but thermocouples of this type must be protected from oxidising atmospheres.

The choice of material for wires on which the samples are suspended presents greater difficulties, since this material must be exposed to the

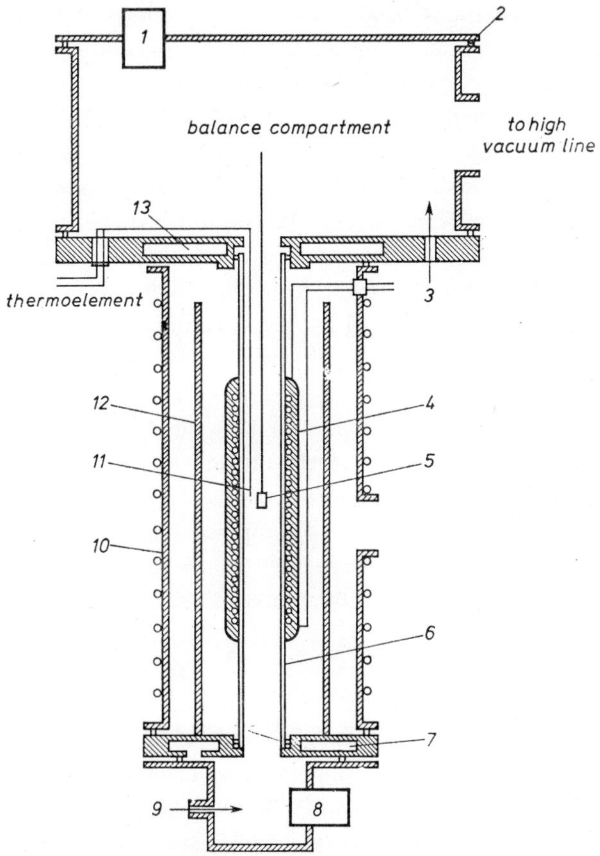

Fig. 161. The thermogravimetric apparatus for studying defect equilibria and kinetics of oxidation of metals at temperatures up to 2000°C (after [250]); *1*—ionisation vacuum meter, *2*—metal welded joints, *3*—gas inlet, *4*—tungsten or molybdenum coil, *5*—sample, *6*—alumina tube, *7*—cooling water, *8*—ionisation vaccuum meter, *9*—gas inlet, *10*—heat screen cooled with water, *11*—thermoelement, *12*—heat screen made of tantalum, molybdenum or alumina.

oxidising atmosphere and must not change its mass. Quartz filaments and platinum wires can be used at temperatures up to 1200°C. At higher temperatures the loss of platinum caused by the formation of its volatile oxides excludes the use of platinum wires. For this reason fibres made of threads of alumina monocrystals or sintered alumina are used at very high temperatures.

The devices described above can only be used on condition that the oxidising medium is a gas (O_2, N_2, Cl_2) or a mixture of gases. In this case the swing balance or the spiral balance can be maintained at room temperature outside the furnace, since the oxidant will not condense on the cold parts of the apparatus. When the oxidant is a solid or a liquid under ambient conditions (S, I, Br), i.e. when the defect equilibria are to be determined for sulphides or halides of metals, the construction of the apparatus must be quite different. In this case the flow system cannot be used and direct measurements of the vapour pressure of the oxidant in the reaction space are very difficult. In order to avoid condensation of the oxidant vapour in any part of the apparatus, the whole space filled with this vapour must be at a temperature higher than that of the vessel containing the solid or liquid oxidant. Under these conditions the vapour pressure of the oxidant in the reaction chamber is constant and is determined by the equilibrium between the liquid or solid oxidant and its saturated vapour. Under these conditions the measurements of the changes of mass of the examined compound MeX are more complicated, since the balance must be placed in the reaction space at an elevated temperature. In this case only quartz spiral balances can be used, since they can work at elevated temperatures and quartz is not corroded by the oxidants even at high temperatures.

The first thermogravimetric apparatus for continuous mesurements of changes of mass of metal sulphides in a sulphur vapour atmosphere was developed in 1952 by Neunhöffer, Hauffe and Rahmel [251, 252]. It was later also used in studies on kinetics of sulphidation, bromination and iodination of metals. The apparatus is shown schematically in Fig. 162. It consists of two tubular electric furnaces placed coaxially one above the other. The upper furnace is fixed, whereas the lower one can be moved along the vertical axis. Inside the furnaces there is an evacuated glass or quartz tube sealed under high vacuum in which the sample of the metal under study is suspended on a quartz spiral. The solid or liquid oxidant is placed on the bottom of the tube.

The length of the reaction tube is so chosen that the sample suspended in the quartz balance is in the zone of the upper furnace and the bottom of the tube containing the oxidant is in the middle of the lower furnace (see Fig. 162). The lower furnace determined the temperature of the solid

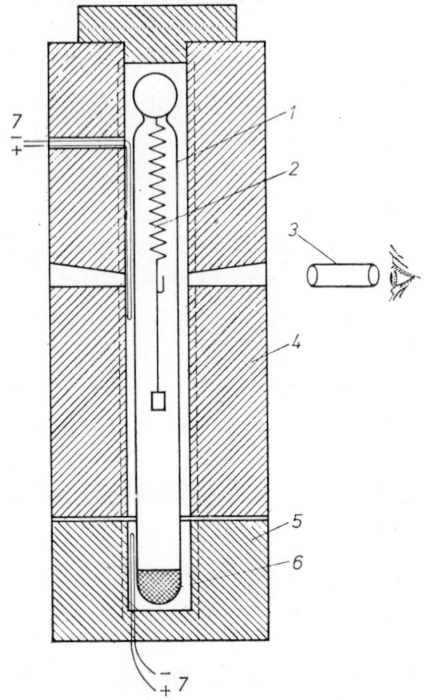

Fig. 162. The scheme of thermogravimetric apparatus for studying the kinetics of oxidation of metals in the oxidant vapours (according to Neunhöffer [251], Hauffe and Rahmel [252]); *1*—reactions tube, *2*—spiral quartz balance, *3* — cathetometer, *4*—upper furnace, *5*—lower furnace, *6*—liquid or solid oxidant, *7*—thermoelements.

or liquid oxidant and consequently the pressure of its vapour in the reaction tube. The upper furnace maintains the required temperature in the reaction chamber.

The changes of mass of the sample are determined, as in other devices of this type, by measuring the elongation of the quartz spiral. Since the spiral is situated in the upper furnace two narrow windows are cut in the furnace at the level of the indicator of the spring. The windows consist of pieces of optical glass which makes it possible to determine the elongations of the spring with a cathetometer.

The apparatus described above is used as follows. The solid or liquid oxidant is placed in the reaction tube. The amount of the oxidant should be much larger than that required for the oxidation of the sample. The tube is evacuated at an elevated temperature in order to remove the gas adsorbed on its walls and absorbed by the oxidant. If the oxidant is a solid under the ambient conditions, it should be melted during the degassing process. Then the bottom of the tube is cooled in liquid air in order to freeze the oxidant and consequently to reduce its vapour pressure in the reaction tube practically to zero. After admitting air to the tube, the spiral, together with the sample suspended on it, are introduced and the apparatus is again degassed and is flushed several times with helium or argon in order to remove the last traces of oxygen. Then the tube is sealed under vacuum and is placed in the furnaces. The outlet of the upper furnace is closed with a ceramic stopper in which there is a metal clamp to hold the tube in the vertical position.

Before starting heating, the lower furnace is placed below the bottom of the reaction tube, in order to preserve the vacuum in the reaction space until both furnaces reach the required temperature. It is advisable to cool the bottom of the tube with liquid air during this period. Then the lower furnace is moved up to the bottom of the upper furnace. This causes a rapid increase of the vapour pressure of the oxidant in the tube. After several minutes the pressure reaches the required value corresponding to equilibrium.

The apparatus developed by Neunhöffer, Hauffe and Rahmel has certain disadvantages which should be discussed. The main limitation of this method results from the necessity of always maintaining the temperature of the lower furnace below the temperature of the reaction space, since only then the sample is under a definite and constant pressure of the oxidant vapour. For this reason the use of arbitrary combinations of oxidant pressure and sample temperature is not possible. It should also be mentioned that determinations of defect equilibria by this method are very tedious, since they require many preliminary operations, particularly sealing of the tube containing the sample and the spiral under a high vacuum, which is very difficult when quartz tubes are used.

In view of these difficulties Geld and Krasowska [253] proposed an interesting modification of the apparatus developed by Neunhöffer, Hauffe and Rahmel. This modification eliminates the need to seal the reaction tube before every measurement. It is shown schematically in Fig. 163.

The apparatus is only slightly different from the classical device developed by Neunhöffer, Hauffe and Rahmel. The sample is suspended on a quartz spiral in the upper furnace. The lower furnace determines the temperature of the molten oxidant and the corresponding pressure of its vapour in the

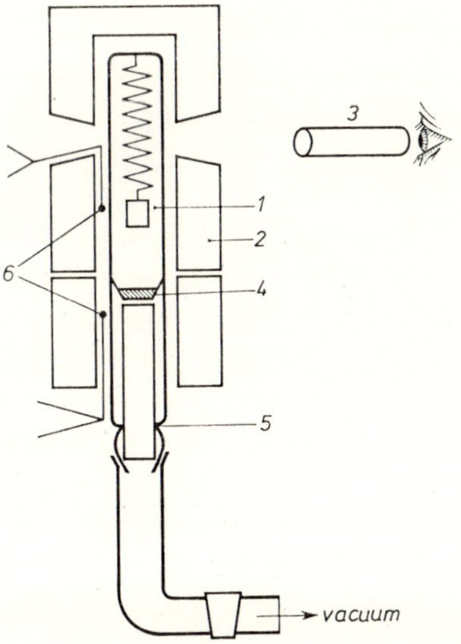

Fig. 163. The scheme of thermogravimetric apparatus for studying the defect equilibria in sulphides and the kinetics of oxidation of metals in sulphur vapours (according to Geld and Krasowska [253]); *1*—reaction tube, *2*—furnaces, *3*—cathetometer, *4*—crucible with sulphur, *5*—vacuum seal formed at the time of solidification of sulphur, *6*—thermo-elements.

reaction space. In Geld and Krasowska's apparatus, unlike the original device, the reaction tube is so long that it protrudes downwards from the furnace and is connected to the high vacuum system. The crucible with the oxidant is situated in the lower part of the tube on a quartz rod. The diameter of the rod is so chosen that it fits the narrow part of the tube. Before starting to heat of the furnaces the tube is evacuated. Then, without breaking the vacuum, the upper furnace is raised to the reaction temperature and heating of the lower furnace is started. The oxidant in the crucible melts, then evaporates and begins to condense on the cold lower part of the tube. Since the gap between the quartz rod and the narrow part of the tube is very small, it soon becomes closed by the condensing oxidant and as a result the reaction chamber becomes disconnected from the high

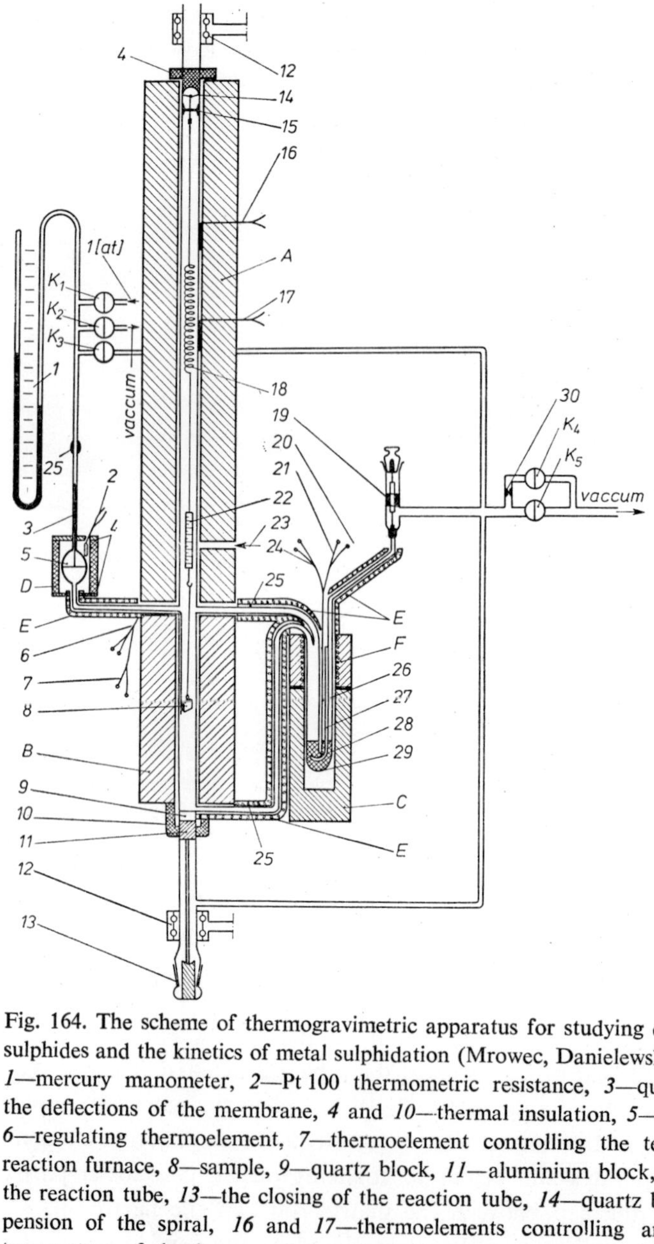

Fig. 164. The scheme of thermogravimetric apparatus for studying defect equilibria in sulphides and the kinetics of metal sulphidation (Mrowec, Danielewski, Stokłosa [255]); *1*—mercury manometer, *2*—Pt 100 thermometric resistance, *3*—quartz rod showing the deflections of the membrane, *4* and *10*—thermal insulation, *5*—quartz membrane, *6*—regulating thermoelement, *7*—thermoelement controlling the temperature of the reaction furnace, *8*—sample, *9*—quartz block, *11*—aluminium block, *12*—clamps fixing the reaction tube, *13*—the closing of the reaction tube, *14*—quartz beam, *15*—the suspension of the spiral, *16* and *17*—thermoelements controlling and regulating the temperature of the furnace of the spiral, *18*—quartz spiral, *19*—vacuum valve, *20*—additional heating element for sulphur, *21* and *24*—thermoelements controlling and

vacuum system. Thus a vacuum seal is formed automatically during the heating of the lower furnace, which eliminates the need to seal the reaction tube before each measurement. After the experiment the narrow part of the tube is heated, which causes the liquidation of the "seal".

The devices described for determining deviations from stoichiometry for sulphides and halides of metals are not as sensitive as those used for measurements in oxygen atmospheres. Their principal disadvantage is the impossibility of direct determination of the pressure of oxidant vapour in the reaction space. As has been mentioned above, this pressure is calculated by assuming that the temperature of the vessel containing the liquid or solid oxidant determines its vapour pressure in the reaction space. Recently Mrowec, Stokłosa and Danielewski [254, 255] developed a new type of apparatus for the determination of defect equilibria in metal sulphides, which not only makes it possible to determine directly the pressure of sulphur vapour in the reaction chamber, but also ensures that the precision in the determination of the changes in mass of the sample is equal to that of modern thermobalances used in studies of the processes taking place in oxygen atmosphere.

This new apparatus is a far reaching modification of the thermobalance developed by Neunhöffer, Hauffe and Rahmel. The pressure of sulphur vapour is directly determined in the reaction space by means of a special membrane manometer. The vessel containing sulphur is situated outside the reaction tube, and the sample can be introduced in the apparatus or removed from it without cutting the reaction tube. A high sensitivity in the measurements is possible due to the use of a long, thermostatted quartz spring of 60 tubs of diameter 30 mm. The apparatus is shown schematically in Fig. 164.

The principal part of the apparatus is a quartz reaction tube. In its upper part there is a quartz beam (14) on which the spiral is suspended. During measurements the temperature in this part of the tube is kept at 480°C in order to prevent the condensation of sulphur vapour. The

regulating the temperature of sulphur, *22*—the scale for reading the elongations of the spiral, *23*—window for observing the elongations of the spiral, *25* and *30*—Pyrex glass–quartz joints, *26*—the screen of the thermocouples, *27*—the screen of the heating element, *28*—sulphur, *29*—sulphur container, K_1–K_5—vacuum taps, *A*—the furnace thermostating the spiral, *B*—the reaction furnace, *C*—the sulphur heating furnace, *D*—the furnace thermostating the membrane, *E–F*—the furnaces thermostatting the sulphur vapour leads.

reaction space around the sample is heated by a separate electric furnace (*B*), which makes it possible to measure the kinetic of sulphidation at temperatures up to 1200°C. The upper and the lower furnaces consist of two semicircular elements, which can be moved away from one another. The container with liquid sulphur, situated near the reaction chamber, is heated with electric furnace (*C*),which ensures the required temperature of liquid sulphur. The sulphur vapour passes from the container to the reaction space through two thermostated glass tubes. The fourth furnace (*D*) maintains the element with the quartz membrane at 400°C. The quartz spiral makes it possible to determine the mass changes of the sample to an accuracy of 10^{-6} g. The elongation of the spiral is observed through a window (*23*) cut in the upper furnace by means of a cathetometer.

The main elements of this apparatus, which ensure its good performance are two vacuum valves (*19* and *9–11*) and the membrane manometer. These are shown in Figs. 165 and 166. The lower valve (Fig. 165), for opening the reaction space, consists of an aluminium block (*4*) with a teflon

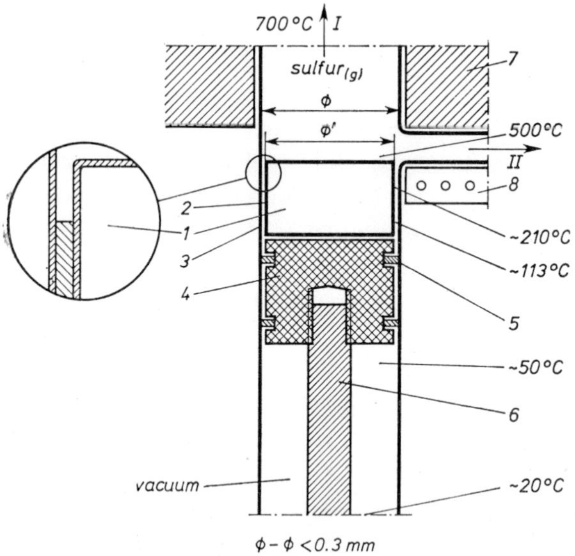

Fig. 165. The scheme of the lower vacuum valve with temperature distribution for sulphur pressure about 1.333 kPa (10 Tr) (after [255]); *1*—quartz block, *2*—the border between liquid and gaseous sulphur, *3*—the border of the solid sulphur, *4*—aluminium piston, *5*—teflon washers, *6*—fixing rod, *7*—reaction furnace, *8*—furnace thermostatting sulphur vapour lead.

seal (*5*) and na empty quartz block (*1*). During the experiment, sulphur penetrates the narrow gap between the reaction tube and the quartz block, where it solidifies automatically creating a vacuum seal. The upper seal (Fig. 166), separating the apparatus from the high vacuum system, works

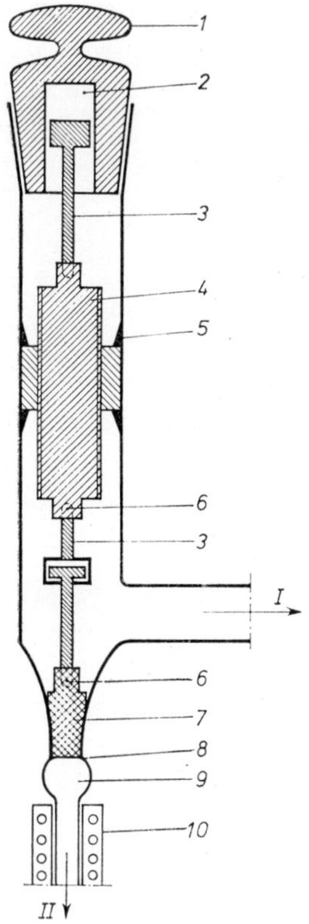

Fig. 166. The scheme of the upper vacuum valve (after [255]); I—to the vacuum system, II—to the sulphur container; *1*—vacuum glass joint, *2*—the lead of the piston, *3-6*—mechanical joints, *7*—conical teflon piston, *8*—solidified sulphur, *9*—sulphur vapour, *10*—the furnace thermostatting the lead to the sulphur container.

on the same principle. After evacuation of the apparatus before the measurement, the conical teflon block (*7*) is lowered by means of screw (*4*) and as a result the apparatus becomes disconnected from the vaccuum line. The seal becomes completely air tight when sulphur begins to condense in the gap between the glass and the teflon block.

Measurements of the kinetics of sulphidation of the metal and of changes in the mass of the resulting sulphide are carried out as follows: The main furnaces *A* and *B* are moved away from one another, the lower valve is opened and the quartz spiral is lowered so that the sample can be suspended on it. The lowering of the spiral is carried out by means of a system of magnets which detach the spiral from the quartz beem (*14*) (the upper hook of the spiral contains a part made of chromium steel which makes

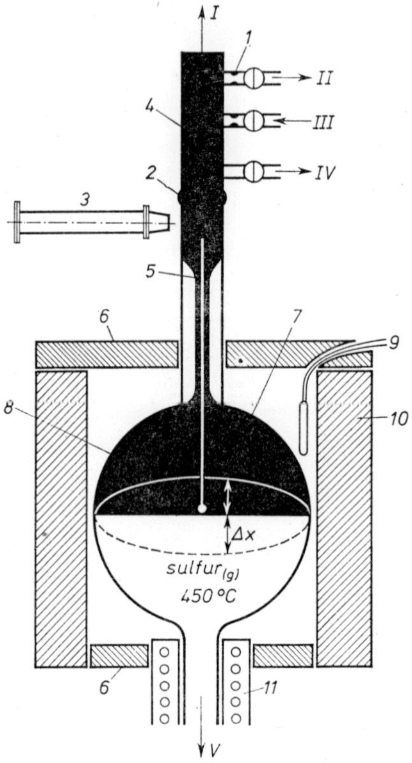

Fig. 167. The scheme of membrane manometer for measuring the pressure of sulphur vapour (after [255]); I—to mercury manometer, II—to vacuum pump, III—air 101.325 kPa (760 Tr), IV—connection to the vacuum system of the microthermobalance, V—connection to the reaction part; *1*—capillaries, *2*—Pyrex glass–quartz joint, *3*—microscope, *4*—quartz rod, *5*—tube leading the quartz rod, *6*—thermal insulation, *7*—quartz vessel, *8*—quartz membrane, *9*—Pt100 thermometric resistance, *10*—the furnace thermostatting the membrane, *11*—the furnace thermostatting the connection to the reaction part.

it possible to use magnets). After checking that the sample is correctly placed in the reaction space and that the spiral is in the right position in the upper part of the apparatus the lower valve and the upper furnace are closed. Then the upper valve is opened and the evacuation of the apparatus and the pressure measuring system is started. At the same time the upper furnace is heated to 480°C, while the lower furnace, all the thermostated glass tubes and the pressure measuring system are heated to the required reaction temperature.

When the pressure in the apparatus falls to a stable value of the order of 1.333×10^{-4} Pa (10^{-6} Tr), and the required temperature is established in both main furnaces, the upper vacuum valve and the tap K_3 to the manometer are closed. At the same time the lower furnace is also closed. When the required temperature is established in the reaction chamber the sulphur container is heated to the temperature correspoding to the pressure of sulphur vapour required. The increasing pressure of sulphur vapour on the membrane of the manometer is compensated by admitting air to the upper part of the system through taps K_1 and K_2 in order to prevent the deformation of the membrane. The position of the membrane can be determined by observing the movements of a quartz rod (*4*) (Fig.164) leaning against the membrane by means of microscope (*3*) (Fig. 167). After the stabilisation of the sulphur vapour pressure in the apparatus and exact compensation of pressures on both sides of the membrane the value of the pressure, shown by the manometer (*1*), (Fig. 164) is recorded. Then the measurements of kinetics of the sulphidation process are started by recording the rate of elongation of the quartz spiral by means of the cathetometer. At the end of the sulphidation process, when the increase of the mass of sample ceases, the temperature or the pressure of sulphur vapour is altered in order to determine the mass of sample corresponding to the next equilibrium state.

The temperature in the reaction chamber is controlled by means of the thermocouple (*7*) situated near the sample. The thermocouple (*6*) maintains the constant temperature in the reaction chamber.

It has been found that the fluctuations of temperature in the reaction chamber do not exceed 5 ± 2°C. The variations of sulphur vapour pressure in the reaction chamber are caused mainly by changes in the temperature of the sulphur container (*29*), but they can be also caused by the consumption of sulphur by the sample when the reaction rate is larger than the rate of diffusion of sulphur vapour to the reaction chamber. For this reason the construction of the apparatus maintains temperature of the sulphur container to within ±0.5°C. The diameters of the container and of the tubes feading sulphur vapour to the reaction chamber are so chosen that the supply of sulphur vapour to the sample is as large as possible.

The membrane manometer (Fig. 167) makes it possible to determine the sulphur vapour pressure with an accuracy of ±26.66 Pa (±0.2 Tr). The required sulphur vapour pressure in the reaction chamber is established within about 5 minutes of melting of the sulphur in the container. The constancy of this pressure during the measurements is of the same order as the sensitivity of the membrane manometer.

After the end of the sulphidation process furnace (*C*) (Fig. 164) is lowered below the level of the sulphur container, which causes a rapid condensation of all the sulphur vapour in the measuring system. When this condensation is complete and the sulphur in the container freezes, both main furnaces are switched off. The upper valve (*19*) is unfrozen by means of a special heater and is opened in order to admit air to the apparatus. Then the lower valve (*11*) is unfrozen and is opened and the spiral is lowered by means of the system of magnets in order to remove the sample from the apparatus.

The results measurements of the deviations of ferrous sulphide $Fe_{1-y}S$ from stoichiometry obtained by means of this apparatus (Fig. 168) [256] indicate that it is very suitable for gravimetric measurements of the con-

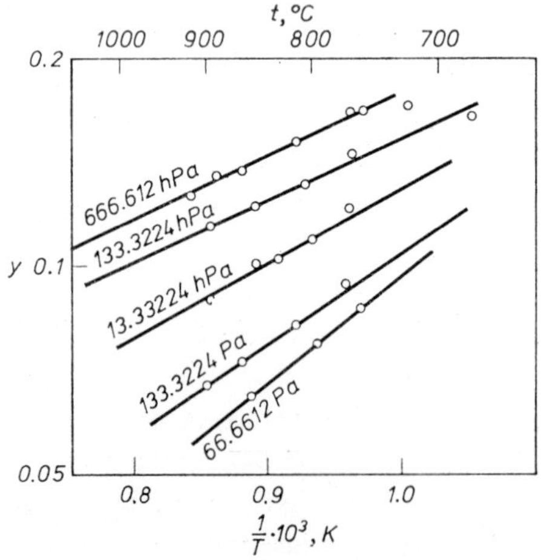

Fig. 168. The temperature dependence of the deviation from stoichiometry y of ferrous sulphide $Fe_{1-y}S$ for several pressures of sulphur vapour (after [765]).

centration of defects in metal sulphides. Unlike other devices of this type it makes it possible to control continuously and directly the sulphur vapour pressure in the reaction chamber during the measurements. The use of the special vacuum valves facilitates access to the reaction chamber and eliminates the necessity of cutting the reaction tube. As a result of the use of spirals, which previously could only employed in oxygen atmosphere, the sensitivity of measurements is very high.

At the begining of the description of gravimetric methods for the determination of defect equilibria we stated that the absolute values of deviation from stoichiometry y can be directly determined from the initial mass of metal sample and the final mass of the resulting compound MeX. However it should be explained that this is only possible in cases when the deviation from stoichiometry is high and consequently the errors resulting from the determination of absolute values of y are low. However in the majority of cases the deviations from stoichiometry are so small that the determination of the value of y with sufficient accuracy requires a different procedure. This problem will be discussed with reference to the results obtained by Mrowec, Godlewski and Stokłosa [202, 257] for cuprous oxide. This work is a good illustration of the difficulties encountered in studying the thermodynamics of point defects in metal oxides.

The deviations from the stoichiometric composition in cuprous oxide are caused by the presence of defects in the form of cation vacancies in the cationic sublattice. The correct stoichiometric formula of cuprous oxide is $Cu_{2-y}O$, where the deviation from stoichiometry y is a direct measure of the concentration of cation vacancies. This concentration, corresponding to the thermodynamic equilibrium state, can be determined directly by gravimetric methods based on measurements of changes of mass of the oxide taking place at high temperatures under the influence of changes in the equilibrium oxygen pressure.

These studies were carried out in the temperature range 900–1100°C and in the pressure range 1.01325–1013.25 hPa (10^{-3}–1 atm). The limits for the temperature and pressure were dictated by the following factors: At pressures lower than 1.01325 hPa (10^{-3} atm) the rate of evaporation of cuprous oxide is so fast that the accurate determination of the deviation from stoichiometry becomes impossible even when allowance for the evaporation is made. The upper pressure limit was dictated by the decomposition pressure of cupric oxide, above which cuprous oxide is oxidised to CuO.

Limitation of the temperature to a relatively narrow range was necessary for many reasons. At temperatures below 950°C the changes in the mass of the sample, caused by the changes in the concentration of defects, were so small that their accurate determination by the gravimetric methods was impossible. Apart from this the homogeneity range of cuprous oxide was very small at these temperatures. At temperatures above 1100°C the rate of evaporation of cuprous oxide was so high that it was not possible to determine the changes of the mass of the sample with a satisfactory reproducibility.

Since the deviations of cuprous oxide from the stoichiometric composition are very small (at 1000°C y is of the order of 10^{-3}), their determination on the basis of the difference between the weight of metal and that of the oxide was not possible even with very sensitive thermobalances (Fig. 159). An additional error was caused by the un-controlled evaporation of the metal during the initial heating of the sample in vacuo. For this reason the deviation from stoichiometry was determined indirectly by measuring the changes of the mass of the oxide caused by changes of oxygen pressure at a constant temperature.

The following procedure was used: A sample of copper was heated in vacuo to a definite temperature. When the temperature became constant the oxidation process was started. During the oxidation the increase of the mass of the sample was measured. When the mass of the resulting cuprous oxide became constant, the oxygen pressure in the apparatus was altered and the corresponding change of the mass of the sample was determined. This operation was repeated many times in order to determine the isothermal dependence of the mass of the sample on oxygen pressure. In addition mass changes of the sample caused by temperature changes at a constant oxygen pressure were determined. Thus the mass changes were determined isothermally and isobarically at least twice. Good agreement between the results proved that all changes in the mass of the sample were due to the system changing from one equilibrium state to another.

All the results were corrected for the effect of evaporation of cuprous oxide. In order to determine the corrections the dependence of the rate of evaporation of cuprous oxide on temperature and pressure was measured in a separate experiment. The results of these studies are shown in the semilogarithmic plot in Fig. 169. It can be seen from the graph that the rate of evaporation of cuprous oxide was relatively high, particularly at low oxygen pressures.

In the calculations of the changes of weight of the sample not only the effect of evaporation of cuprous oxide, but also various other factors were taken into account. Particular attention was paid to the dependence of the gas buoyancy effect on temperature and oxygen pressure. For this

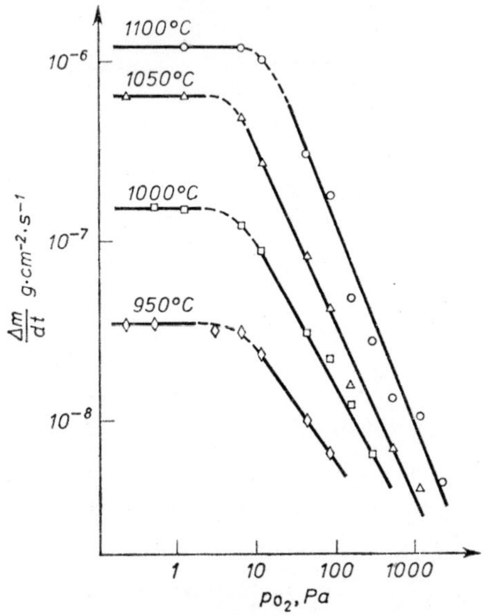

Fig. 169. The dependence of the rate of evaporation of cuprous oxide $Cu_{2-y}O$ on temperature and oxygen pressure shown in the double logarithmic plot (after [202]).

reason the calibration of the apparatus was carried out by means of an alumina monocrystal sample having the same shape and dimensions as the cuprous oxide sample.

It is of interest that the corrections for gas buoyancy were of the same order, and in some cases even larger, than the observed changes of the mass of the cuprous oxide sample. The same applies to the corrections for the effect of evaporation of cuprous oxide. The determination of all the corrections made it possible to determine not only the changes of the mass of the sample examined but also the total amount of oxygen present in the sample. It was required later for calculating the absolute values of deviations from stoichiometry on the basis of the changes in the mass of the sample.

Figure 170 shows the results of measurements of changes in the mass of the sample caused by changes of oxygen pressure at several temperatures. They are the mean values of the results obtained for several samples and are shown as the double logarithmic plots. The changes of mass Δw shown in the graph were calculated with respect to the mass of oxygen present

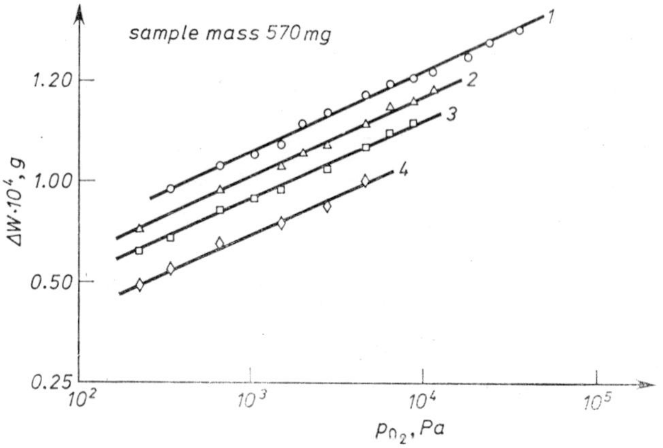

Fig. 170. The dependence of the changes of the mass of cuprous oxide on the equilibrium oxygen pressure shown in the double logarithmic plot (after [202]). Curves *1–4* correspond to 1100, 1050, 1000 and 950°C respectively.

in the sample at the lowest pressure used in the investigation. It follows from the graph that with a double logarithmic plot the dependence is linear and that the slope does not depend on temperature. Hence the dependence of the changes of the mass of the oxide on equilibrium oxygen pressure can be expressed by the following empirical equation:

$$w = \text{const}\, p_{O_2}^{1/4.2} \tag{3.2.4}$$

In order to determine the absolute values of the deviation from stoichiometry from experimentally determined changes of the mass of cuprous oxide it is necessary to define a reference state. Results of studies carried out by various authors [258, 259] indicate that good results are obtained when it is assumed that the deviation of an oxide, with a deficit of metal or an excess of oxygen (p type semiconductors), from the stoichiometric composition tends to zero when the equilibrium oxygen pressure tends to zero. On this assumption, the dependence of the changes of the mass of the oxide on oxygen pressure can be expressed by the following equation:

$$m = m_0 + \text{const}\, p_{O_2}^{1/n} \tag{3.2.5}$$

where m and m_0 are the mass of the oxide at pressure p_{O_2} and the mass of the stoichiometric oxide at $p_{O_2} = 0$. It follows from this expression that the plot of the changes of the mass of a sample of the oxide against the n-th root of the oxygen pressure for several temperatures should be a set of straight lines having different slopes and converging at the point corresponding to the stoichiometric composition of the oxide. By using this plot it is possible to find the value of m_0 and hence to calculate the absolute values of the deviation from stoichiometry.

In the case of oxides (or sulphides) showing an excess of metal or a deficit of oxygen (n type semiconductors)the dependence of the changes of mass on oxygen (or sulphur) pressure is inverted:

$$m = m_0 + \text{const}\, p_{O_2}^{-1/n} \quad (3.2.5a)$$

It follows from equation (3.2.4) that, in the case of cuprous oxide, the exponential coefficient in the general equation (3.2.5) has a value close to 1/4. The correct value of this exponent can be obtained by means of mathematical analysis of the dependence of the mass changes on the n-th root of the oxygen pressure. Such an analysis, carried out by a trial and

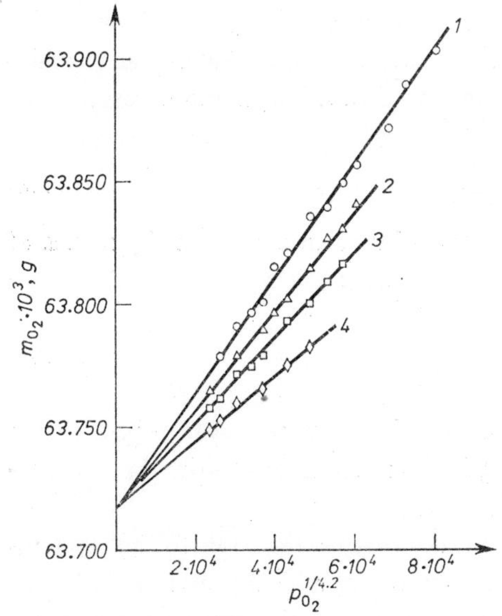

Fig. 171. The dependence of the changes of the mass of cuprous oxide sample on the equilibrium oxygen pressure raised to the power 1/4.2 (after [202]). Curves *1–4* correspond to 1100, 1050, 1000 and 950°C respectively.

error method, has shown that the best correlation of the results is obtained when the exponent $1/n$ is equal to 1/4.2. The plot of changes of the mass of the sample against oxygen pressure raised to the power 1/4.2 is shown in Fig. 171. It follows from this plot that the results fall on straight lines for all the temperatures and converge at a point which, according to equation (3.2.5), determines the mass of oxygen present in the compound when it has the stoichiometric composition.

In order to calculate the absolute values of y, the following equation can be used:

$$\frac{[\mathrm{Me}]_{\mathrm{total.}}}{[\mathrm{X}]_{\mathrm{total.}}} = \frac{a-y}{b} \tag{3.2.6}$$

where $[\mathrm{Me}]_{\mathrm{total.}}$ and $[\mathrm{X}]_{\mathrm{total.}}$ are the concentrations of the metal and the oxidant respectively, expressed in gram-atoms per unit volume, in a compound having the general formula $\mathrm{Me}_{a-y}\mathrm{X}_b$. In the case of cuprous oxide this expression has the following form:

$$\frac{[\mathrm{Cu}]_{\mathrm{total.}}}{[\mathrm{O}]_{\mathrm{total.}}} = 2-y \tag{3.2.7}$$

The total concentration of oxygen in the sample $[\mathrm{O}]_{\mathrm{total.}}$ can be regarded as the sum of the oxygen concentration in the compound having the stoichiometric composition $[\mathrm{O}]_{\mathrm{stoich.}}$ and the concentration of the excess oxygen $[\mathrm{O}]_y$. Since

$$2[\mathrm{O}]_{\mathrm{stoich.}} - [\mathrm{Cu}]_{\mathrm{total.}} = 0 \tag{3.2.8}$$

equation (3.2.7) can be transformed to obtain an expression by means of which the value of y can be directly calculated:

$$y = \frac{2[\mathrm{O}]_y}{[\mathrm{O}]_{\mathrm{total.}}} = \frac{2\Delta m}{m_{\mathrm{O}}^{\mathrm{s}} + \Delta m} \tag{3.2.9}$$

where Δm is the increase of the mass of the sample with respect to the stoichiometric composition and $m_{\mathrm{O}}^{\mathrm{s}}$ is the mass of oxygen in the stiochiometric compound.

The values of y calculated by means of this equation for many temperatures and oxygen pressures are shown graphically in Fig. 172 with a double logarithmic plot. The broken line shows the upper limit of homogeneity of Cu_2O phase in the range of temperature and oxygen pressure examined.

In order to illustrate the temperature dependence of y, the above mentioned results are again shown in Fig. 173 in the Arrhenius system of coordinates for several oxygen pressures. It can be seen from the plot that the temperature dependence of y plotted in this system of coordinates has, as expected,

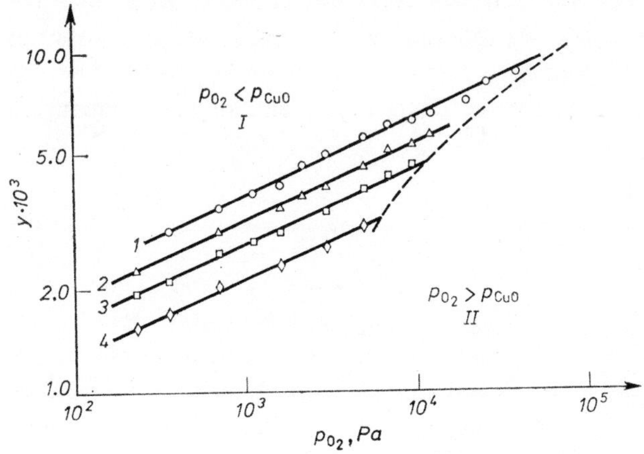

Fig. 172. Changes of deviation from stoichiometry of cuprous oxide as a function of equilibrium oxygen pressure shown in the double logarithmic plot (after [202]). Curves *1–4* correspond to 1100, 1050, 1000 and 950°C respectively. The broken line shows the limit of the homogeneity range of $Cu_{2-y}O$ phase.

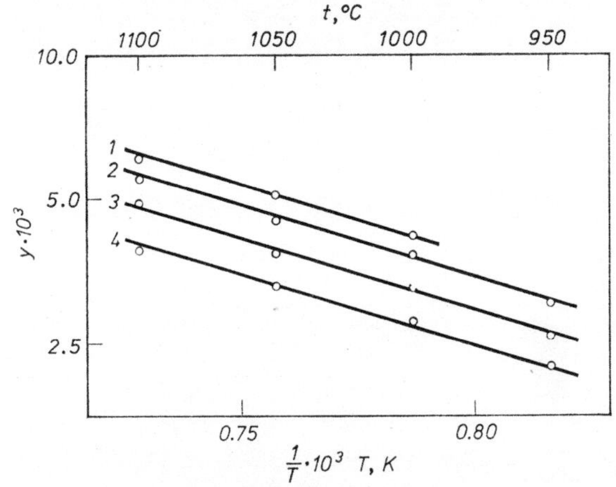

Fig. 173. The temperature dependence of the deviation from stoichiometry of cuprous oxide for several oxygen pressures (after [202]). Curves *1–4* correspond to 6.941 kPa (6.85×10^{-2} atm), 4. 256 kPa (4.20×10^{-2} atm), 2.999 kPa (2.96×10^{-2} atm), 1.601 kPa (1.58×10^{-2} atm) respectively.

a linear character and that the slope of the straight lines does not depend on oxygen pressure within the limits of experimental error. Hence the enthalpy of formation of point defects in cuprous oxide does not depend on equilibrium oxygen pressure and is equal to $\Delta H_f = 50.24 \pm 4.1868$ kJ/mole (12 ± 1 kcal/mole). Since it was shown at the same time that the character of the dependence of y on equilibrium oxygen pressure does not depend on temperature (Fig. 172), the results can be presented in the form of a single empirical equation expressing the dependence of y on temperature and equilibrium oxygen pressure:

$$y = 0.886 p_{O_2}^{1/4.2} \exp\left(\frac{-12{,}000}{RT}\right) \tag{3.2.10}$$

3.2.2 Volumetric methods

The concentration of defects in metal oxides can be also determined very accurately by volumetric methods consisting in measurements of the volume of oxygen bound by the sample. The interpretation of the results is analogous to that used in the case of the gravimetric methods. The only difference results from the need to calculate the mass of the reacted oxygen from its volume. Volumetric methods are not suitable for studies of the equilibria

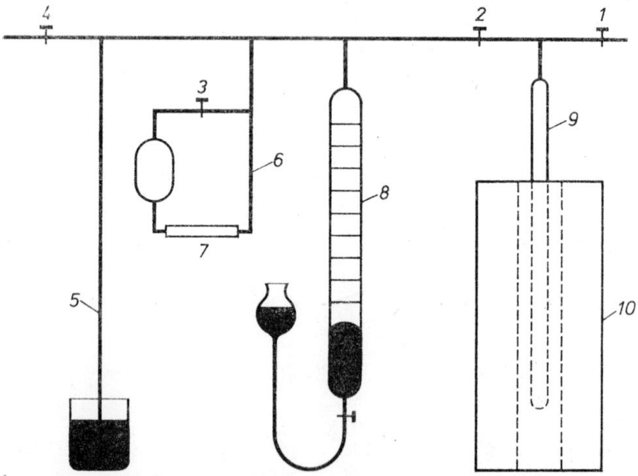

Fig. 174. The scheme of volumetric apparatus for studying the kinetics of oxidation of metals (after Engell–Hauffe–Ilschner [260]); *1–4* vacuum taps, *5*—mercury manometer, *6*—tensiometer, *7*—capillary with the scale, containing a drop of liquid of low vapour pressure, *8*—gas burette, *9*—reaction tube, *10*—furnace.

of defects in sulphides and halides of metals, since they are applicable only to gaseous oxidants but not to vapours.

In view of the high sensitivity of the volumetric methods several ingenious devices for studies of the equilibria of defects in metal oxides and for measurements of the kinetics of high temperature oxidation of metals as well as chemisorption of gases on solids have been developed.

Figure 174 shows the apparatus developed by Engell, Hauffe and Ilschner [260]. The principal element of this apparatus is a differential manometer (tensiometer) which makes it possible to determine the volume of the gas consumed at a constant pressure in the reaction chamber. It consists of an perfectly horizontal capillary situated between the apparatus and the reference container. The capillary is calibrated and has a suitable scale. It contains a drop of liquid having very low vapour pressure (e.g. diethyl phthalate). Before the reaction the pressures of the gas in the apparatus and in the reference vessel are the same. In this situation the drop of liquid in the capillary is in the zero position. During the reaction the drop moves to the right as a result of the decrease of the amount of gas in the reaction chamber. When the drop reaches the end of the scale the loss of gas in the apparatus is compensated by means of a microburette. This causes the return of the drop to the zero position.

The principle of the measurements is as follows: a sample of the metal to be examined is placed in the reaction tube, tap *2* is closed and the tube is evacuated by means of a high vacuum pump. Then the reaction space is heated to the required temperature without breaking the dynamic vacuum. During this time the remaining part of the apparatus is thoroughly degassed and is filled with the oxidant through tap *4*. During the filling of the apparatus, tap *3* is open. In this way the required pressure of the oxidant, measured by a mercury gange, is established everywhere in the apparatus with the exception of the reaction space. Then the high vacuum line is disconnected from the reaction space by closing tap *1* and opening tap *2* at the same time. In order to ensure complete equilibration of pressures in the apparatus tap *4*, controlling the flow of the gas to the apparatus, should be closed 5–10 seconds after opening tap *2*. Since the measurement starts only after closing tap *4*, the reaction cannot be controlled during the first few seconds. This is one of the disadvantages of this method. However it should be stressed that this method is very sensitive, since it makes it possible to determine changes in the volume of oxygen corresponding to changes in the mass of the sample of the order of 10^{-6} g. Since the sensitivity of the measurements is so high, both the reaction zone and the remaining part of the apparatus, which is at room temperature, must be very thoroughly thermostated, in order to eliminate the very pronounced effect of temperature changes on the readings of the tensiometer and the microburette.

Figure 175 shows a more detailed diagram of the volumetric apparatus developed by Jenkins [261]. The principle of measurements is analogous

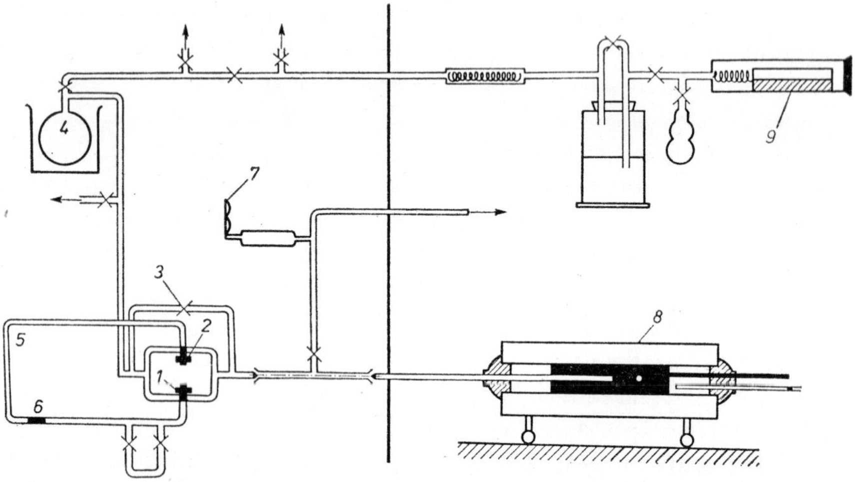

Fig. 175. The scheme of the volumetric apparatus for studying the kinetics of oxidation of metals (after A. Jenkins [261]); *1-3*—vacuum taps, *4*—oxygen container, *5*—capillary with the scale, containing the indicator drop, *6*, *7*—manometers, *8*—furnace *9*—device for generating oxygen by decomposition of potassium permanganate.

to that of Hauffe's apparatus, but in the case of Jenkin's device there is no need to compensate for the loss of gas in the reaction chamber by means of the microburette, which makes it possible to carry out continuous measurements. The system of taps *1*, *2* and *3* connects the differential manometer to the reference vessel and the reaction chamber in such a way that the drop in the capillary can move to the right and to the left. When the reaction chamber is connected to the lower part of the capillary through tap *1* the upper part of the capillary is connected to the reference container through tap *2*, while tap *3* is closed, the consumption of oxygen in the reaction chamber causes a shift of the drop in the capillary to the right. When the drop reaches the end of the scale, tap *3* is opened. As a result the drop stops and a fresh aliquot of oxygen is admitted to the reaction chamber. Then the lower part of the capillary is connected to the reference container by turning tap *1* and the upper part of the capillary is connected to the reaction vessel by turning tap *2*. In this situation, when tap *3* is closed, the drop begins to move to the left.

According to Jenkins, a 1 mm shift of the drop in a capillary having a diameter of 1 mm corresponds to the consumption 1 μg of oxygen in the reaction chamber at pressures of the order of 1013.25 hPa (1 atmosphere). This proves that the sensitivity of this type of apparatus is very high.

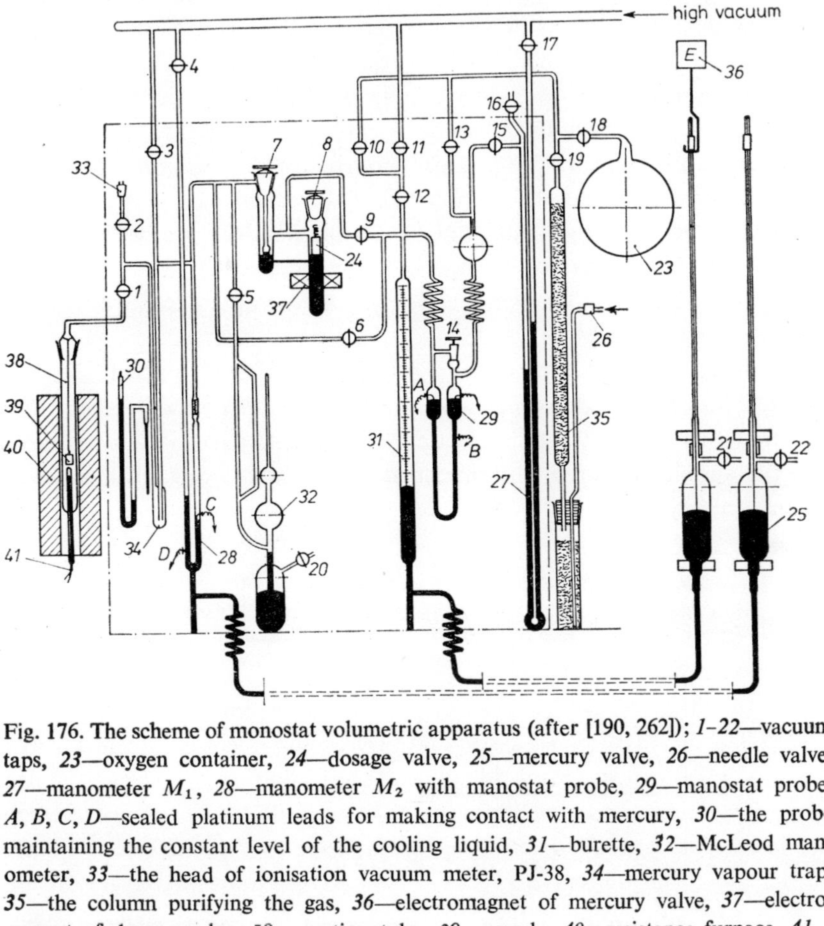

Fig. 176. The scheme of monostat volumetric apparatus (after [190, 262]); *1–22*—vacuum taps, *23*—oxygen container, *24*—dosage valve, *25*—mercury valve, *26*—needle valve, *27*—manometer M_1, *28*—manometer M_2 with manostat probe, *29*—manostat probe, *A, B, C, D*—sealed platinum leads for making contact with mercury, *30*—the probe maintaining the constant level of the cooling liquid, *31*—burette, *32*—McLeod manometer, *33*—the head of ionisation vacuum meter, PJ-38, *34*—mercury vapour trap, *35*—the column purifying the gas, *36*—electromagnet of mercury valve, *37*—electromagnet of dosage valve, *38*—reaction tube, *39*—sample, *40*—resistance furnace, *41*—thermoelement Pt-PtRh, *42*—heat screen separating the reaction furnace from the measuring part of the apparatus.

Recently Mrowec and Przybylski have [192, 262] developed a new type of volumetric apparatus, which, in view of its advantages, deserves to be discussed.

This apparatus contains a special system of manostats, which makes it possible to measure very accurately the volumes of oxygen consumed by the sample under isothermal and isobaric conditions. During these measurements all operations, with the exception of the reading of mercury level in the burette, are made automatically. So far this apparatus has been used for measurements of pressures in the range 1.333–7.999 × 10^4 Pa (10^{-2} to 600 mm Hg). However this range will be considerably widened as a result of improvements now being made.

The apparatus is shown schematically in Fig. 176. It consists of two principal parts, in which the pressures p_1 and p_2 are different. The two parts are separated by tap *24*. The first part consists of the reaction tube *38*, the manostat probe *28* and the manometer. The second part consists of the burette *31* connected to the manometer *27*, the manostat probe *29* and the gas container *23*. Both the burette and manostat probe *28* are connected to mercury valves *25*. The construction of the valves is described in detail in the original publication.

A separate part of the apparatus consists of an electronic system (Fig. 177) coupled with electromagnets *36* and *37* and platinum contacts *AB* and *CB*, which automatically maintains a set pressure in both parts of the apparatus. The pressure is measured with manometers *27* and *28* using a cathetometer to an accuracy of 133.322 × 10^{-2} Pa (0.01 mm Hg).

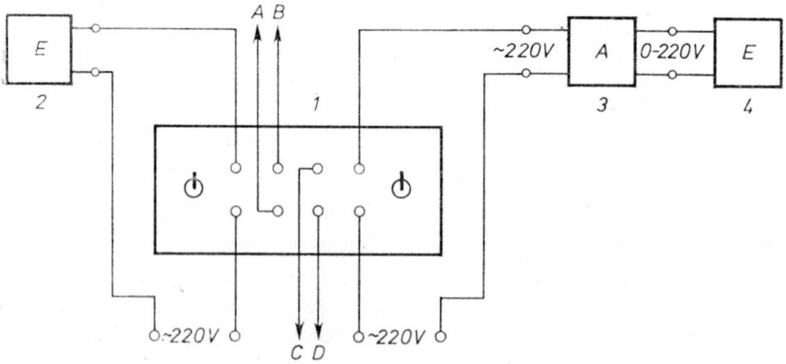

Fig. 177. The scheme of the electrical circuit (after [190]); *1*—electronic relay, *2*—electromagnet of the dosage valve, *3*—autotransformer, *4*—electromagnet of the mercury valve, *A*, *B*, *C*, *D*—platinum contacts.

The measurements consist of the folowing operations: After the introduction of the sample in the reaction zone, the pressure in the apparatus is reduced to 133.322×10^{-6} Pa (10^{-6} mm Hg). Then, without disconnecting the vacuum and keeping valves *1* and *3* open, the reaction zone and the sample are heated to the required temperature. After a suitable time of heating and after checking the pressure in the reaction tube *38*, vacuum valves *1* and *3* are closed. Then oxygen from container *23* is admitted to the reamining part of the apparatus also under vacuum and degassed through valves *10*, *12* and *18* in such a way that two manostats reach the required pressures p_1 and p_2. For this purpose burette *31* is filled with oxygen under pressure p_2, measured with manometer *27*, and the electronic relay *1* (Fig. 177) is switched on. Pressure p_1 is established using valve *25* and by raising the mercury level in manometer *28* to a suitable height. The electrical system of valve *24* is then switched on keeping vacuum valves *7* and *9* open.

Valve *24* consists of two vessels connected to one another and filled with mercury. In the left hand vessel, there is a capillary with porous glass sealed from underneath, connected to vacuum tap *7*. The right hand vessel, surrounded by electromagnet *37*, contains a metal rod. When the electromagnet is switched on, the rod rises and as a result some of the mercury flows from the left hand vessel to the right. The porous glass becomes uncovered and as a result oxygen from burette *31* passes through it to manometer *28* until the mercury level falls below platinum contact *C*. After closing vacuum tap *14* in monostat probe *29*, the initial mercury level in the burette is re-established by means of electromagnet *36*.

The measurement starts at the moment that valve *1* is opened. At first a slight decrease of the pressure in the apparatus takes place as a result of the increase in volume. This pressure change is immediately eliminated as a result of the automatic opening of valve *24*. During the reaction, the consumption of oxygen by the sample causes a slight decrease of oxygen pressure in the reaction chamber and consequently the mercury level in manometer *28* immediately rises. When this level touches contact *C*, the electronic system switches on electromagnet *27*, which automatically opens valve *24*. The introduction of an aliquot of oxygen to the reaction chamber causes a decrease of the pressure in manostat probe *29* and a rise of the mercury level to contact *A*. At the same time electromagnet *36* is switched on. This opens mercury valve *25* and causes a stepwise rise of the mercury level in burette *31* until the pressure in probe *29* is equilibrated. At this time the electronic system is switched off until the pressere in the reaction chamber again decreases. Contact probe *30* is used to avoid disturbances resulting from the changes of the level of liquid used for cooling certain parts of the apparatus. Probe *30* maintains the constant level of the cooling liquid in freezer *34*.

The apparatus makes it possible to carry out continuous measurements without any limitation on the amount of oxygen that can be added to the burette. When the mercury level reaches the upper limit of the burette, electromagnet *36* and the electronic control system for valve *24* are switched on. The vacuum pump is then connected to valve *21*. The opening of the mercury valve causes a flow of mercury from the burette to the valve container. A suitable quantity of oxygen is admitted to the burette through valves *12, 10* and *18* until the mercury level reaches platinum contact *A*. This switches on electromagnet *36* which automatically adjusts the pressure in the measuring part to the value, p_2, which existed there before the filling of the burette with the fresh aliquot of oxygen. At the same time the original level is re-established in the burette.

The measurement is continued by switching on the electrical control system of valve *24*. The compensation for the amount of gas in the burette takes 3–4 minutes. This operation does not disturb the reaction and the measurement of the amount of oxygen consumed.

In order to determine the sensitivity of the apparatus at various oxygen pressures in the reaction chamber, the difference, Δh_1, between the mercury levels in manometer *28* near contact *C* before valve *24* is switched on and after switching it off was measured. This difference is, on the average, 6.666 Pa (0.05 mm Hg), which in the case of a reaction taking place at a pressure of 666.61 Pa (5 mm Hg) represents only 1% of the total pressure. Obviously variations of the pressure decrease with increasing pressure in the reaction chamber, i.e. the constancy of pressure in the reaction chamber increases with increasing oxygen pressure.

The volume of oxygen added to the reaction zone is $\Delta V_1 = \Delta h_1 \cdot S$, where S is the cross sectional area of manometer *28* near contact *C*; e.g. during a reaction of oxidation of a metal under constant pressure p_1, the volume oxygen introduced at each addition is given by the following relationship:

$$\Delta V_1 \cdot p_1 = \Delta h_2 \cdot a \cdot p_2 \tag{3.2.11}$$

where a is the calibration constant of the burette.

The measurement of the aliquot of oxygen added consists in the determination of the difference between the mercury levels in the burette Δh_2 caused by the introduction of oxygen at constant pressure p_2 in the reaction chamber. It follows from equation (3.2.11) that the value of Δh_2 is affected by the ratio of pressures, p_1/p_2, in the two parts of the apparatus. The choice of the most favourable ratio of pressures depends on the rate of the oxidation process. This ratio decreases with decreasing rate of oxidation of the sample. Pressure p_2 must be so chosen that the amount of oxygen in the burette is sufficient for the fastest stage of the process, i.e. so that gas is only added to the burette during the later stages of the reaction, when it is usually slower than at first. Excessively high pressures p_2 should not be used, since this can lead to errors resulting from additions of large amounts of gas. On the other hand when pressure p_2 is too low, the introduction of the required amount of gas implies that the mercury in the burette is raised very high (large value of Δh_2) and that valve *24* has to be used too frequently. In this case the mercury level in the burette would rise practically continuously, which would affect the precision of the measurement.

The most favourable value of the pressure ratio, p_1/p_2, determined in special studies varies from 1/4 to 1/7. At this ratio the introduction of an aliquot of oxygen corresponds to a difference between the mercury levels in the burette of about 0.5–1.0 mm. In the pressure range 1.333–1.999 $\times 10^4$ Pa (10^{-2}–600 mm Hg), the shift of the mercury level in the burette with calibration constant 4.4542 ml/cm corresponds to a mass of oxygen of the order of 10^{-7} g.

It follows from the above discussion that the apparatus ensures an extremely high accuracy in the measurement of the amount of oxygen reacting with the sample and that the operation of the apparatus is very simple.

3.2.3 The electrochemical method

Engell [263] developed an original electrochemical method which makes it possible to determine very accurately the excess or the deficit of metal in compounds of the MeX type. The method consists of measurements of electrical current flowing in an external circuit during electrolytic dissolution of the compound examined in a suitable solution. In the case of *p*-type semiconductors the sample examined is placed on the cathode, which causes neutralisation of the excess of positive charges (electron holes) during electrolysis. In the case of *n*-type semiconductors the dissolution takes place on the anode and as a result the excess of negative charges (quasi-free electrons) is lost to the current source. Thus the current flowing in the external circuit is a direct measure of the excess charges (positive or negative) and hence of the deviation from stoichiometry.

The high sensitivity of the method is due to the possibility of direct determining the deviation from stoichiometry. In the determination of this deviation by chemical analysis of the compound examined, the total amount of metal is determined, although the excess or deficit of this component represents only a small fraction of the total amount.

In spite of its advantages Engell's method is not widely used, since it is difficult to find a suitable electrolyte. However it should be stressed that studies carried out by this method on ferrous oxide $Fe_{1-y}O$ provided the first experimental data about the distribution of cation vacancies in the oxide scale on iron [264].

3.2.4 The redox method

The redox method developed by Bunsen and Rupp [265] and improved by Dereń *et al.* [266, 267] is much more widely used in studies of the deviations from stoichiometry in semiconductors. It consists of the determination of excess positive or negative electric charge, connected with deviation from stoichiometry, by means of a suitable redox reaction. In the case of the reactions of semiconductors of the *p*-type with hydrochloric acid, the excess positive charge causes oxidation of an equivalent number of chloride ions to free chlorine. When the resulting solution is treated with a potassium iodide solution, chlorine displaces iodine from its salt and the free iodine can be determined volumetrically. In the case of semiconductors of the *n*-type the excess of metal is determined analogously. However in this case the excess negative charge causes the reduction of potassium permanganate which can be determined colourimetrically. An iodine solution can be used instead of the permanganate solution.

It follows from the above description that this method is based on classical chemical determinations. However, since only the excess or the deficit of metal are determined, the sensitivity of the method is at least two orders of magnitude higher than that of direct chemical analysis. Dereń and his collaborators have shown that the excess of metal in compounds of the ZnO type can be determined with an accuracy of the order of 10^{-5} g per g of the sample. The deficit of metal in compounds of the NiO type can be determined with the same accuracy. However the redox method has certain disadvantages, which limit its application, e.g. the method is limited to substances soluble in very dilute hydrochloric acid.

3.2.5 Rosenburg's kinetic method

In 1961 Rosenburg [268] proposed a new, extremely promising method for determining equilibrium concentrations of point defects in crystal lattices of compounds of the MeX type. The great advantage of this method is the fact that it makes it possible to determine the total concentration of defects not only in crystals with deviations from stoichiometry but also in stoichiometric compounds. The gravimetric, chemical and electrochemical methods used up to the present time only make it possible to determine the concentrations of point defects indirectly, on the basis of either deviations from stoichiometry or the equivalent number of electronic defects. However it is known that in compounds of the MeX type the total concentration of defects does not always correspond to the deviation from stoichiometry, even in the case of semiconductors of the n or p type (see Chapter 1 of the present work). In the case of solid electrolytes the concentration of lattice defects is not related at all to the deviations from stoichiometry and hence with the concentration of electron defects, so that the methods discussed above are completely useless.

Another significant advantage of Rosenburg's method consists in the fact that it makes it possible to determine not only the concentration of point defects, but also the chemical diffusion coefficient $\tilde{D}$ (see Chapter 2.1) at the same time.

The method is based on measurements of the kinetics of oxidation of the relevant metal at a constant temperature and under a definite oxidant pressure, under conditions for which the equilibrium concentration of defects C_D and chemical diffusion coefficient of the oxidation product (compound MeX) are to be determined. The method can be used under the following conditions [189, 269].

1. The oxidation of the metal must take place according to the parabolic rate law and the layer of scale of the oxidation product must consist of a single phase adhering to the metallic core. Only in this case is the diffusion of defects in the scale the rate determining step for the overall oxidation process.

2. The growth of scale must be caused by diffusion of one reacting species only, i.e. the metal or the oxidant.

3. The type of defect present in the crystal lattice of the oxidation product must be known, since only then it is possible to say which reacting species participates in the transport of matter through the layer of scale on MeX.

4. The concentration distribution of lattice defects in the oxidation product during its formation should be linear (the defect self-diffusion coefficient does not depend on the concentration).

Such an oxidation process takes place according to Tammann's parabolic rate law:

$$\frac{dx}{dt} = \frac{k'_p}{x} \tag{3.2.12}$$

where x is the thickness of the scale at time t and the parabolic reaction rate constant k'_p is related to the concentration of defects and their mobility by the following equation:

$$k'_p = \tilde{D} N_d \tag{3.2.13}$$

where $\tilde{D}$ is the chemical diffusion coefficient and N_d is the concentration in mole fractions of defects at the phase boundary where they are formed.

Thus the parabolic Tammann constant is the product of two important quantities, which can not be calculated separately on the basis of the results of classical kinetic studies of the oxidation of metals. Rosenburg has shown that, by performing the oxidation process in a special way, it is possible to determine the values of $\tilde{D}$ and N_d from a single measurement. This discovery considerably extends the range of physico-chemical methods used in studies on properties of solids.

In Rosenburg's method the following procedure is used: The metal under study is oxidised for a definite time, after which the thickness of the reaction product has the certain value x_0. A preliminary estimate of this thickness must made for the given system and reaction conditions on the basis of Rosenburg's theory. (This will be explained later in this book). Then the oxidation is terminated by removing the oxidant from the reaction chamber. Usually, it is not possible to decrease the oxidant pressure to a value equal to the dissociation pressure of the scale, since these pressures are of the order of 1.033×10^{-3}–1.0133×10^{-25} Pa (10^{-8}–10^{-30} atm) or even lower. However it is enough to decrease the pressure in the apparatus to a value of the order of 101.325×10^{-3} Pa (10^{-6} atm) and to close the apparatus tightly, since then reaction in the system decreases the partial pressure of component X_2 to the equilibrium value. The oxidation continues until the system Me–MeX–X_2 reaches thermodynamic equilibrium, i.e. until the time when the partial pressure of oxidant X_2 in the reaction zone reaches a value equal to the dissociation pressuie of compound MeX.

Under these conditions the concentration of lattice defects in the entire scale layer is uniform and corresponds to equilibrium with the metal phase. In the case of compounds with a deficit of metal or an excess of the oxidant ($Me_{1-y}X$ or MeX_{1+y}), this is the lowest concentration of defects at the given temperature and, in the case of compounds with an excess of metal or a deficit of the oxidant ($Me_{1+y}X$ or MeX_{1-y}), it is the highest possible concentration of defects. This situation is shown schematically in Fig. 178.

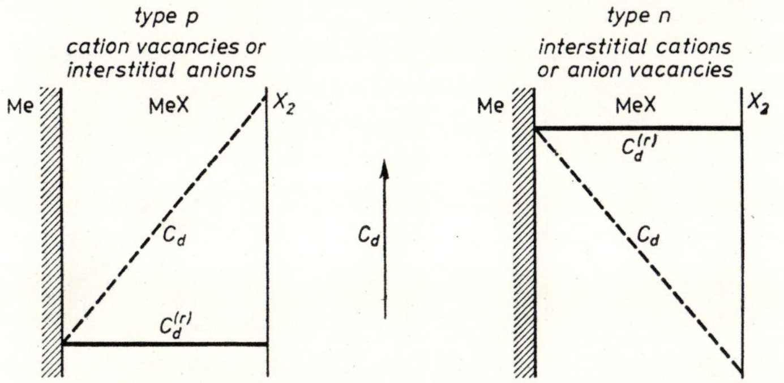

Fig. 178. The distribution of concentration of defects in the growing scale layer of the *p* or *n* type after the establishment of the thermodynamic equilibrium in the Me–MeX–X_2 system (after [269]). The broken line shows the distribution of defects in the reaction product before the interruption of the oxidation process.

Obviously the establishment of this state requires a definite time of heating, which also can be estimated for specified conditions on the basis of Rosenburg's theory.

After the homogenisation process, the oxidant is again introduced to the reaction chamber and the measurement of the rate of increase of the thickness or the mass of the product is started. Immediately after the second introduction of the oxidant, the process of increase of the thickness of the scale takes place in a different way from that during the period preceding the break in the reaction. This is because, during the homogenising heating, the scale was completely saturated with the metal, which, as has already been mentioned, in the case of semiconductors of the *p* type, corresponds to the minimum concentration of cation vacancies or interstitial anions and in the case of semiconductors of the *n* type — to the maximum concentration of interstitial cations or anion vacancies. When the oxidant is reintroduced in the reaction zone, the concentration of metal in the surface layer of the scale rapidly decreases as a result of the increase in the concentration of cation vacancies or interstitial anions (type *p*) or as result of the decrease in the concentration of interstitial cations or anion vacancies (type *n*).

During the reaction, the change in the concentration of defects extends more deeply and more deeply in the scale, as a result of diffusion processes, finally leading to the reestablishment of a linear concentration gradient in the scale, as existed before the interruption of the oxidation process. Figure 179

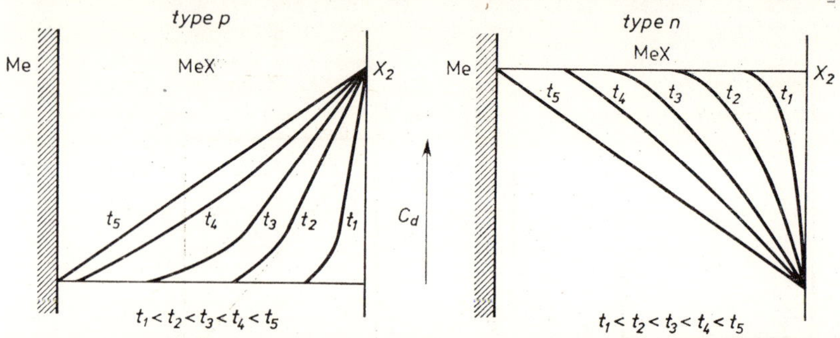

Fig. 179. Changes of the distribution of defects in the *p* or *n* type scale after the reintroduction of the oxidant to the reaction chamber (after [269, 766]). The horizontal lines show the distribution of defects before the introduction of the oxidant.

shows schematically the successive stages of changes in the distribution of concentration of defects in the scale after the reintroduction of the oxidant in the reaction chamber.

From the point of view of the theory of diffusion, the growth of the scale after restarting the oxidation reaction can be divided into two stages. In the initial stage the diffusion processes take place only in a part of the scale layer, while thermodynamic equilibrium exists on the Me–MeX phase boundary. This is the case of diffusion in a semi-infinite system. It should be mentioned that, during this stage of the reaction, only the scale layer is "oxidised" and the metallic base does not participate in the process. In the second stage, when the linear concentration gradient of defects is re-established in the scale, the steady state diffusion begins to operate (if the increase in the thickness of the scale during this stage in comparison with its original thickness, x_0, is neglected. Hence the condition: $x \ll x_0$ must be fulfilled).

Since it was assumed that coefficient $\tilde{D}$ is independent of the concentration of defects and that defects are formed and disappear only on phase boundaries, the changes of concentration of defects taking place in the scale after the reintroduction of the oxidant to the reaction chamber are related to the reaction time and to the position coordinate according to the following form of Fick's second law:

$$\frac{\partial C(x)}{\partial t} = \tilde{D}\frac{\partial^2 C(x)}{\partial x^2} \tag{3.2.14}$$

where $C(x)$ is the local concentration of defects at a distance x from the surface of the metal, t is the reaction time measured from the moment of reintroduction of the oxidant to the reaction chamber.

The description of diffusion processes, and consequently the course of the oxidation reaction, requires the solution of equation (3.2.14) for the following boundary conditions:

1. when $t = 0$, $C(x) = C_0$ for all values of x in the range from $x = 0$ to $x = x_0$ (thermodynamic equilibrium in the system Me–MeX–X);
2. when $t > 0$, $C(x) = C_0$ for $x = 0$ and $C(x) = C_1$ for $x = x_0$.

The solution of Fick's second law for these boundary conditions leads to the following expression:

$$C(x) = C_0 - \frac{(C_0 - C_1)^x}{x_0} +$$

$$+\frac{2}{\pi}\sum_{1}^{\infty}\frac{C_1\cos n\pi - C_0}{n}\sin\left(\frac{n\pi x}{x_0}\right)\exp(-D'n^2t)+$$

$$+\frac{4C_0}{\pi}\frac{1}{(2m+1)}\sin(2m+1)\frac{\pi x}{x_0}\exp[-D'(2m+1)^2t] \tag{3.2.15}$$

where $D' = \tilde{D}\pi^2/x_0^2$.

On differentiating function $C(x)$ with respect to x and calculating for $x = x_0$, Rosenburg obtained, after integration, the following expression for the relation between the reaction time and the number of defects which are formed or annihilated on unit surface area of the scale:

$$\Delta = \text{const}\left(\frac{\Delta m}{A}\right) = \tilde{D}\int_0^t\left(\frac{C}{x}\right)_{x=x_0}dt = \frac{\tilde{D}(C_1 - C_0)t}{x_0} +$$

$$+\frac{2x_0}{\pi^2}\sum_{1}^{\infty}[C_1 - C_0(-1)^n]\left[\frac{1}{n^2}\right]\exp(-D'n^2t - 1)+$$

$$+\frac{4C_0x_0}{2}\sum_0^\infty \frac{1}{(2m+1)^2}\{\exp[-D'(2m+1)^2t]-1\} \tag{3.2.16}$$

where $\left(\frac{\Delta m}{A}\right)$ is the mass of the oxidant combining with unit surface area of the scale.

When the process takes place at pressures much greater than the dissociation pressure of the scale, the concentration of defects at the phase boundary where they are formed is many times greater than their concentration in the region where they disappear, i.e. $C_1 \ll C_0$. In this case equation (3.2.16) has the following form:

$$\Delta = \frac{2x_0C_0}{\pi^2}\sum_1^\infty \frac{(-1)^2}{n^2}+\frac{4x_0C_0}{\pi^2}\sum_0^\infty \frac{1}{(2m+1)^2}+$$

$$+\frac{x_0C_0D't}{\pi^2}-\frac{2x_0C_0}{\pi^2}\sum_1^\infty \frac{(-1)^n}{n^2}\exp(-n^2D't)-$$

$$-\frac{4x_0C_0}{\pi^2}\sum_0^\infty \frac{1}{(2m+1)^2}\exp[-(2m+1)^2D't] \tag{3.2.17}$$

Equation (3.2.17) describes the course of the oxidation process after the reintroduction of the oxidant in the reaction chamber, since the number of defects which are created or annihilated on the outer surface of the scale

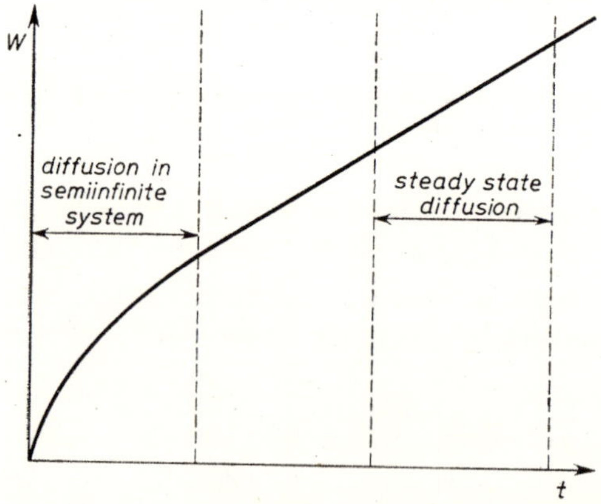

Fig. 180. The scheme of the course of the oxidation process after the reintroduction of the oxidant to the reaction chamber (after [269]). (W stands for the mass of oxidant combining with the unit surface area of the sample).

is directly proportional to the amount of the oxidant combining with this surface $\left[\Delta = \text{const}\left(\frac{\Delta m}{A}\right)\right]$. The numerical solution of this equation for assumed values of the parameters D and C_0 leads to a relationship between $\Delta m/A$ and t shown schematically in Fig. 180.

It follows from the figure that, in the initial stage of the reaction, the growth of the scale layer is parabolic, but in later on it becomes linear. The paralinear part of the curve is understandable in view of the assumptions made in the solution of Fick's second law. In the initial stage the scale layer behaves as a semi-infinite diffusion system and since the chemical diffusion coefficient for defects is independent of their concentration, the relation between the mass of the combined oxidant and the reaction time can be expressed by the following equation:

$$\frac{\Delta m}{A} = \text{const} = \text{const}'\sqrt{\tilde{D}t} \tag{3.2.18}$$

Later on, the concentration gradient of defects in the scale becomes practically constant, since the increase of the scale thickness becomes negligibly small in comparison with the value of x_0 ($\Delta x \ll x_0$). This means that the course of the reaction is linear. In agreement with this conclusion, Rosenburg has shown that his general equation (3.2.17) has two simple limiting solutions:

$$\left(\frac{\Delta m}{A}\right)_1 = (1.128\,C_0\sqrt{\tilde{D}})\sqrt{t}, \quad \text{when } t \ll \frac{x_0^2}{D} \tag{3.2.19}$$

and

$$\left(\frac{\Delta m}{A}\right)_2 = \left(\frac{\tilde{D}C_0}{x_0}\right)t + \frac{x_0 C_0}{3}, \quad \text{when } t > \frac{x_0^2}{2D} \tag{3.2.20}$$

Equation (3.2.19) describes the initial parabolic section of the curve shown in Fig. 180 with an accuracy of about 1%, when the condition $t \ll x_0^2/D$ is fulfilled. Equation (3.2.20) characterises the linear course of the reaction for times longer than $x_0^2/2D$.

These equations can be expressed in a slightly different form, which is more suitable for graphical interpretation:

$$\left(\frac{\Delta m}{A}\right)_1 = k_p\sqrt{t} + c_p \tag{3.2.21}$$

$$\left(\frac{\Delta m}{A}\right)_2 = k_1 t + c_1 \tag{3.2.22}$$

where the parabolic rate constant k_p is the slope of the straight line obtained by plotting $(\Delta m/A)_1$ against $\sqrt{t}$ and the linear constant k_l is the slope of the straight line obtained by plotting $(\Delta m/A)_2$ against t. Constants c_l and c_p are the intercepts on the axis of ordinates of extrapolated plots of $(\Delta m/A)_2$ against t and $(\Delta m/A)_1$ against $\sqrt{t}$ respectively.

When the growth of the scale layer after the re-introduction of the oxidant takes place from the very beginning according to the parabolic rate law, the plot of $(\Delta m/A)_1$ against $\sqrt{t}$ passes through the origin and $c_p = 0$. However, as a rule, various factors disturb the parabolic course of the reaction at the beginning. Such disturbances are caused by the still unstabilised partial pressure of the oxidant on the MeX–X_2 phase boundary, chemisorption processes, etc. In this case the plot of $(\Delta m/A)_1$ against $\sqrt{t}$ does not pass through the origin and as a result the constant, c_p, in equation (3.2.21) has appreciable positive or negative values depending on the factors disturbing the course of the reaction in the initial stage. These values are taken into account in further calculations.

Since k_p and k_l, as well as c_p and c_l, can be determined by means of one measurement, the two unknown quantities $\tilde{D}$ and C_d can be calculated from the following equations:

$$\tilde{D} = \left[\frac{1.128 k_l x_0}{k_p}\right]^2 \qquad (3.2.23)$$

and

$$C_d = \frac{\left[\dfrac{k_p}{1.128}\right]^2}{k_l x_0} \qquad (3.2.24)$$

The concentration of defects can be also calculated by means of the following equation:

$$C_d = \frac{3(c_l - c_p)}{x_0} \qquad (3.2.25)$$

where C_d is the equilibrium concentration of defects on the phase boundary at which they are formed. In the case of semiconductors of the p type, C_d is the concentration of cation vacancies or of interstitial anions in the crystal remaining in equilibrium with the oxidant at the pressure during the second oxidation. In the case of semiconductors of the n type, C_d is the concentration of interstitial cations or anion vacancies in equilibrium with the metallic phase.

In certain cases, the value of C_d calculated in this way requires correction. When the crystal contains cation vacancies or interstitial cations, the equilibrium concentration of defects is equal to the calculated value of C_d only if each atom of combined oxidant leads to formation of one vacancy or the elimination of one interstitial cation, i.e. only in the case of compounds of the Me_1X_1 type (NiO, CoO, ZnO etc.), since the concentration of defects is calculated from the amount of oxidant that reacts with unit surface area of the scale. In the case of compounds such as Cu_2O or Al_2O_3, this condition is not fulfilled, since in the case of cuprous oxide the incorporation of one oxygen atom causes the formation of two cation vacancies and in the case of alumina three oxygen atoms cause the elimination of two interstitial cations. Hence the calculated value of C_d should be multiplied by the coefficient resulting from the stoichiometric formula of the scale compound. In general, for compounds of the Me_aX_b type this coefficient is equal to a/b.

When the defects are present in the anion sublattice, the calculated value of C_d does not require any corretions irrespective of the stoichiometric formula of the compound and is equal to the real concentration of defects. In this case, either an anion vacancy becomes filled with an atom (or an ion) of the oxidant, or this atom (or ion) becomes incorporated in an interstitial position. Thus one atom of the oxidant results in the formation or the annihilation of one lattice defect irrespective of the stoichiometric formula of compound MeX. The final conclusions can be expressed in the following form:

1. When the defects are formed at the scale/gas interface in the case of cation vacancies ($Me_{1-y}X$), the concentration of defects $C_d = a/b\, C'_d$ and in the case of interstitial anions (MeX_{1+y}), $C_d = C'_d$, where C'_d is the value of the concentration of defects resulting directly from the measurements.

2. When the defects are formed on the metal/scale interface then in the case of interstitial cations ($Me_{1+y}X$) the concentration of defects $C_d = a/b\, C'_d$ and in the case of compounds MeX_{1-y}, $C_d = C'_d$.

In order to check the correctness of the calculations of C_d and $\tilde{D}$ by means of the approximate equations (3.2.21) and (3.2.22) the values of C_d and $\tilde{D}$ should be substituted in the generalised form of Rosenburg's equation in order to calculate the function $(\Delta m/A) = f(t)$. If the values of $(\Delta m/A)$ obtained in this way as a function of t are in agreement with the experimental curve, the values of C_d and $\tilde{D}$ can be regarded as having been calculated correctly. Lack of agreement indicates that the values of k_p and k_l were determined with incorrect time intervals.

The solution of the general equation (2.3.17) for each measurement of the values of C_d and $\tilde{D}$ would be very tedious. For this reason Rosenburg solved his general equation numerically and obtained the numerical values shown in Table 13.

It should be stressed that in the check calculations, it is necessary to take into account the correction c_p, since Rosenburg's solution was derived for the undisturbed reaction. Finally it should be pointed out that the

TABLE 13

The results of the numerical solution of Rosenburg's general equation

$\frac{D_d t}{x_0^2}$	$\left(\frac{D_d t}{x_0^2}\right)^{1/2}$	$\frac{\Delta\pi^2}{x_0 D_d}$
0.02026	0.1422	0.159
0.03040	0.1744	0.195
0.04053	0.2011	0.226
0.05066	0.2250	0.253
0.06079	0.2466	0.276
0.07092	0.2664	0.299
0.08115	0.2845	0.320
0.09118	0.3020	0.339
0.1013	0.3183	0.357
0.1114	0.3338	0.375
0.1216	0.3485	0.392
0.1317	0.3629	0.408
0.418	0.3765	0.423
0.1621	0.4026	0.453
0.1824	0.4272	0.480
0.2026	0.4501	0.507
0.2229	0.4720	0.582
0.2432	0.4930	0.556
0.2837	0.5325	0.603
0.3242	0.5694	0.647
0.3647	0.6038	0.691
0.4053	0.6366	0.733
0.5066	0.7117	0.837
0.6079	0.7795	0.939
0.7092	0.8422	1.041
0.8105	0.9002	1.142
0.9118	0.9549	1.243
0.0132	1.0065	1.344

determination of correct values of C_d and $\tilde{D}$ by this method is possible only when the following conditions are fulfilled:

1. The layer of the oxidation product must be compact in the whole cross-section and must adhere to the metallic core during all the successive stages of the measurements.

2. During reaction in steady state conditions local thermodynamic equilibrium must exist at both phase boundaries.

3. The growth of the scale layer must follow the parabolic rate law.

4. The surface of the layer must be microscopically smooth (the roughness coefficient $f \cong 1$).

5. The time that the metal sample is heated in vacuo after preliminar oxidation, in order to homogenise it, must exceed $x_0^2/\tilde{D}$.

6. The determination of the parabolic rate constant k'_p must be carried out during a time much shorter than $x_0^2/\tilde{D}$ and the determination of the linear constant k_l during a time longer than $x_0^2/2\tilde{D}$. The studies can therefore be planned rationally only, when the order of magnitude of $\tilde{D}$ is known. When this value cannot be estimated by any method, the corresponding thickness x_0 must be determined on the basis of preliminary measurements.

7. The value of $C_d x_0/3$ must be appreciably greater than the sensitivity of the measurements, since it follows from equation (3.2.20) that this expression corresponds to the intercept of the extrapolated linear part of function $(\Delta m/A) = f(t)$ on the y axis (see Fig. 180). Thus the time of the parabolic course of the reaction increasses with increasing thickness of the preliminary scale layer x_0 and increasing concentration of defects. This leads to the conclusion that the thickness x_0 must be always chosen so that it is possible to determine the value of k_p with the required accuracy. On the other hand the parabolic course of the reaction should not last too long, since the increase of the scale thickness of layer during the entire kinetic run must be much smaller than the initial thickness x_0.

Rosenburg has pointed out that with present measuring techniques, the concentration of defects can still be determined when their number is 10^{15} per cm^3, i.e. of the order of 10^{-4} at.%. These values are at least two orders of magnitude smaller than those determinable by other methods.

Another important advantage of Rosenburg's method results from the possibility of repeating the measurement of the values of C_d and $\tilde{D}$ several times on the same sample without removing it from the apparatus. After the first determination of the values of k_p and k_l at a given thickness x_0

the oxidant can be removed from the reaction chamber and after a suitable period heating to homogenese the sample, the whole measuring procedure can be repaeted taking into account the inceased value of x_0. In this way the measurement can be repeated several times. This gives a whole family of curves as shown schematically in Fig. 181. In the initial part they form

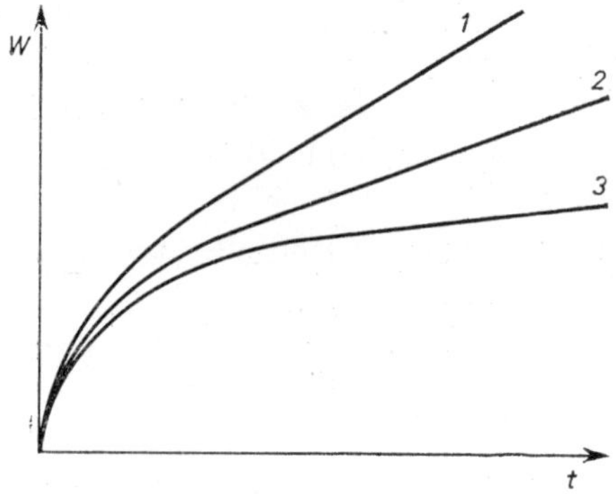

Fig. 181. The course of the oxidation after the reintroduction of the oxidant to the reaction chamber (after [269]). Curves *1*, *2*, and *3* correspond to different thicknesses of primary layer x_0.

one curve, which characterises the course of the change of concentration of defects in the surface part of the layer, which is obviously independent of the thickness x_0. When the initial thickness of the scale x_0 is larger, the time of the parabolic course of the reaction is longer, since the scale behaves like a semi-infinite system for a longer time. When the steady state is reached, the course of the reaction becomes linear, but the larger the thickness of the layer of scale the slower the rate is and as a result the angle between the straight lines and the time axis becomes smaller (Fig. 181).

In order to avoid misunderstandings it should be mentioned that Rosenburg did not use the chemical diffusion coefficient $\tilde{D}$ in his work but the coefficient of defect diffusion D_d. This was wrong, since D_d corresponds to conditions without gradient, whereas in the growing scale a concentration gradient of defects is obviously present and the diffusion process takes place under the influence of this gradient. It was shown in Chapter 2.6 that, under these conditions, the mobility of defects is determined by the chemical diffusion coefficient which is larger that the coefficient of diffusion of defects, since defects have, as a rule, definite effective charges (see equation (2.6.4)).

Rosenburg's method has been applied several times to various systems, but the results have not been compared with those obtained by classical methods. Recently this was done by Fryt, Mrowec and Walec [187, 188, 269] for the Co–$Co_{1-y}O$–O_2 system. This system was chosen, since cobaltous oxide is the only material in which the structure and the concentration of defects as well as their mobility have been thoroughly studied. It shows a relatively large deviation from stoichiometry, which is caused by defects in the cationic sublattice ($Co_{1-y}O$). For this reason the concentration of defects and its dependence on temperature and equilibrium oxygen pressure could be determined with a relatively high accuracy by the thermogravimetric method. The chemical and self-diffusion coefficients were also determined with equally high accuracy as a function of temperature and oxygen pressure. This was possible because on one hand the mobility and concentration of cation vacancies in $Co_{1-y}O$ are relatively high and on the other the very long half life time of the radioactive cobalt isotope ^{60}Co facilitates the accurate determination of D_{Co}.

The studies were carried out in a simple thermogravimetric device, which made it possible to measure continuously the increases of the mass of the oxidised sample with an accuracy of the order of 10^{-4} g. The required partial pressure of oxygen in the reaction chamber was obtained by means of a suitable mixture of argon and oxygen which flowed through the apparatus under the pressure of 1013.25 hPa (1 atm) at the rate of 20 dm^3/h. The studies were carried out in the temperature range 1000–1300°C and oxygen pressure range 303.975 Pa–101.325 kPa (3×10^{-3}–1 atm).

The choice of the temperature and oxygen pressure ranges ensured the formation of a homogeneous scale consisting exclusively of the $Co_{1-y}O$ phase.

In agreement with the results obtained by other authors it was found that in these conditions oxidation of cobalt took place according to the parabolic rate law and the growth of the compact, homogeneous scale took place as a result of outward diffusion of metal (the marker situated at the metal/scale interface).

The thickness of the primary oxide layer x_0 was estimated on the basis of the values of $\tilde{D}$ determined by Price and Wagner [270]. It varied, depending on the reaction conditions, from 0.055 to 0.083 cm. The scale layers of the required thickness x_0 were obtained by oxidation of a sample of metal at 1100 or 1200°C in air. Then purified argon was passed through the reaction chamber maintained at the temperature of the intended measurement. The partial pressure of oxygen in the purified argon was of the order of the dissociation pressure of CoO. In these conditions oxidation was completely stopped and the mass of the sample was constant.

After homogenisation a mixture of argon and oxygen, in which the partial pressure of oxygen had the required value, was passed through the reaction chamber and the measurements of the kinetics of further growth of the scale were started. After one cycle of measurements the system was again equilibrated and the measurements were repeated under another oxygen pressure. This procedure was repeated over the whole temperature range, using the same sample of metal. The experiment was repeated several times using different samples.

Figure 182 illustrates a set of curves obtained at 1200°C for several oxygen pressures. Analogous results were obtained at other temperatures.

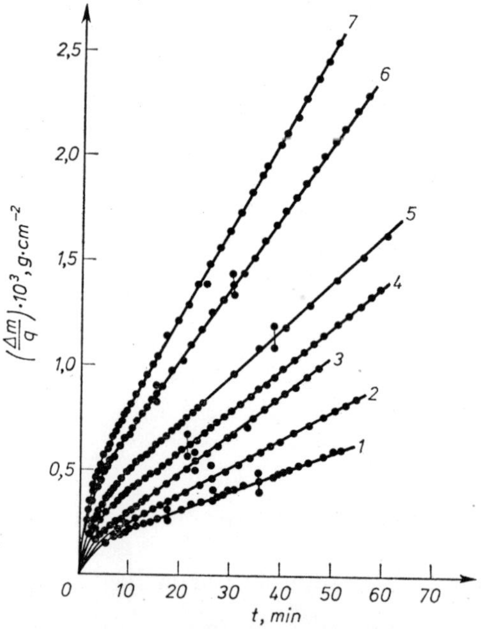

Fig. 182. The kinetics of the growth of $Co_{1-y}O$ layer after the reintroduction of oxygen to the reaction chamber at 1200°C (after [187]). The curves correspond to the following oxygen pressures: 101.325 kPa–354.64 Pa (1–3.5×10^{-3} atm), 202.65 kPa–820.733 Pa (2–8.1×10^{-3} atm), 303.975–3.101 kPa (3–3.06×10^{-2} atm), 405–8.196 kPa (4–8.0×10^{-2} atm), 506.63–21.278 kPa (5–2.1×10^{-1} atm), 607.95–50.663 kPa (6–5.0×10^{-1} atm), 709.275–101.325 kPa (7–1.0 atm). The end of the parabolic course of the reaction is marked on each curve with ↕ and the beginning of the linear course with ↔.

It follows from the graph that, in accordance with expectation, the curves are paralinear. The values of the constants k_l were calculated from the slopes of the linear parts of the plots and the values of the constants C_l from the intercepts of the extrapolated straight lines on the ordinate axis.

In order to calculate the values of k_p and C_p, the results of the measurements obtained in the initial oxidation period were plotted using parabolic coordinates ($\Delta m/A$ vs $\sqrt{t}$).

It follows from the graph shown in Fig. 183 that with these coordinates the course of the oxidation is linear. This makes it possible to determine

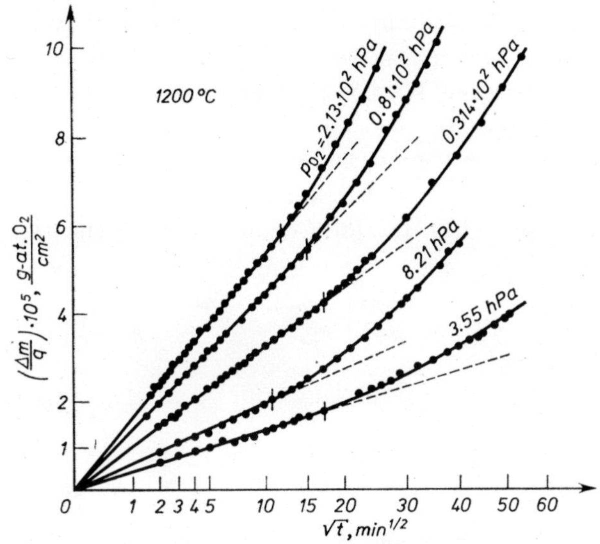

Fig. 183. The course of oxidation of cobalt at 1200°C after reintroduction of oxygen to the reaction chamber shown in the parabolic system (the initial stage) (after [183]).

the value of k_p (in this case $C_p = 0$). In later stages of the oxidation the experimental points deviate from linearity as a result of the gradual change of the reaction kinetics from parabolic to linear.

The values of k_p and k_l determined in this way, as well as the values of C_p and C_l, were used for the calculation of the dependence of the concentration of cation vacancies and chemical diffusion coefficients on temperature and equilibrium oxygen pressure. It was found that the concentration of defects was the following function of temperature and equilibrium oxygen pressure:

$$N_{V_{Co}} = 1.26 p_{O_2}^{1/3.85} \exp\left(\frac{12500}{RT}\right) \tag{3.2.26}$$

The chemical diffusion coefficient is independent of oxygen pressure and is the following function of temperature:

$$\tilde{D}_{CoO} = 6.6 \times 10^{-3} \exp\left(-\frac{24400}{RT}\right) [\mathrm{cm}^2/\mathrm{s}] \tag{4.2.27}$$

Fisher and Tannhauser [38, 39], Eror and Wagner [191] and Koel and Gellings [271] have determined the dependence of the equilibrium concentration of cation vacancies in cobaltous oxide on temperature and oxygen pressure on the basis of the deviation of this oxide from stoichiometry by the gravimetric method. Price and Wagner [270] and Koel and Gellings [271] have determined the temperature dependence of the chemical diffusion coefficient of this oxide by the thermogravimetric method and on the basis of the results of measurements of electrical conductivity. The results of these studies can be expressed in the form of the following empirical equations:

$$N_{Co} = 0.94 \times 10^{-1} p_{O_2}^{1/3.86} \exp\left(-\frac{7200}{RT}\right) \tag{3.2.28}$$

(Tannhauser and Fisher)

$$N_{Co} = 0.95 \times 10^{-1} p_{O_2}^{1/3.8} \exp\left(-\frac{6500}{RT}\right) \tag{3.2.29}$$

(Error and Wagner)

$$N_{Co} = 1.45 \times 10^{-1} p_{O_2}^{1/4.0} \exp\left(-\frac{7800}{RT}\right) \tag{3.2.30}$$

(Koel and Gellings)

$$D_{CoO} = 4.33 \times 10^{-3} \exp\left(-\frac{24000}{RT}\right) [\mathrm{cm}^2/\mathrm{s}] \tag{3.2.31}$$

(Price and Wagner)

$$D_{CoO} = 9.3 \times 10^{-3} \exp\left(-\frac{24000}{RT}\right) [\mathrm{cm}^2/\mathrm{s}] \tag{3.2.32}$$

(Koel and Gellings)

Comparison of equation (3.2.27) with equations (3.2.31) and (3.2.32) indicates that the results of the determination od chemical diffusion coef-

ficients obtained by Rosenburg's kinetic method are in excellent agreement with those obtained by the classical methods. In the case of determinations the concentration of defects, this agreement is limited to the effect of oxygen concentration on the concentration of defects (the exponents at the oxygen pressure in equations (3.2.26) and (3.2.28)–(3.2.30) are practically identical). The enthalpy of formation of defects determined by the kinetic method (equation (3.2.26) is considerably higher than the enthalpy determined by the classical methods. This is due to the fact that Rosenburg's method gives the total concentration of defects, whereas the other methods give only the concentration of the predominant defects, which give rise to the deviation from stoichiometry.

It has been shown in Chapter 2.6 that the self-diffusion coefficient of metal or oxygen in compounds MeX is related to the chemical diffusion coefficient and the concentration of defects according to simple equation (2.6.5). In the case discussed this equation can be expressed in the following form:

$$D_{\mathrm{Co}} = \frac{D_{\mathrm{CoO}} N_{\mathrm{Co}}}{(1+p)} \qquad (3.2.33)$$

In the range of temperature and oxygen pressure examined the cationic vacancies in cobaltous oxide are singly ionised (the oxygen pressure exponent in equations (3.2.26) and (3.2.28)–(3.2.30) is close to 1/4). This means that $p = 1$. Equation (3.2.33) is therefore reduced to the following form:

$$D_{\mathrm{Co}} = \tfrac{1}{2}\tilde{D}_{\mathrm{CoO}} N_{\mathrm{Co}} \qquad (3.2.34)$$

On introducing the empirical equations (3.2.26) and (3.2.28) in equation (3.2.34) we obtain the following equation expressing the dependence of the self-diffusion coefficient of cobalt in cobaltous oxide on oxygen pressure and temperature:

$$D_{\mathrm{Co}} = 8.3\times 10^{-3} p_{\mathrm{O_2}}^{1/3.85} \exp\left(-\frac{37800}{RT}\right) [\mathrm{cm^2/s}] \qquad (3.2.35)$$

As it has already been mentioned, the self-diffusion coefficient of cobalt in $Co_{1-y}O$ was very exactly determined in a wide range of temperature and oxygen pressure by Carter and Richardson by means of the radioisotopic method and recently also by Chen, Peterson and Reeves. Since Chen and his collaborators carried out their measurements at only one oxygen pressure equal to 212.78 hPa (0.21 atm), the comparison the results

obtained by the kinetic method with those obtained by the radioisotopic method was limited to this oxygen pressure. In this case the empirical equation (3.2.35) is reduced to the following form:

$$D_{Co} = 5.2\times10^{-3}\exp\left(-\frac{37800}{RT}\right)\ [\mathrm{cm}^2/\mathrm{s}] \tag{3.2.36}$$

(Mrowec and collab., Rosenburg's kinetic method).

The results obtained by Chen and collab. and by Carter and Richardson for the same oxygen pressure [p_{O_2} = 21.278 kPa (0.21 atm)] can be expressed in the following form:

$$D_{Co} = 5.0\times10^{-3}\exp\left(-\frac{38400}{RT}\right)\ [\mathrm{cm}^2/\mathrm{s}] \tag{3.2.37}$$

(Chen and collab., radioisotope method)

$$D_{Co} = 5.2\times10^{-3}\exp\left(-\frac{38000}{RT}\right)\ [\mathrm{cm}^2/\mathrm{s}] \tag{3.2.38}$$

(Carter and Richardson, radioisotope method).

The equation expressing the temperature dependence of D_{Co} for oxygen pressure equal to 21.278 kPa (0.21 atm) derived by the Fueki–Wagner method (Chapter 3.3.2) should also be mentioned here:

$$D_{Co} = 5.0\times10^{-3}\exp\left(-\frac{37800}{RT}\right)\ [\mathrm{cm}^2/\mathrm{s}] \tag{3.2.39}$$

(Mrowec and Przybylski, the kinetic Fueki–Wagner method).

Comparison of equation (3.2.27) with equations (3.2.31) and (3.2.32) as well as equations (3.2.36) and (3.2.39) with equations (3.2.37) and (3.2.38) leads to the conclusion that the results concerning chemical diffusion coefficient and self-diffusion coefficient obtained by the kinetic methods are in surprisingly good agreement with those obtained by the classical methods. This agreement exists not only to the activation energy of diffusion of cation vacancies and cobalt ions in the oxide, but also to the absolute values of the coefficients $\tilde{D}_{CoO}$ and D_{Co} over the whole range of temperature and oxygen pressure (the numerical values of preexponential factors in equations (3.2.36)–(3.2.39) are practically identical). Since the kinetic methods are incomparably simpler than the radioisotopic ones and in addition provide more accurate results, it is obvious that they should be widely used.

3.2.6 The coulometric titration method

The concentration of point defects in crystals of MeX compounds can be in some cases determined with great accuracy by the coulometric titration method [274]. This method can be used when the predominant lattice disorder is connected with deviations from stoichiometry and when solid electrolyte suitable for the system under study is available. The method consists in the electrochemical determination of the amount of oxidant admitted to the space occupied by the compound under study or the amount of oxidant removed from this space and simultaneous determination of its activity (pressure) in this space. The method requires a system for solid state electrolysis (coulometric titration), which could also act as a solid state cell. Since compounds showing deviations from stoichiometry are electronic semiconductors, the construction of such a system requires blocking of the flow of electron with ionic semiconductor (solid electrolyte). In the case of oxides the zirconium dioxide doped with calcium oxide is usually used. This solid solution shows practically pure anionic conductivity over a wide range of temperature and oxygen pressure (it is a solid electrolyte). In the case of sulphides suitable anionic semiconductors have not yet been found. Only solid electrolytes containing defects in the cation sublattice, i.e. cationic semiconductors are known. Such electrolytes, unlike anionic semiconductors, cannot block electrons in all sulphide cells, but only those where the solid electrolyte and the sulphide under study have common cations.

In the case of oxides, the cell can be represented by the following scheme:

$$\mathrm{Pt}|p'_{O_2}|\mathrm{MeO}|\mathrm{ZrO_2(CaO)}|p''_{O_2}|\mathrm{Pt} \qquad (3.2.40)$$

$$\xrightarrow{\hspace{6em}}$$
$$O^{2-}$$
$$\xleftarrow{\hspace{6em}}$$

The construction on this system must ensure the complete isolation of the two electrode compartments from one another, in order to make it possible to use any oxygen pressures p'_{O_2} and p''_{O_2} on the two sides of the cell. One of the electrodes in this cell is the compound, MeO, in equilibrium with oxygen at pressure p'_{O_2} and the other electrode is oxygen at pressure p''_{O_2}. Since the electrodes are separated from one another by the solid electrolyte, the potential of the cell is related to oxygen pressure according to the following equation:

$$E = -\frac{RT}{nF}\log\left(\frac{p''_{O_2}}{p'_{O_2}}\right) \qquad (3.2.41)$$

where F is the Faraday constant, n is the number of electrons exchanged

in the electrode processes (in this case $n = 4$), R is the gas constant and T is the absolute temperature.

The pressure of the gas reference electrode p''_{O_2} is fixed [e.g. 1013.25 hPa (1 atm)] and hence the measurement of the potential of the cell makes it possible to determine the pressure p'_{O_2} which is established in the left hand electrode compartment, when the system MeO–O_2 reaches thermodynamic equilibrium. Obviously such measurements must be carried out at a sufficiently high temperature at which the establishment of the equilibrium is possible.

It is convenient to start the investigation by making the pressures of oxygen equal ($p'_{O_2} = p''_{O_2}$). This can be done by short circuiting the cell for a short time, which causes spontaneous transport of oxygen ions throught the solid electrolyte and electrons in the outside circuit until the pressures of oxygen in the two electrode zones become equal. This is equivalent to the decrease of the cell potential to zero.

Then the cell is connected to an external source of direct current. As a result electrolysis (coulometric titration) takes place. If the positive pole is connected to the gas electrode (on the right), the oxygen in the left hand electrode compartment surrounding compound MeO is reduced on the surface of the solid electrolyte and migrates through anion vacancies of the solid electrolyte in the form of ions to the gas electrode, where it is reoxidised to O_2 on the platinum contact. Consequently the amount of oxygen and hence its pressure in the space surrounding the compound under study is decreased. When the flow of current through the system is interrupted, after some time the oxygen pressure becomes stabilised at a definite level corresponding to the new equilibrium in the MeO–O_2 system. The establishment of this equilibrium is manifested by a constant value of the potential of the cell, which also makes it possible to calculate the oxygen pressure at the left hand electrode, corresponding to the new defect equilibrium.

When the time during which the current flows and its intensity are known, the amount of oxygen titrated from the space of the left hand electrode can be calculated:

$$\Delta n_{O_2} = \frac{IT}{4F} \tag{3.2.42}$$

where Δn_{O_2} is the number of moles of oxygen, I is the current intensity and t is the time of its flow.

In order to calculate the amount of oxygen liberated from the oxide as a result of the establishment of the new equilibrium of defects, it is necessary to subtract from Δn_{O_2} the number of moles of oxygen, which left the cathode compartment and caused the decrease of oxygen pressure Δn_{O_2} (gas). This value can be calculated when the volume V occupied by the gas in the cathode compartment and the difference of pressures are known:

$$\Delta n_{O_2}(\text{gas}) = \left(\frac{V}{RT}\right) \Delta p \tag{3.2.43}$$

where Δp is the difference between oxygen pressures existing before and after the titration. Thus the number of moles of oxygen released by the oxide Δn_{O_2} (sample) is:

$$\Delta n_{O_2}(\text{sample}) = \Delta n_{O_2} - \Delta n_{O_2}(\text{gas}) \tag{3.2.44}$$

This value can be used for calculating the change of the stoichiometric composition of the oxide by means of the following equation:

$$\Delta y = \left[\frac{2M\Delta n_{O_2}(\text{sample})}{m}\right] \tag{3.2.45}$$

where M is the molecular weight of oxygen, m is the mass of the sample and y is the deficit or excess of oxygen with respect to the stoichiometric composition, which depends on the type of defect present in the lattice ($Me_{1-y}O$, MeO_{1+y} or $Me_{1+y}O$, MeO_{1-y}).

In order to determine the absolute values of y, it is necessary to carry out a series of measurements for obtaining the dependence of this value on the equilibrium oxygen pressure. It was shown in Chapter I of this work that the deviation from stoichiometry can increase or decrease with increasing oxygen pressure according to the following equation:

$$y = \text{const}\, p_{O_2}^{\pm 1/n} \tag{3.2.46}$$

where the exponent $1/n$ is positive in the case of compounds of the $Me_{1-y}O$ or MeO_{1+y} type and is negative in the case of compounds of the $Me_{1+y}O$ or MeO_{1-y} type. The absolute value of this coefficient determines the degree of ionisation of the defects.

The measured value, Δy, is the difference between the two values of y corresponding to the different equilibrium oxygen pressures, one of which (the initial one) is the pressure at the reference electrode. By using equation (3.2.46) we can therefore express Δy by the following equation:

$$\Delta y = \text{const}[p_{O_2}^{\pm 1/n} - p_{O_2}^{\pm 1/n}(B)] \tag{3.2.47}$$

where $p_{O_2}(B)$ is the oxygen pressure at the reference electrode.

In order to calculate the values of the exponent $1/n$ in equation (3.2.47), it is necessary to carry out a series of measurements of Δy for various oxygen pressures and to determine the dependence of Δy on oxygen pressure by the method of successive approximations. The determination of the numerical value of this coefficient is equivalent to the determination of the degree of ionisation of defects. The proportionality constant in equation (3.2.47) can be calculated from the slope of the straight lines obtained in the relation $\Delta y = f[p_{O_2}^{\pm 1/n} - p_{O_2}^{\pm 1/n}(B)]$.

Introducing this value in equation (3.2.46) we can finally calculate the dependence of the absolute value for deviation from stoichiometry (the concentration of defects) on oxygen pressure. By repeating the above procedure for several temperatures it is possible to determine also the entropy and enthalpy of formation of defects.

The above method was used for investigating the thermodynamics of defects in nickel oxide, cobaltous oxide, ferrous oxide and uranium di-

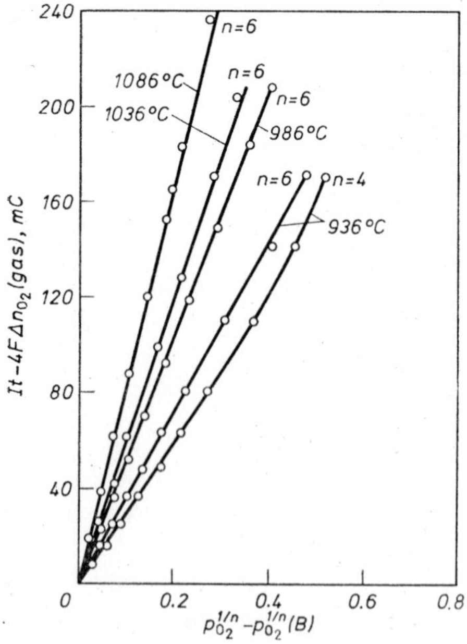

Fig. 184. The results of coulometric titration of nickel oxide as a function of oxygen pressure for several temperatures (after [275]); [n is the denominator of the exponent $\frac{1}{n}$ in equation (3.2.46)].

oxide. Figure 184 illustrates the results of coulometric titration of nickel oxide at several temperatures [275]. It follows from the graph that only in the case when the exponent $1/n$ is equal to $1/6$ is the function linear. This means that in nickel oxide over the range of temperature and oxygen pressure examined doubly ionisedcation vacancies predominate.

The sensitivity of this method is very high, on condition that the measurements are carried out in suitable conditions, in particular in an apparatus where the reaction chamber containing the sample is hermetically sealed, since only then does the coulometric titration make it possible to determine the real amount of oxygen released or absorbed by the sample. The sample must not touch the solid electrolyte, since a chemical reaction between the two solid phases can affect the results of the determinations.

Figures 185, 186 and 187 show schematically three modern devices for the coulometric titration of metal oxides, developed by Steel [276], Schmalzried [277] and Rapp [275]. They ensure the fulfillment of the above conditions. In all these devices the sample compartment is made air tight by

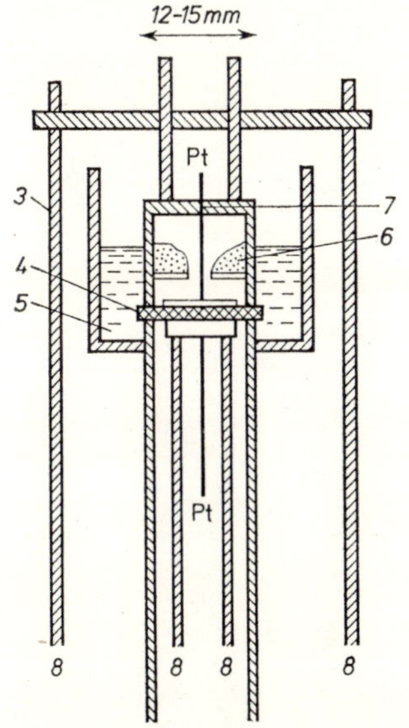

Fig. 185. The scheme of the apparatus for measuring the deviations from stoichiometry of metal oxides by the coulometric titration method (after [276]). *1*—the reference electrode, *2*—Pt-O_2 electrode, *3*—alundum tube, *4*—solid electrolyte, *5*—liquid electrolyte (e.g. sodium silicate), *6*—sample, *7*—metal-ceramic washer, *8*—pressing springs.

means of sealed glass joints. In all of them the sample is out of direct contact with the solid electrolyte. The simplest and the best seems to be the device developed by Rapp (Fig. 187). In this apparatus the oxide sample (in finally crystalline form) is placed in a platinum net situated at a contact made of the same metal. The space surrounding the sample is minimised by means of a special vacuum tight alundum tube, which is closed at the bottom by the solid electrolyte. The contact between the tube and the solid electrolyte is made air tight by means of Pyrex glass which makes perfect joints with these materials. Since the sample is powdered and the electrode space is small, the equilibrium state in the system oxide–oxygen is established in a relatively short time. This apparatus was used

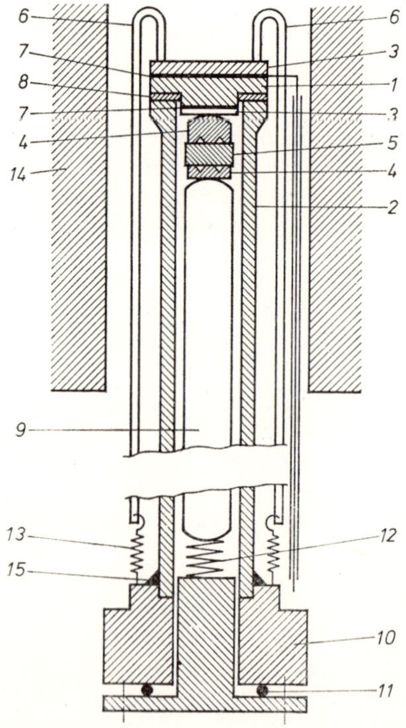

Fig. 186. Apparatus for studying deviations from stoichiometry of metal oxides by the coulometric titration method (after [277]); *1*—solid electrolyte, *2*—quartz tube, *3*—alundum disc, *4*—alundum clamps for the sample, *5*—sample, *6*—quartz rods, *7*—platinum contacts, *8*—layer of high melting glass, *9*—quartz tube closed on both ends, *10*—metal block, *11*—rubber ring (shock absorber), *12*—spring, *13*—fixing springs, *14*—resistance furnace, *15*—welding material.

in the investigation of the thermodynamics of defects in nickel oxide shown in Fig. 184.

All the methods of determination of concentrations of defects described above consist in direct or indirect determination of deviations from stoichiometry. The only exception is Rosenburg's kinetic method. It was shown

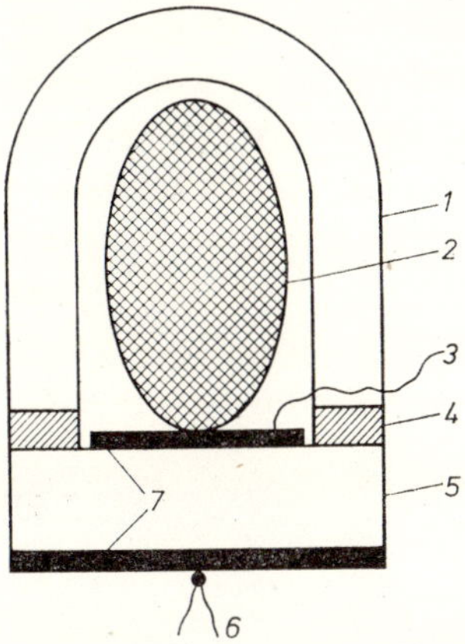

Fig. 187. An element of the apparatus for measuring deviations from stoichiometry of metal oxides by the coulometric titration method (after [275]); *1*—alundum cylinder, *2*—powdered sample in platinum net, *3*—platinum contact (I), *4*—Pyrex glass washer, *5*—solid electrolyte, *6*—thermoelement acting also as the second contact (II), *7*—platinum electrodes.

in Chapter 2.9 that the concentration of defects in compounds MeX not necessarily equal to the degree of deviation from stoichiometry y, but can be higher. For this reason Rosenburg's method is very important.

There is also a possibility of indirect determination of the total concentration of defects in the compounds discussed on the basis of the values of self-diffusion and chemical diffusion coefficients [278]. This method was used for calculating the total concentration of defects in cobaltous oxide ($Co_{1-y}O$). It was found that at temperature above 1000°C this concentration is higher than the deviation from the stoichiometric composition, y (see Chapter 2.9).

The concentration of defects is related to the above diffusion coefficients as follows:

$$[\mathrm{def}] = \frac{D(1+p)}{\tilde{D}} \tag{3.2.48}$$

where [def] is the total concentration of defects in the given sublattice, p is the degree of ionisation of defects and D is the self-diffusion coefficient of metal or oxidant, depending on the type of disorder in the crystal lattice of the examined compound MeX.

The degree of ionisation of defects p can be determined on the basis of measurements of electrical conductivity (see Chapter 1.3), self-diffusion (see Chapter 2.1) or deviations from stoichiometry (see Chapter 1.2.3), since it is directly related to the exponent $1/n$ of the oxygen pressure in equations (1.3.7), (1.3.8), (1.2.56), (1.2.58), (2.1.17) and (2.1.18). This relationship has the following form:

$$\frac{1}{n} = \frac{1}{2(1+p)} \tag{3.2.49}$$

Equation (3.2.48) can therefore be expressed in the following form:

$$[\text{def}] = \frac{D}{2\tilde{D}\left(\frac{1}{n}\right)} \tag{3.2.50}$$

The chemical diffusion coefficient can be determined directly from measurements of electrical conductivity or changes of the mass of sample caused by changes of oxidant pressure (see Chapter 3.4). The self-diffusion coefficient is often determined by means of radioactive isotopes (tracers) (see Chapter 3.3). In order to obtain the value of D, the diffusion coefficient of tracer D_t obtained by this method should be divided by the correlation factor (see equation 2.2.1). Equation (3.2.50) therefore becomes:

$$[\text{def}] = \frac{D_t}{2\tilde{D}\left(\frac{1}{n}\right)f} \tag{3.2.51}$$

The above method of indirect determination of the concentration of defects has two important advantages. Firstly, it provides information regarding the total concentration of defects in the given crystal sublattice, irrespective of the degree of deviation from stoichiometry. Secondly, it gives more accurate results than those obtained by direct determination of deviations from stoichiometry, since coefficients D and $\tilde{D}$ can be determined with a relatively high degree of accuracy. This was shown in Chapter 3.2.5 for the example of cobaltous oxide.

3.3 Methods of determination of self-diffusion coefficients

It has already been stated that the term "self-diffusion" means the random walk of defects in crystals which are in thermodynamic equilibrium with their surroundings. Studies on self-diffusion are carried out in order to determine the rate of this movement, the measure of which is the self-diffusion coefficient, and to determine the dependence of this rate on temperature and external pressure. In view of the great importance of these studies many methods of measuring these phenomena have been developed. In this chapter we will briefly discuss only two of them, namely the classical and still the most commonly used tracer method and the new, very promising method based on the kinetics of oxidation of metals.

3.3.1 The tracer method

This method includes several techniques using of tracer atoms, both radioactive and stable isotopes. In the most frequently used technique the tracer atoms of metal or oxidant are introduced in the surface layer of the crystal and the crystal is heated under defined conditions. The distribution of concentration of tracer atoms in the crystal is then determined by various methods. The heating period must be such that the atoms of the tracer do not reach the opposite surface of the crystal. Figure 188 shows a typical distribution curve for the concentration of a tracer in crystal. From the

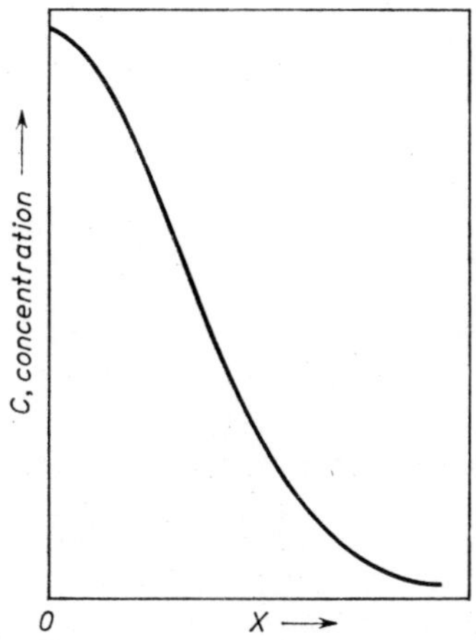

Fig. 188. The schematic plot of the distribution of the tracer concentration in MeX crystal.

point of view of the theory of diffusion it is a semi-infinite system. In this case the solution of Fick's second law gives the following simple equation:

$$c = \frac{c_0}{2\sqrt{\pi D t}} \exp\left(-\frac{x^2}{4Dt}\right) \tag{3.3.1}$$

where c is the concentration of the tracer at a distance x from the surface of the crystal, c_0 is the concentration of the tracer on the surface before diffusion annealing and t is the time of annealing.

Taking logarithms of equation (3.3.1), we obtain

$$\log c = \log\left(\frac{c_0}{2\sqrt{\pi D t}}\right) - \frac{x^2}{2.303 \cdot 4Dt} \tag{3.3.2}$$

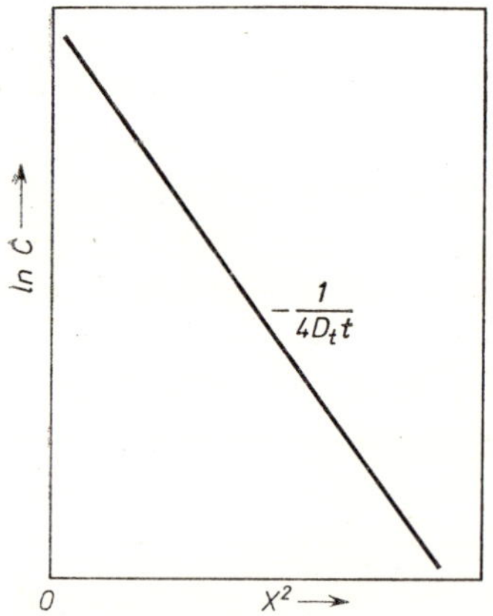

Fig. 189. The schematic plot of the distribution of the tracer concentration in MeX crystal ($\ln c$ is plotted against x^2).

It follows from equation (3.3.2) that the plot of $\log c$ against x^2 is a straight line. Its slope makes it possible to calculate the self-diffusion coefficient D (Fig. 189):

$$\left(\frac{\partial \log c}{\partial x^2}\right)_t = -\frac{1}{9.212\, Dt} \tag{3.3.3}$$

The determination of the absolute values of the concentration of the tracer in the sample would be very difficult, but it is not necessary. Since only the variation of the concentration of the tracer need be known, it is possible to measure a quantity which is directly proportional to this concentration. Such a quantity is the intensity of radiation of individual layers of the crystal, which can be successively collected from its surface by any method.

It should be mentioned that the linear dependence of the logarithm of concentration (activity) of the tracer on the square of the distance coordinate x is a proof for the lattice mechanism of diffusion. Fisher [274] has shown that when diffusion along grain boundaries predominates, the logarithm of the concentration of the tracer is a linear function of x and not of x^2:

$$\ln c = -\left\{\frac{2(D_v/D_g)d^{1/2}}{(\pi D_v t)^{1/4}}\right\}x + \text{const} \qquad (3.3.4)$$

where D_v and D_g are the coefficients of lattice diffusion and intergranular diffusion respectively and d is the "thickness" of the grain boundary.

In the studies on tracer diffusion coefficients, the principal problem consists in obtaining the layer containing tracer atoms, which should be as thin as possible and should adhere perfectly to the surface of the crystal. In the case of metals this problem does not present great difficulties, since the radioactive isotope of the metal can be introduced in the surface layer of the sample by means of ion exchange or electrolysis; e.g. when a sample of silver is immersed in a solution of $AgNO_3$ containing radioactive silver atoms, some of the atoms in the surface layer of the crystal lattice of the sample are replaced by radioactive atoms and as a result a monolayer of tracer atoms is obtained. If the required concentration of tracer atoms in the surface layer of the sample cannot be obtained by ion exchange, the electrolytic deposition of the isotope on the surface is used.

In the case of compounds of the MeX type, i.e. oxides, sulphides and halides of metals, the introduction of the isotope in the surface of the crystal is more difficult. As a rule it is not possible to obtain the required concentration of tracer atoms by ion exchange and consequently this simplest and best method is practically inapplicable in studies on self-diffusion in ionic crystals. Therefore the deposition of radioactive isotopes of the metals is usually carried out by evaporation, precipitation or electrolysis. The deposited metal is then converted to the compound with the oxidant. It is essential that the layer deposited on the surface of the crystal is in intimate contact with the surface and does not show any structural differences from the rest of the crystal.

When the self-diffusion of the oxidant is studied, there is no need to deposit a suitable layer of the compound on the surface of the sample, since it is enough to heat the crystal in an atmosphere of the oxidant containing tracer atoms. However in this case the boundary conditions for the solution of Fick's second law are slightly different, since the concentration of the isotope on the surface of the sample does not change during the diffusion annealing (assuming that the amount of gas is large in comparison with the mass of the sample). When this fact is taken into account, the following relationship is obtained:

$$c = C_0\left\{1-\operatorname{erf}\left[\frac{x}{2(Dt)^{1/2}}\right]\right\} \tag{3.3.5}$$

where C_0 is the concentration of tracer atoms on the surface of the crystal and erf is the error function. The values of this function have been tabulated. Some of them are given in Table 14.

TABLE 14

Selected values of the error function

z	erf (z)	z	erf (z)	z	erf (z)
0	0	0.55	0.5633	1.3	0.9340
0.025	0.0282	0.60	0.6039	1.4	0.9523
0.05	0.0564	0.65	0.6420	1.5	0.9661
0.10	0.1125	0.70	0.6778	1.6	0.9763
0.15	0.1680	0.75	0.7112	1.7	0.9838
0.20	0.2227	0.80	0.7421	1.8	0.9891
0.25	0.2763	0.85	0.7707	1.9	0.9928
0.30	0.3286	0.90	0.7970	2.0	0.9953
0.35	0.3794	0.95	0.8209	2.2	0.9981
0.40	0.4284	1.0	0.8427	2.4	0.9993
0.45	0.4755	1.1	0.8802	2.6	0.9998
0.50	0.5205	1.2	0.9103	2.8	0.9999

Figure 190 shows schematically the distribution of concentration of tracer atoms of the oxidant in the crystal for the case discussed (C_0 = const).

It was shown in the second part of this book that the self-diffusion coefficient of atoms or ions in the crystal is directly proportional to the concentration of defects. Since this concentration is a function of both the temperature and the equilibrium oxidant pressure, the measurements of self-diffusion coefficients must be carried out as functions of both these parameters. On the basis of the results of studies on the dependence of the coefficient, D, on the equilibrium oxidant pressure, it is possible to come to significant conclusions regarding the mechanism of diffusion and the degree of ionisation of defects. It should be stressed that the great majority of the results of studies on self-diffusion carried out up to the present time is not suitable for such interpretation. As a rule, in these studies the effect of pressure on the value of coefficient D was not taken into account. Only the effect of

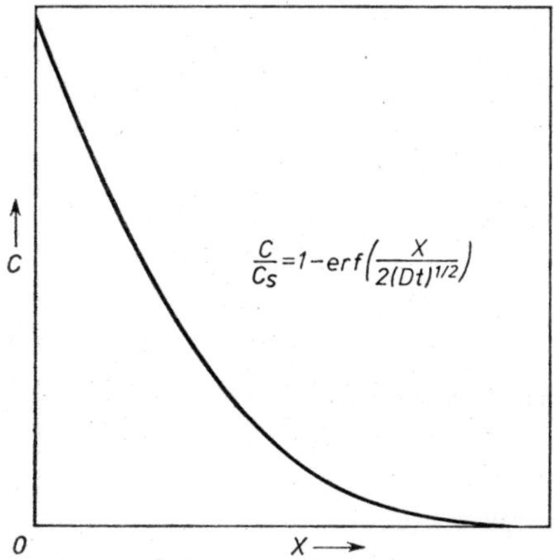

Fig. 190. The schematic plot of the distribution of the tracer concentration in MeX crystal in the case when the tracer concentration on the surface of the sample is constant

temperature was studied and many authors gave no information regarding the atmosphere in which the diffusion annealing was carried out. The results of these studies have no scientific value and from the practical point of view they can be used only for estimating the transport properties of the materials investigated. Only in recent years has a systematic increase of the number of studies, in which the effects of both temperature and pressure on self-diffusion rate are taken into account, been observed.

There are three principal methods of determination of the distribution of radioactive tracers in samples subjected to diffusion annealing. One of them, called the layer method, consists in collecting successive layers of the sample parallel to the original surface on which the radioactive isotope had been deposited before the annealing. Then the radioactivity fractions of the sample is determined and the dependence of the concentration of tracer atoms on the distance from the surface of the sample is plotted. The plot is used for calculating the self-diffusion coefficient. The accuracy of the results obtained by this method depends on the precision in the collection the successive thin layers of the sample.

The fractionation of the sample can be carried out either chemically or mechanically.

The chemical method consists in etching of the sample in a suitable solution and determination of the radioactivity of the solution. If the rate of dissolution of the compound under study is known, the time of etching can be suitably chosen and very thin layers can be collected. However, in this case, particular attention should be paid to the possibility of non-uniform etching of the sample. Since the rate of dissolution of a crystalline substance usually depends on the orientation of the individual crystals, the formation of an uneven surface may occur, which becomes rougher during the etching process. For this reason the chemical method is not widely used.

In the mechanical method the layers are collected by means of an accurate lathe or a suitable grinding machine. Modern machines of this type are very precise and make it possible to collect relatively thin layers. However this method can be used only in the case of substances which are not very brittle, i.e. mainly metals. For this reason Letaw, Slifkin and Portnoy [280] developed a special grinding machine which makes it possible to collect parallel layers of brittle substances a few micrometers thick. This precision in collecting the layers of sample makes it possible to use much shorther periods of diffusive annealing and to measure the self-diffusion coefficients at relatively low temperatures.

Figure 191 shows a new device for the accurate collection of thin layers of brittle substances by manual grinding. The device was developed by Bruin and Clark [281]. The sample (*B*) is placed in a conical groove cut in a teflon washer (*C*), which is fixed in the head of the instrument (*D*), parallel to its base (*E*). The axial position of the sample in the device is determined with an accuracy of $\pm$ 1μm by means of a micrometer fixed to the side of the machine. The device containing the sample is placed in lead (*F*), which has a slit cut in the middle. The width of the slit is equal to the diameter of the sample. The slit contains a suitable abrasive material with properties which depend on the properties of the sample. By sliding the system of rings in lead (*F*) backwards and forwards manually it is possible to remove a defined thickness from the sample, which is determined by the micrometer. By means of this device the self-diffusion coefficients were determined for very brittle substances, such as beryllium oxide BeO etc.

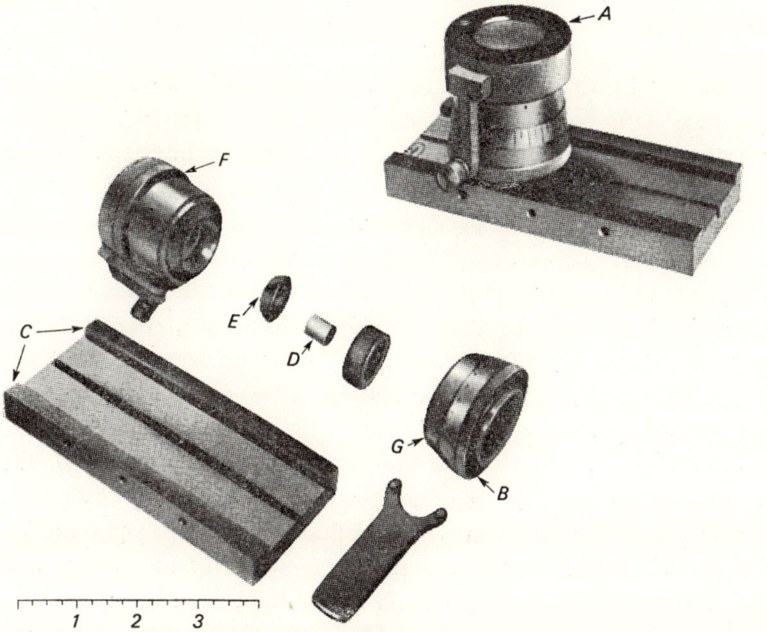

Fig. 191. The device for mechanical collection of thin layers of sample of a brittle material (oxides, sulphides) (after [281]); *A*—the device ready for polishing with the sample fixed inside, *B*—sample, *C*—conically cut teflon sample holder, *D*—main head, *E*—base of the head, *F*—leading rail of the head.

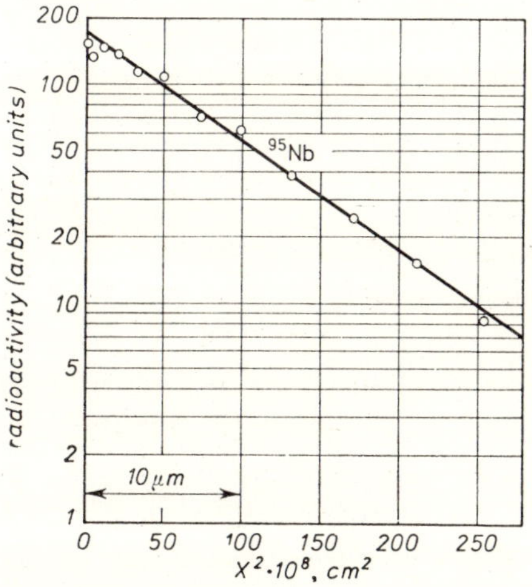

Fig. 192. The distribution of radioactivity in a sample of niobium after diffusive annealing at 1607°C (after [281]).

The plot in Fig. 192 shows the distribution of radioctivity in a sample of niobium, determined by the above method. A layer of radioactive niobium isotope ^{95}Nb with a thickness of a few atomic monolayers was deposited on the surface of the sample, which was then heated at 1607°C. It follows from the graph that the thickness of the layers collected mechanically was about 1 μm. The linear dependence of the logarithm of the radioactivity on the square of the depth of penetration of the isotope clearly indicates that, in these conditions, volume diffusion predominates.

Another method of determining the distribution of the tracer concentration in ionic crystals is the autoradiographic technique. After diffusion annealing the sample is cut perpendicularly to the surface on which the isotope had been deposited before annealing. After polishing, the sample is placed on a photographic film sensitive to the radiation of the isotope. After development, the film shows the distribution of the tracer in the sample, since the darkening of the film depends on the intensity of radiation. By analysing this darkening by means of a suitable microspectrophotometer it is possible to determine the concentration gradient of the isotope and hence to calculate the self-diffusion coefficient of the tracer.

Since the degree of darkening of the film is directly proportional to the concentration of the radioactive isotope, the distribution of this isotope in the sample can be expressed by the following equation:

$$P = P_0 \exp\left(-\frac{x^2}{4Dt}\right) \tag{3.3.6}$$

where P is the degree of darkening of the film at distance x from the surface of the sample coated with a layer of the isotope before annealing and P_0 is the degree of darkening on this surface. When $\log P$ is plotted against x^2, the plot is a straight line having a slope of $-1/4\,Dt$. Since the time of diffusion annealing t is known, the diffusion coefficient D can be calculated.

This method is simpler and less time consuming than the layer method, but it can only be used for determining relatively large values of diffusion coefficient. The resolving power of autoradiograms is not very high, which makes it necessary to use a relatively deep penetration of the isotope to the sample in order to obtain sufficiently accurate results. The depth of the penetration obviously depends on the time of annealing, but at relatively low values of D the preparation of sufficiently thick diffusion layers takes so long that the autoradiographic method becomes practically useless. It should be also mentioned that this method can only be used, if the energy of the radiation of the isotope is so low that its range is by at least one order of magnitude shorter than the thickness of the diffusion layer. This is another limitation of the use of this method in studies on self-diffusion.

The third method consists in measurements of radioactivity of the surface of the sample before and after diffusion annealing. During the annealing the radioactivite isotope, which was originally present only on the surface of the sample, diffuses to the bulk of the sample and as a result the radioactivity of the surface decreases in proportion to the diffusion rate. Hence the self-diffusion coefficient can be calculated when the decrase of radioactivity of the surface and the annealing time are known.

In this case the solution of Fick's second law gives the following relationship:

$$\frac{I}{I_0} = \exp(\mu^2 Dt)[1-\text{erf}(\mu\sqrt{Dt})] \tag{3.3.7}$$

where I is the radioactivity of the surface of crystal after annealing for time t, I_0 is the initial activity and μ is the coefficient of absorption of the radiation.

In order to calculate D it is necessary to plot I/I_0 against $\mu^2 Dt$ and to compare the plot with the experimental curve in order to calculate the numerical value of $\mu^2 Dt$ and hence the diffusion coefficient. Since the absorption coefficient μ and the annealing time, t, are known, the diffusion coefficient D can be calculated.

Although this method appears to be simple, it is fairly tedious, since special conditions of measurement must be observed and certain supplementary determinations must be carried out accurately. Thus the absorption of the radiation of the isotope by the sample must be exactly known. Comparative studies have shown that even when these special conditions are observed during the measurements of the intensity of radiation and when the self-absorption of the substance examined is known, the metod based on the measurements of radioactivity of the surface does not give reliable results. For this reason Kofstad [282] and recently Lundy [283], in their discussion of the methods of determination of self-diffusion in solids, have come to the correct conclusion in the author's opinion, that the layer method gives the most reliable results.

However Askill has pointed out that the best results can be obtained only, if the method selected is the most suitable for the problem to be solved. Although there is no doubt that the layer method gives the best results, it is not a universal method. It gives the best results only in the cases when the radioactive isotope emits γ radiation. When it emits soft β radiation the usefulness of the method is very limited. On the other hand the autoradiographic method gives the best results with soft radiation, but it cannot be used when the tracer emits hard γ radiation. The method based on measurements of the decrease of intensity of radiation can be used for both β and γ radiation, but the correctness of the results depends on exact values of absorption coefficients being known as well as rates of evaporation of the isotope from the surface of the sample etc., which considerably limits the applicability of this method.

3.3.2 The kinetic method

In recent years, a new method of determination of self-diffusion coefficients in compounds of the MeX type has been developed. It is based on measurements of the kinetics of oxidation of metals and, like Rosenburg's method, it consists in determination of the rate of growth of a compact, single phase layer of MeX on the surface of the metal under isobaric and isothermic conditions.

The growth of this layer takes place according to the parabolic rate law (see Chapter 3.2.5), which means that the diffusion of one or both reagents in the scale is the rate determining step of the oxidation process.

In 1951, when C. Wagner [242] was developing his theory of oxidation of metals, he already derived the general relation between the parabolic rate constant for the oxidation of metal and the self-diffusion coefficients of the reagents in a growing, homogeneous compact scale layer:

$$k'_p = \int_{a'_X}^{a''_X} \left(\frac{z_1}{|z_2|} D_{Me} + D_X \right) d \ln a_X \tag{3.3.8}$$

where a'_X and a''_X are the activities of the oxidant on the metal/scale and scale/oxidant interfaces respectively, z_1 and z_2 are the valencies of cations and anions in the compound MeX, forming the scale, D_{Me} and D_X are the self-diffusion coefficients of the metal and the oxidant and k'_p is the parabolic rate constant of the scale growth according to equation (3.2.3).

The determination of the values of k'_p is troublesome, since it requires tedious and not very accurate determinations of the thickness of the scale as a function of its time of formation. In order to obtain a single kinetic curve it is necessary to oxidise many samples at various time intervals and, after preparing suitable sections, to determine the thickness of the scale by means of a microscope.

However the value of this constant can be determined very accurately on the basis of the results of investigation of the kinetics of oxidation of the metal obtained by the thermogravimetric or the volumetric methods which were described in the last chapter. These methods consist of continuous measurements of weight gains of the sample during the reaction, i.e. of the mass of the oxidant combining with the sample. The plot of the square of the weight gain against the oxidation time is a straight line having a slope equal to the parabolic rate constant k_p'' for the formation of scale according to Pilling and Bedworth's equation:

$$(\Delta m/A)^2 = k_p'' t \tag{3.3.9}$$

where $(\Delta m/A)$ is the weight gain (Δm) per unit surface area A. The parabolic rate constant k_p'' is related to the Tammann constant k_p' following equation:

$$k_p' = \frac{1}{2}\left(\frac{\tilde{V}|z_2|}{A_X}\right)^2 k_p'' \tag{3.3.10}$$

where $\tilde{V}$ is the volume of one gram-equivalent of the scale, z_2 is the valency of the oxidant and A_X is the atomic weight of the oxidant.

Wagner's general equation is valid on condition that the rates of self-diffusion of the metal and the oxidant in the scale are of the same order of magnitude (i.e. when point defects are simultaneously present in both sublattices). Such cases are exceptional. As a rule the crystal lattice disorder of MeX compounds in practice is limited to one sublattice (cationic or anionic). When the defects are present in the cationic sublattice (compounds of the $Me_{1-y}X$ or $Me_{1+y}X$ type), the growth of scale takes place as a result outward diffusion of the metal, since $D_{Me} \gg D_X$ and Wagner's general equation becomes:

$$k_p' = \int_{a_X'}^{a_X''} \frac{z_1}{|z_2|} D_{Me}\, d\ln a_X \tag{3.3.11}$$

When the defects occur almost exclusively in the anionic sublattice (compounds of the MeX_{1-y} or MeX_{1+y} type), the scale growth takes place as a result of the inward diffusion of the oxidant, since $D_X \gg D_{Me}$. In this case general equation (3.3.8) becomes:

$$k_p' = \int_{a_X'}^{a_X''} D_X\, d\ln a_X \tag{3.3.12}$$

In order to calculate coefficients D_{Me} or D_X from equations (3.3.11) or (3.3.12), these equations must be integrated. It has been shown in the second chapter of this book that self-diffusion coefficients depend on the concentration of defects and hence on the oxidant pressure [see equations (2.1.26) and (2.1.27)]. The integration therefore requires a knowledge of the dependence of D_{Me} or D_X on a_X or p_{X_2}. Since this function cannot be predicted (this would require the knowledge of the degree of ionisation of defects and other parameters), Wagner's equation was not used for the above calculations until recently.

Only in 1965 Pettit [284] as well as Fueki and Wagner [285] showed that in the case of compounds of the $Me_{1-y}X$ or MeX_{1+y} type the solution of this problem is possible by means of the inverse procedure, i.e. by differentiation of equation (3.3.8). These authors observed that this equation consists of two terms. One of them is a function of a''_X and the other, a function of a'_X. At a constant temperature, the activity of the oxidant on the metal/scale interface (a'_X) is constant and independent of the external pressure, since it is assumed in Wagner's theory (confirmed experimentally) that local thermodynamic equilibrium exists at this phase boundary.

Hence equation (3.3.8) can be differentiated with respect to a''_X:

$$\left(\frac{dk'_p}{d\ln a''_X}\right)_{a'_X} = \frac{z_1}{|z_2|}D_{Me}+D_X \tag{3.3.13}$$

In this equation the activity of the oxidant can be replaced by its pressure according to the following relationship:

$$a\ln a_X = \tfrac{1}{2}d\ln p_X \tag{3.3.14}$$

which leads to the following general solution:

$$\left(\frac{dk'_p}{d\ln p''_{X_2}}\right)_{p'_{X_2}} = \frac{z_1}{2|z_2|}D_{Me}+D_X \tag{3.3.15}$$

This equation can be simplified to the following forms for the compounds $Me_{1-y}X$ and MeX_{1+y} respectively:

$$\left(\frac{dk'_p}{d\ln p''_{X_2}}\right)_{p'_{X_2}} = \frac{z_1}{2|z_2|}D_{Me} \tag{3.3.16}$$

and

$$\left(\frac{dk'_p}{d\ln p''_{X_2}}\right)_{p'_{X_2}} = \frac{1}{2}D_X \tag{3.3.17}$$

It follows from equations (3.316) and (3.3.17) that the coefficients D_{Me} and D_X can be determined on the basis of measurements of the dependence of the oxidation rate of the metal on the oxidant pressure. In order to determine function $dk_p'/d\ln p_{X_2}''$ it is necessary to carry out a series of measurements of the parabolic rate constant for various oxidant pressures and to plot k_p' against $\ln p_{X_2}$. The plot is a curve, which on graphical differentiation directly gives the numerical values of the differential $dk_p'/d\ln p_{X_2}''$ for any pressure in the range of p_{X_2} examined and hence the values of D.

The authors of this method have made a first attempt to apply it to Ni–NiO–O_2 system in order to determine the self-diffusion coefficients of nickel D_{Ni} in nickel oxide which, as was shown in Chapter 3.2, is a p-type metal deficit semiconductor ($Ni_{1-y}O$). Figure 193 shows, as an example,

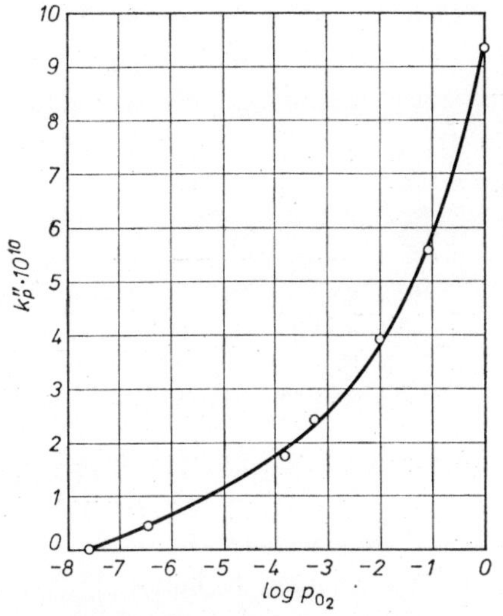

Fig. 193. Parabolic dependence of the rate of oxidation of nickel on oxygen pressure at 1190°C (after [285]) (the semilogarithmic plot).

the results of kinetic studies on oxidation of nickel at 1190°C and various oxygen pressures. Fueki and Wagner carried out analogous measurements at 1000–1400°C. On the basis of the results of these studies the dependence of the coefficient D_{Ni} on oxygen pressure and temperature was determined over wide temperature and pressure ranges.

The results of these studies are shown graphically in Fig. 194. Up to the present, they are the only values of self-diffusion of nickel in nickel oxide obtained experimentally over such a wide range of oxygen pressure. It follows from the graph that the degree of ionisation of defects (cation vacancies) in $Ni_{1-y}O$ strongly depends on both temperature and oxygen pressure. This fact has not yet been satisfactorily explained in spite of

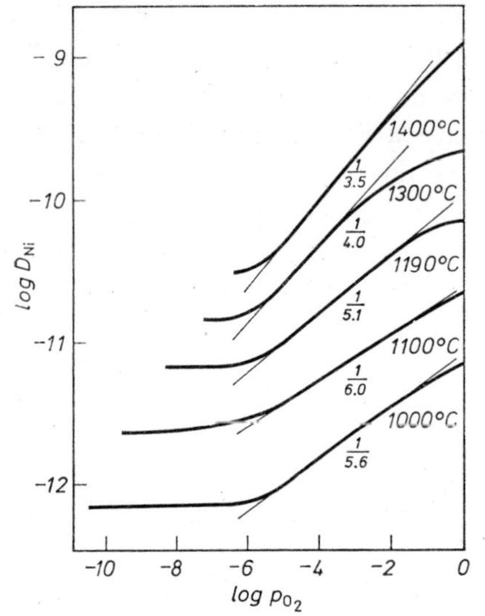

Fig. 194. The dependence of the self-diffusion coefficient of nickel in nickel oxide on oxygen pressure for various temperatures (after [285]). The figures at the curves stand for the slopes of tangents, which in a certain range of pressures approximately characterise the above dependence.

many attempts to do so. Apart from this, it also follows from the graph that in the range of oxygen pressures lower than about 10.1325×10^{-2} Pa (10^{-6} atm) the intrinsic ionic Frenkel type disorder [equation (1.2.1)] predominates in the cationic sublattice of nickel oxide, since in this range the self-diffusion coefficient of nickel is independent of oxygen pressure. This result could have been predicted, since the concentration of cation vacancies resulting from the quasi-chemical surface reaction decreases with decreasing equilibrium oxygen pressure (see the first part of this book) and as a result the intrinsic ionic disorder becomes predominant at oxygen pressures lower than a certain critical value.

The results of the above studies are a good illustration of the importance of investigations of self-diffusion in the characterisation of transport and semiconducting properties of ionic crystals.

The Fueki-Wagner method was checked by Mrowec and Przybylski [188, 189] for the cobalt–cobaltous oxide–oxygen system. This system was chosen for the same reasons as those in the case of checking Rosenburg's method (see Chapter 3.2.4).

Kinetic studies on the oxidation of cobalt were carried out at 950–1300°C in oxygen pressure range 1.01325×10^{-2}–1013.25 hPa (10^{-5}–1 atm) by the thermogravimetric and the volumetric methods. The studies were carried out on flat discs of spectrally pure cobalt with dimensions such that during the entire determination of the kinetic curve the scale was compact and tightly adhering to the core.

The parabolic rate constants of the oxidation were determined by Mrowec and Stokłosa's method [286, 287] taking into account the changes of the

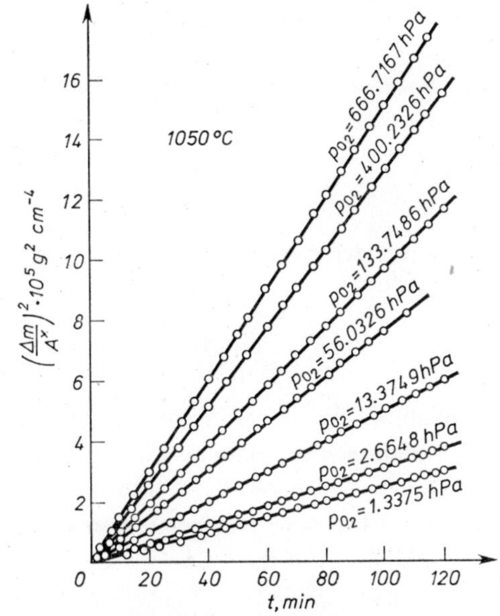

Fig. 195. Oxidation of cobalt at 1050°C under various oxygen pressures (after [190]) (parabolic plots).

surface area of the core resulting from metal consumption and thermal expansion. Figure 195 shows as an example the parabolic region for oxidation of cobalt at 1000°C at several oxygen pressures. The dependence of the parabolic rate constant on the logarithm of oxygen pressure at the same temperature is shown in Fig. 196.

Graphical differentiation of the curves obtained at several temperatures made it possible to calculate coefficient D_{Co} as a function of temperature and oxygen pressure by means of equation (3.3.16). Selected results of these

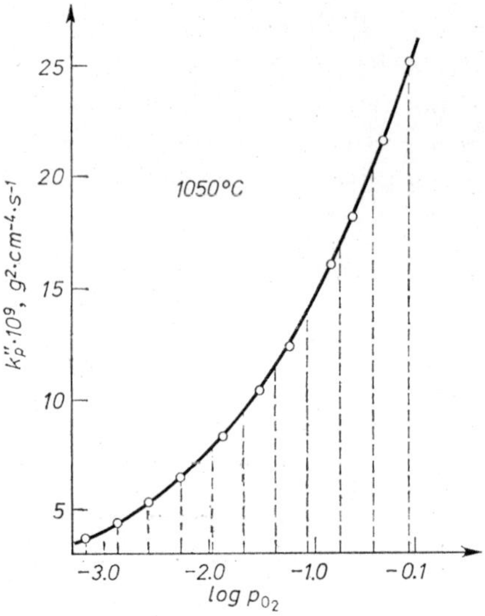

Fig. 196. The dependence of the parabolic rate constant of oxidation of cobalt on oxygen pressure shown in the semilogarithmic plot (after [190]).

calculations are shown graphically in Fig. 197 on an Arrhenius plot at several oxygen pressure. For comparison, the results of radioisotopic studies carried out by Chen and collab. [273] as well as by Carter and Richardson [272], corrected for the correlation effect, have also been plotted in Fig. 197. The tracer diffusion coefficients D^t_{Co} determined by the above authors by means of radioactive cobalt ^{60}Co were recalculated in order to obtain self-diffusion coefficients D_{Co} by means of the following equation:

$$D_{Co} = D^t_{Co} \frac{1}{f} \tag{3.3.18}$$

assuming that the correlation coefficient, f, is equal to 0.78.

It is clear from the graph that the values of self-diffusion coefficients of cobalt in cobaltous oxide obtained indirectly by the Fueki–Wagner method are in excellent agreement with the results obtained directly by means of the radioisotopic measurements. In order to give a better illustration of this agreement the results of the kinetic and the radioisotopic studies are

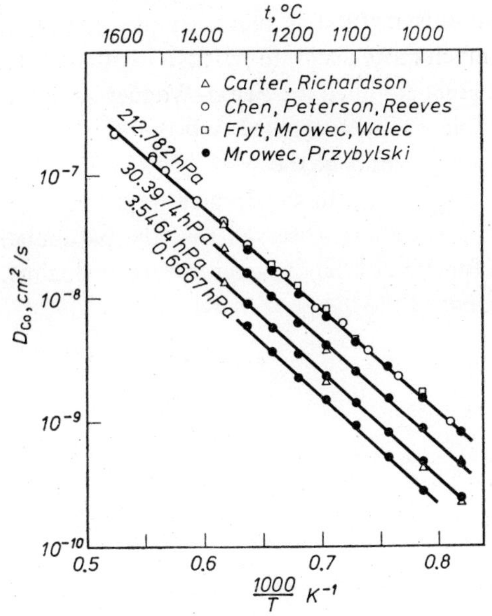

Fig. 197. The temperature dependence of the diffusion coefficient of cobalt in $Co_{1-y}O$ in the Arrhenius system (after [190]). The results obtained by Carter and Richardson (Δ) and by Chen and coworkers (○) were recalculated according to equation (3.3.18); □—results obtained by Fryt, Mrowec and Walc by means of the kinetic Rosenburg's method; ●—results obtained by Mrowec and Przybylski.

shown below in the form of empirical relationships between the coefficient D_{Co} and temperature at an oxygen pressure of 20.265 kPa (0.2 atm).

$$\frac{D^{t}_{Co}}{f} = D_{Co} = 5.0\times 10^{-3}\exp\left(-\frac{38\,400}{RT}\right)\ [cm^2/s] \qquad (3.3.19)$$

(Chen and collaborators)

$$\frac{D^{t}_{Co}}{f} = D_{Co} = 5.2\times 10^{-3}\exp\left(-\frac{38\,000}{RT}\right)\ [cm^2/s] \qquad (3.3.20)$$

(Carter and Richardson)

$$D_{Co} = 5.0\times 10^{-3}\exp\left(-\frac{37\,800}{RT}\right)\ [cm^2/s] \qquad (3.3.21)$$

(Mrowec and Przybylski).

The above agreement proves that the Fueki–Wagner kinetic method can be successfully used for determining self-diffusion coefficients in metal oxides. It should be stressed that this method is much simpler than the radioisotopic methods, but gives much more accurate results.

Recently Mrowec and Stokłosa [288] modified the Fueki–Wagner method. This modification makes it possible to calculate the self-diffusion coefficients of metals and oxidants in compounds of the $Me_{1-y}X$ and MeX_{1+y} types from the results of kinetic studies in a much simpler way, but with equal or even better accuracy. These authors observed that the parabolic rate constant for scale growth is a function of the oxygen pressure analogous to that of the self-diffusion coefficient of the metal or oxidant:

$$k_p' = \text{const}\, p_{X_2}^{1/n} \tag{3.3.22}$$

and

$$D_{Me} = D_0\, p_{X_2}^{1/n} \quad (\text{for } Me_{1-y}X) \tag{3.3.23}$$

or

$$D_X = D_0\, p_{X_2}^{1/n} \quad (\text{for } MeX_{1+y}) \tag{3.2.24}$$

Using this analogy it is possible to replaced coefficients D_{Me} and D_X (which are dependent on the oxidant pressure) in equations (3.3.11) and (3.3.12) with the product $D_0 p_{X_2}^{1/n}$ and to carry out the integration after expressing the activity of the oxidant as a pressure. This procedure leads to the following relationships:

$$k_p' = \frac{z_1}{|z_2|} D_0 n[(p_{X_2}'')^{1/n} - (p_{X_2}')^{1/n}] \tag{3.3.25}$$

for compounds of the $Me_{1-y}X$ type and

$$k_p' = D_0 n[p_{X_2}'')^{1/n} - (p_{X_2}')^{1/n}] \tag{3.3.26}$$

for compounds of the MeX_{1+y} type.

These equations can be simplified when the measurements of k_p' are carried out at pressures much higher than the equilibrium oxidant pressure at the metal/scale interface, i.e. the dissociation pressure of the MeX compound at the given temperature. In this case $p_{X_2}' \ll p_{X_2}''$ and consequently equations (3.3.25) and (3.3.26) become:

$$k_p' = \frac{z_1}{|z_2|} D_0 n(p_{X_2}'')^{1/n} \tag{3.3.27}$$

and

$$k_p' = D_0 n(p_{X_2}'')^{1/n} \tag{3.3.28}$$

Taking logarithms of these expressions we obtain the following relationships:

$$\log k_p' = \log\left(\frac{z_1}{|z_2|}\right) D_0 n + \frac{1}{n} \log p_{X_2}'' \tag{3.3.29}$$

and

$$\log k_p' = \log D_0 n + \frac{1}{n} \log p_{X_2}'' \tag{3.3.30}$$

It follows from these equations that in order to find the required values of $1/n$ and D_0 it is necessary to determine the function $\log k_p' = f(\log p_{X_2})$. This problem is much simpler than that encountered in the case Fueki and Wagner's method, since the above function is linear. For this reason coefficients $1/n$ and D_0 can be simply and accurately determined by the graphical or analytical method. After introducing the values of these coefficients in equations (3.3.23) or (3.3.24) we can calculate the self-diffusion coefficients of the metal or the oxidant in the MeX compound examined for any oxidant pressure.

Mrowec and Stokłosa [288] have used this method for determining the self-diffusion coefficient of copper in cuprous oxide ($Cu_{2-y}O$) as a function of temperature and oxygen pressure. In Fig. 198 the results of these studies for an oxygen pressure of 13.1723 Pa (1.3×10^{-4} atm) are compared with

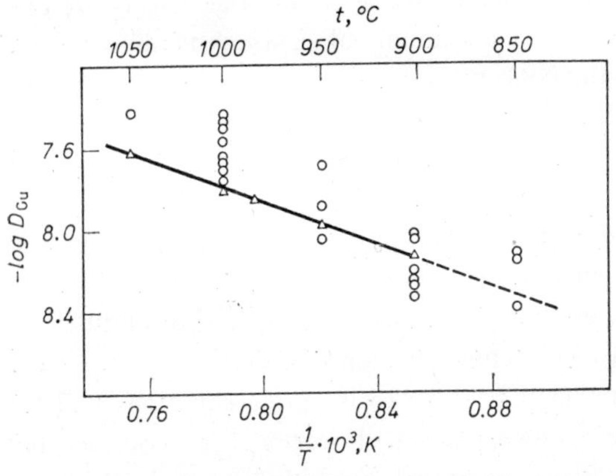

Fig. 198. The temperature dependence of the self-diffusion coefficient of copper in cuprous oxide at oxygen pressure equal to 10.1325 Pa (10^{-4} atm) (after [288]); ○ and Δ correspond to the results obtained by the radioisotopic (tracer) method and the kinetic method, respectively (the size of signs Δ shows the maximum scatter of the results).

the results obtained in the same conditions by Moore and Selikson [289], who used the tracers method. The graph clearly shows that the accuracy of the determination of D by the kinetic method is many times higher than that of the classical tracer method.

It has already been mentioned that the applicability of the kinetic method is limited, since it can be used only in the case of compounds showing a real deficit of metal ($Me_{1-y}X$) or a real excess of the oxidant (MeX_{1+y}), i.e. semiconductors of the p-type. In these cases during the scale growth, the defects (cation vacancies or interstitial anions) are formed on its outer surface and disappear on the inner phase boundary. For this reason the maximum concentration of defects is established on the outer surface of the scale and is proportional to the oxidant pressure. Consequently the parabolic rate constant for the oxidation and the equilibrium concentration of defects are analogous functions of oxygen pressure, which makes it possible to calculate the self-diffusion coefficients by the differential method of Fueki and Wagner or by the integral method proposed by Mrowec and Stokłosa.

When the scale consists of a compound of the $Me_{1+y}X$ or MeX_{1-y} type, i.e. n-type semiconductor, the distribution of defects (interstitial cations or anion vacancies) in the scale is inverted. In this case the defects are formed at the inner phase boundary, where their concentration reaches the maximum value corresponding to the dissociation pressure of the scale at the given temperature. On the outer surface, where the defects are annihilated, their concentration has a minimum value, which is almost zero. In this case the variation in the concentrations of defects, which determines the rate of growth of the scale, is practically independent of the external oxidant pressure and Wagner's general equation cannot be solved by any of the above methods.

However Kröger [15] and Kofstad [247] pointed out that this equation can be solved for any case, where a linear concentration gradient of defects in the scale is assumed. This assumption implies that the chemical diffusion coefficient $\tilde{D}$ is independent of the concentration of defects over the whole depth of the scale, which makes it possible to integrate equations (3.3.11) and (3.3.12) after suitable transformation. For pressures considerably exceeding the dissociation pressure of the scale a simple relationship between the parabolic rate constant for oxidation and the self-diffusion coefficient of the metal or the oxidant in the scale is finally obtained:

$$k'_p = (p+1)D = N_d\tilde{D} \tag{3.3.31}$$

TABLE 15

Self-diffusion coefficients of metals in $Cu_{2-y}O$ and $Co_{1-y}O$ and in $Fe_{1-y}S$ calculated from oxidation or sulphuration kinetics by means of equations (3.3.16), (3.3.29) and (3.3.31)

P'_{O_2} hPa	P'_{S_2} hPa	D_{Me}, cm^2/s								
		D_{Cu} in $Cu_{2-y}O$			D_{Co} in $Co_{1-y}O$			D_{Fe} in $Fe_{1-y}S$		
		temperature 1000°C			temperature 1000°C			temperature 800°C		
		eqn. (3.3.16)	eqn. (3.3.29)	eqn. (3.3.31)	eqn. (3.3.16)	eqn. (3.3.29)	eqn. (3.3.31)	eqn. (3.3.16)	eqn. (3.3.29)	eqn. (3.3.31)
6.687	0.066	4.6×10^{-8}	4.3×10^{-8}	4.2×10^{-8}				7.8×10^{-8}	1.0×10^{-7}	9.2×10^{-8}
40.023	0.659	7.3×10^{-8}	6.8×10^{-8}	6.9×10^{-8}				1.5×10^{-7}	1.8×10^{-7}	1.7×10^{-7}
80.553	6.586	8.0×10^{-8}	8.1×10^{-8}	8.2×10^{-8}				2.8×10^{-7}	3.1×10^{-7}	3.0×10^{-7}
101.325	65.861	8.3×10^{-8}	8.6×10^{-8}	8.7×10^{-8}				5.1×10^{-7}	5.9×10^{-7}	5.4×10^{-7}
40.54	658.613				9.8×10^{-10}	1.0×10^{-9}	9.0×10^{-10}	9.4×10^{-7}	1.0×10^{-6}	9.6×10^{-7}
70.928					1.2×10^{-9}	1.2×10^{-9}	1.1×10^{-9}	1.2×10^{-6}	1.2×10^{-6}	1.1×10^{-6}
111.458					1.4×10^{-9}	1.4×10^{-9}	1.3×10^{-9}			
435.698					2.3×10^{-9}	2.3×10^{-9}	2.2×10^{-9}			

where D is the self-diffusion coefficient of the metal or the oxidant (depending on the type of lattice disorder) and p is the degree of ionisation of cation or anion defects.

It follows from equation (3.3.31) that in this case, in order to determine the self-diffusion coefficient it is necessary to know the degree of ionisation of defects under the given conditions. The validity of this equation has been checked by Mrowec and his collaborators for the following systems: Fe–$Fe_{1-y}S$–S_2 [290], Co–$Co_{1-y}O$–O_2 [291, 192] and Cu–$Cu_{2-y}O$–O_2 [288]. Selected values of self-diffusion coefficients of copper in Cu_2O, cobalt in CoO and iron in FeS calculated from kinetic measurements, by means of equations (3.3.16), (3.3.29) and (3.3.31), are compared in Table 15. It is seen that the agreement between the data obtained by the three methods discussed above is completely satisfactory.

However it should be mentioned that equation (3.3.31) can be used as the basis of correct calculations only if the mobility of ionic defects, which is the measure of the chemical diffusion coefficient $\tilde{D}$, does not depend on their concentration. The results of studies carried out by Pettit [284] and Mrowec [288, 290, 291, 198] indicate that this condition is fulfilled in the case of compounds MeX in which the concentration of defects is low. This offers wide possibilities of applying equation (3.3.31) to calculations of self-diffusion coefficients in compounds of the MeX type on the basis of kinetic studies on the oxidation of metals.

3.4 Methods for determination of chemical diffusion coefficients

It has been shown in the second chapter of this book that knowledge of the chemical diffusion coefficient $\tilde{D}$ is necessary for characterisation of many important properties of solids, particularly the homogenisation time and the reactivity in processes governed by diffusion of reagents in the solid phase. The chemical diffusion coefficient is a measure of the mobility of defects, when a concentration gradient exists in the crystal, i.e. in non-equilibrium conditions.

In view of the importance of chemical diffusion coefficients, many methods for their determination have been developed. They are mainly based on kinetic measurements of changes in the mass of the sample or of changes in its electrical conductivity accompanying the passage of the system from one equilibrium state to another. Colour changes of the substance caused by changes in the concentration of defects are also utilised. In this case the chemical diffusion coefficient is determined by measuring the rate of movement of the boundary of the colour change in the crystal. In some cases the change in thermodynamic conditions may lead to the inversion of the predominant type of lattice disorder (see the first part of the book), which is manifested as a change in semiconducting properties from type n to type p or vice versa. This is also utilised in the studies discussed. In this case the rate of movement of the boundary of the inversion in the crystal, i.e. the rate of migration of the p-n junction, is measured. Finally coefficient $\tilde{D}$ can be determined by measuring the kinetics of electrostatic potential changes in suitable solid state cells. In this chapter the most important of these methods, consisting of measurements of changes of mass, electrical conductivity and relaxation time in solid state cells will be outlined. (Rosenburg's kinetic method was discussed in Chapter 3.2.5).

3.4.1 The thermogravimetric method

This method, like Rosenburg's method, gives the most reliable results, since the changes of mass are directly related to the equalisation of the differences of concentrations of defects in the crystal. However the application of this method is limited to cases in which the changes of the mass of the sample, caused by changes in the concentration of defects, are so large that they can be determined gravimetrically with sufficient accuracy. In practice it means that the method is limited to compounds with considerable deviations from stoichiometry.

In this method the following procedure is used: The sample is placed in the microthermogravimetric apparatus described in Chapter 3.2.1., which makes it possible to determine accurately the changes in the mass of the sample during the time in which it is in the reaction zone. The sample is then homogenised at a given temperature under a definite oxidant pressure. When the mass of the sample becomes constant, the system is in equilibrium and the concentration of defects is constant. Then the oxidant pressure in the reaction chamber is rapidly altered and the kinetics of the changes of the mass accompanying the passage of the system to the new equilibrium state is measured.

The changes of the mass of the sample take place as a result of the release or the binding of the oxidant by the surface. If the initial equilibrium state corresponds to the higher oxidant pressure, then, after the rapid drop of the pressure, the sample releases the oxidant to the surroundings until the new equilibrium state is reached (reduction takes place). In the case of compounds of the $Me_{1-y}X$ or MeX_{1+y} type this corresponds to a decrease of the concentration of cation vacancies or interstitial anions respectively and the corresponding decrease of the number of electron holes. In the case of compounds of the $Me_{1+y}X$ or MeX_{1-y} type this is equivalent to an increase of the concentration of interstitial cations or anion vacancies and the corresponding increase of the number of quasi-free electrons. Obviously in the reverse case, i.e. when the initial equilibrium state corresponds to the lower oxidant pressure, after the rapid increase of the pressure the concentration of defects changes in the opposite direction as a result of the binding of the oxidant by the surface of the sample (oxidation takes place).

The choice of the initial equilibrium is not important from the point of view of interpretation of the results, since the boundary conditions in the solution of Fick's second law are identical for both cases. The formation or annihilation of defects takes place on the surface of the crystal, and as a result a difference between the concentrations of defects on the surface and in the bulk of the crystal appears and then gradually disappears as a result of the diffusion process.

From the point of view of solid state diffusion theory the changes of the concentration of defects can be divided into two stages. Initially they are limited to the surface of the sample. For this reason the system can be regarded as semi-infinite. In this case the solution of Fick's second law for a flat rectangular sample leads to the following simple equation [162]:

$$\tilde{D} = \left(\frac{\Delta m_t}{A}\right)^2 \frac{\pi}{4t(\Delta C)^2} \tag{3.4.1}$$

where Δm_t is the change of the mass of the sample after time t, A is the surface area of the sample and ΔC is the total change of concentration of defects.

It follows from equation (3.4.1) that, in the initial stage of the reaction, the squares of the changes of the mass of the sample should be a linear function of the reaction time. Hence the chemical diffusion coefficient can be calculated from the slope of the straight line plotted on these coordinates.

In the later stages of the reaction the changes in the concentration of defects reach deeper and deeper parts of the sample and finally they spread right through it. In this case the solution of Fick's second law leads to the following exponential equation:

$$\left[1-\left(\frac{\Delta m_t}{\Delta m_k}\right)\right] = \frac{8}{\pi^2}\exp\left(-\frac{\tilde{D}^2 t}{4l^2}\right) \tag{3.4.2}$$

where Δm_k is the total change of the mass of the sample when the new equilibrium state is established and l is one half of the thickness of the sample.

Taking logarithms of this equation we obtain:

$$\log\left[1-\left(\frac{\Delta m_t}{\Delta m_k}\right)\right] = \log\left(\frac{8}{\pi^2}\right) - \left(\frac{\tilde{D}\pi^2 t}{2.303\cdot 4l^2}\right) \tag{3.4.3}$$

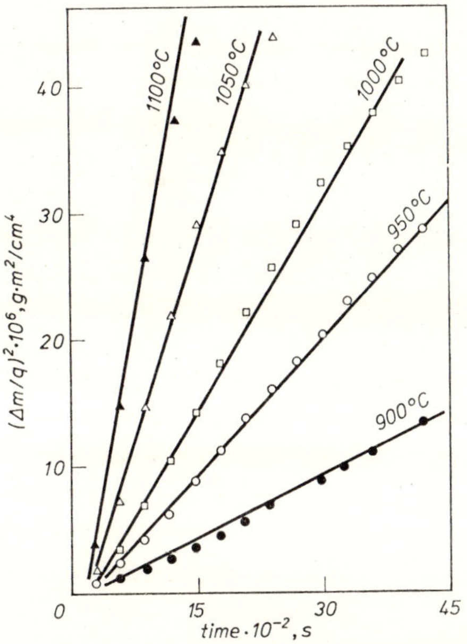

Fig. 199. Parabolic plot of the changes of the mass (reduction) of $Fe_{1-y}O$ sample caused by changes of partial oxygen pressure in the reaction chamber (the initial stage). The composition of $Fe_{1-y}O$ before the measurement: $Fe_{0.8888}O$; after the reduction: $Fe_{0.9500}O$ (after [162]).

It follows from equation (3.4.3) that, in this stage of the reaction, the logarithm of the expression $[(1-\Delta m_t)/\Delta m_k]$ is a linear function of time and that the chemical diffusion coefficient can be calculated from the slope of the straight line plotted with these coordinates. Figures 199 and 200 illustrate the results of the determination of the chemical diffusion coefficient of iron in wustite ($Fe_{1-y}O$) obtained by the thermogravimetric method. It follows from the graph in Fig. 199 that the changes of the mass of the sample in the initial stage of the reaction are a linear function of time with a parabolic plot and in the later stage they are a linear function

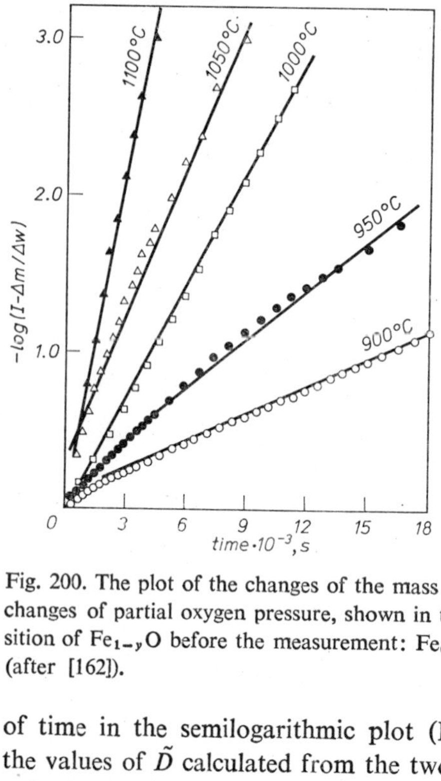

Fig. 200. The plot of the changes of the mass (reduction) of $Fe_{1-y}O$ sample caused by changes of partial oxygen pressure, shown in the semilogarithmic system. The composition of $Fe_{1-y}O$ before the measurement: $Fe_{0.8888}O$; after the reduction: $Fe_{0.9500}O$ (after [162]).

of time in the semilogarithmic plot (Fig. 200). The agreement between the values of $\tilde{D}$ calculated from the two graphs is completely satisfactory.

It should be mentioned that this method makes it possible to check the results internally by calculating $\tilde{D}$ for different stages of the reaction.

3.4.2 *Electrical conductivity method*

It has been shown in the first part of this book that the concentration of ionic defects in compounds of the MeX type showing deviations from stoichiometry is strictly connected with the concentration of electronic defects. Thus, by studying the changes of the electrical conductivity of the sample caused by changes in thermodynamic conditions, it is possible to obtain information about the mobility of ionic defects. However while the thermogravimetric method gives direct information on this mobility, the method based on measurements of electrical conductivity requires the assumption that the mobility of electron defects is independent of their concentration, since the electrical conductivity is given by the product:

$$\sigma = B_i n_i q_i \tag{3.4.4}$$

where B_i is the mobility of the current carriers, n_i is their concentration and q_i is their charge.

It follows from this equation that the electrical conductivity of the crystal is proportional to the concentration of the charge carriers on condition that their mobility does not depend on their concentration. It is also necessary to assume that the degree of ionisation of ionic defects does not change in the concentration range examined. Since only then the ratio of the ionic to the electronic defects remains constant. In the case of small deviations from stoichiometry, i.e. in the case of a low concentration of defects, both these conditions are usually fulfilled, which means that the method can be used.

The procedure used in this method is essentially the same as in the thermogravimetric method. The electrical conductivity of the sample is measured as a function of time when the system changes to a new equilibrium state. The results of these measurements are then used for calculating the chemical diffusion coefficient by means of a modified equation (3.4.3) in which the changes of the mass of the sample are replaced by the changes of its electrical conductivity [162]:

$$\log\left[1-\left(\frac{\Delta\sigma_t}{\Delta\sigma_k}\right)\right] = \log\left(\frac{8}{\pi^2}\right)-\left(\frac{\tilde{D}\pi^2 t}{2.303\cdot 4l^2}\right) \tag{3.4.5}$$

where $\Delta\sigma_t$ and $\Delta\sigma_k$ are the changes of the electrical conductivity at time t and after the new equilibrium state is reached, respectively.

3.4.3 *Solid state cell potential method*

The methods of determination of coefficients $\tilde{D}$ described consist of measurements of changes of mass or electrical conductivity of the sample caused by changes in the concentration of defects in the crystal. The changes in the concentration of defects take place under the influence of rapid changes of the partial pressure of the oxidant in the atmosphere surrounding the crystal. An analogous effect can be obtained by passing electric current for a short time through a specially constructed system containing a solid electrolyte. By measuring the time dependence of the potential changes of such a solid state cell it is possible to determine chemical diffusion coefficients [293].

In this method the following procedure is used: The sample of a disc of the MeX compound is prepared with a suitable thickness, is placed between an analogous tablet made of a solid electrolyte and the reference electrode. In the case of oxides the solid electrolyte is usually calcium oxide stabilised zirconia [$ZrO_2(CaO)$]. This solid solution has practically pure anionic conductivity over a wide range of temperature and oxygen pressure.

In the case of sulphides, the solid electrolyte is silver iodide with practically pure cationic conductivity. The low melting point of this compound (540°C) is a disadvantage, seriously limiting the application of this method for determining chemical diffusion coefficients in metal sulphides.

The use of solid electrolytes for electron blocking is necessary, since compounds with deviations from stoichiometry are almost pure electron conductors (see the first part of this book). For this reason information on the mobility of ionic defects can be obtained by measuring the electrical conductivity of the crystal on condition that the electronic component of this conductivity is eliminated.

In the case of oxides the following solid cell can be constructed:

$$\mathrm{Pt|Ni|NiO|ZrO_2(CaO)|MeO|Pt} \qquad \begin{matrix} \xrightarrow{\qquad\qquad} \\ O_2^{-} \\ \xleftarrow{\qquad\qquad} \end{matrix} \tag{3.4.6}$$

The MeO compound is placed between the platinum electrode and the solid electrolyte. Nickel oxide mixed with metallic nickel is a source of oxygen, from which oxygen ions can be transported through the solid electrolyte to MeO, or can be taken from it by the same route. Obviously, instead of Ni–NiO other systems, e.g. Fe–FeO, Cu–Cu_2O or oxygen gas, can be used.

The system shown in scheme (3.4.6) is heated to a definite temperature, keeping the external circuit closed. As a result the concentration of defects in the disc of oxide MeO is established at a definite level corresponding to the oxygen activity in the system Ni–NiO. Then a definite amount of oxygen is introduced in the disc by passing an electric current from the external source from the right to the left:

$$(-)|Pt|Ni|NiO|ZrO_2(CaO)|MeO|Pt(+) \tag{3.4.7}$$

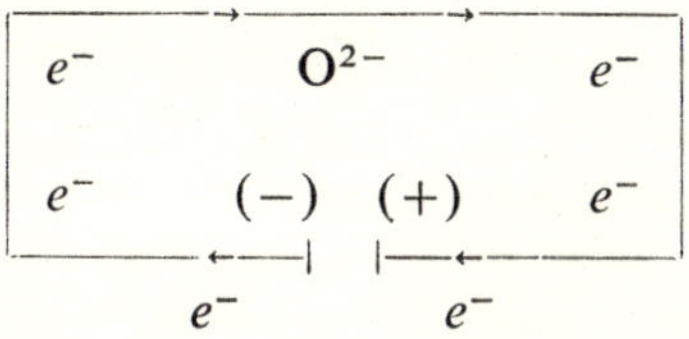

As a result of this process the activity of oxygen in the MeO compound at the phase boundary with the solid electrolyte increases, which is equivalent to a change in the concentration of defects in the surface layer of MeO (an increase of the concentration of cation vacancies or interstitial anions in the case of compounds of the $Me_{1-y}O$ or MeO_{1+y} type or a decrease of the concentration of interstitial cations or anion vacancies in the case of compounds of the $Me_{1+y}O$ or MeO_{1-y} type). The flow of the current through the system shown in scheme (3.4.7), used for causing a definite change of the activity of the oxidant (or metal) in the compound examined is called coulometric titration. When the time of the flow of current through the system is much shorter than that required for the disappearance of the differences in the concentration of defects in the MeO tablet, the rate of equalisation of the concentration of defects in the tablet can be measured by determining the time dependence of the potential changes of the cell after disconnecting the external current source. When the MeO disc is sufficiently thick, the system can be regarded as semi-infinite and the chemical diffusion coefficient can be calculated by means of the following equation [293]:

$$\Delta V_t = E\left(\frac{8}{\pi^2}\right)\exp\left(-\frac{\pi^2 \tilde{D} t}{4L^2}\right) \tag{3.4.8}$$

where ΔV_t is the change of potential of the cell during time t counted from the moment of interruption of current flow, E is the potential of the cell during the period preceding the coulometric titration and L is the thickness of the MeO disc.

Taking logarithms of equation (3.4.8) we obtain:

$$\log(\Delta V_t) = \log E\left(\frac{8}{\pi^2}\right) - \frac{\pi^2 \tilde{D} t}{2.303 \cdot 4L^2} \tag{3.4.9}$$

It follows from this equation that the plot of the logarithm of potential changes against time is linear and consequently coefficient $\tilde{D}$ can be simply calculated graphically or analytically.

In order to illustrate the above methods, selected values of chemical diffusion coefficients in several oxides and sulphides of metals are shown in Table 16.

TABLE 16

Chemical diffusion coefficients in some MeX-type compounds

Compound	Temperature °C	Atmosphere	Method*	D cm²/s
CoO	1000	$Ar–O_2$	tg	
CoO	1000	$Ar–O_2$	pe	3.2×10^{-7}
NiO	1000	$Ar–O_2$	pe	2.0×10^{-7}
FeO	1000	$CO–CO_2$	tg	1.4×10^{-6}
FeO	1100	$CO–CO_2$	tg	7.8×10^{-6}
Cu_2O	1000	O_2	pe	3.0×10^{-4}
MnO	1030	$CO–CO_2$	pe	1.2×10^{-6}
ZnO	600	O_2	pe	4.0×10^{-6}
FeS	800	$H_2–H_2S$	tg	7.0×10^{-6}
FeS	900	$H_2–H_2S$	tg	1.7×10^{-5}
Cu_2S	440	Ar	ec	8.0×10^{-2}
Ag_2S	200	Ar	ec	6.0×10^{-1}

* pe — electrical conductivity method
tg — thermogravimetric method
ec — solid cell electromotive force method

4. Selected Values of Parameters of Self-diffusion and Heterodiffusion in Metals, Alloys and Metal Oxides, Sulphides and Halides

This chapter is a supplement to the descriptive material in previous chapters. It contains selected results of studies on the rates of self-diffusion and heterodiffusion in metals, alloys and metal oxides, sulphides and halides. Only the data which can be used to calculate the energy of activation of diffusion E_D and the frequency factor D_0 in the Arrhenius equation:

$$D = D_0 \exp - \frac{E_D}{RT}$$

have been included. For this reason the tables only contain the numerical values of D_0 and E_0 for the temperature range given. These data will make it possible for the reader to calculate the diffusion coefficient D at any temperature on the basis of the Arrhenius equation.

The numerical data are divided into 6 paragraphs. Paragraph 4.1. contains data concerning self-diffusion in pure metals. In paragraph 4.2 data characterising the rates of diffusion of dopants in metals are assembled. In paragraph 4.3 the quantities for diffusion in alloys are given. Finally, paragraphs 4.4, 4.5 and 4.6 contain the results of studies on self-diffusion, heterodiffusion and chemical diffusion in oxides, sulphides and halides of metals.

4.1 Self-diffusion in metals

Material	Tracer	Temperature range °C	Activation energy E_D		Frequency factor D_0 cm^2/s	Ref.
			kJ/mole	kcal/mole		
Aluminium	^{26}Al	400–610	144.4	34.5	—	[294]
	Al	130–200	123.5 ± 3.9	1.28 ± 0.04eV	0.19 ± 0.06	[295]
Antimony	^{124}Sb	500–630	201.0	48.0	56.0	[296]
Beryllium	^{7}Be	565–1065	157.4	37.6	0.52	[297]
		565–1065	165.0	39.4	0.62	[297]
Cerium	^{141}Ce	719–771	87.5	20.9	8.5×10^{-3}	[298]
		745–791	84.6	20.2	7×10^{-3}	[298]
		528–692	153.7	36.7	5.5×10^{-1}	[298]
Cadmium	^{115}Cd	130–280	80.0	19.1	0.10	[299]
		180–300	86.2	20.6	0.68	[300]
		240–300	85.8	20.5	0.031	[301]
		220–300	82.1	19.6	0.012	[301]
Chromium	^{51}Cr	1200–1600	306.5	73.2	0.28	[302]
		1030–1545	308.6	73.7	0.2	[303]
		1000–1750	339.1	81.0	1.6	[304]
β-Cobalt	^{60}Co	772–1048	273.8	65.4	0.50	[305]
		1192–1297	260.4	62.2	0.17	[305]
		1047–1311	287.6	68.7	1.66	[306]
Copper	^{64}Cu	760–1060	211.4	50.5	0.78	[307]
	^{64}Cu, ^{67}Cu	700–1060	211.3 ± 0.1	2.19 ± 0.01	0.78	[308]
	^{64}Cu	863–1057	243.7	58.2	0.33	[309]
		780–890	260.0	49.2	2.32	[310]
Germanium	^{71}Ge	730–916	290.6	69.4	10.8	[311]
		730–916	324.5	77.5	44.0	[311]

4.1 *(cont.)*.

Material	Tracer	Temperature range °C	Activation energy E_D		Frequency factor D_0 cm^2/s	Ref.
			kJ/mole	kcal/mole		
Gold	^{198}Au	700–900	176.3	42.1	0.117	[312]
		850–1050	176.94	42.26	0.107	[313]
		706–1010	191.3	45.7	0.15	[314]
Indium		130–149	88.8	21.2	8.0	[315]
α-Iron	^{55}Fe, ^{59}Fe	700–790	251.2	60.0	2.0	[316]
		790–900	239.4	57.2	1.9	[316]
		860–900	239.9	57.3	2.0	[315]
		638–768	254.1	60.7	27.5	[317]
		808–884	240.7	57.5	2.01	[317]
γ-Iron		900–1200	284,3	67.9	1.3	[318]
		1064–1395	270.0	64.5	0.18	[316]
		950–1200	283.4	67.7	2.5	[319]
		1156–1350	268.0	64.0	0.22	[315]
		1075–1340	283.9	67.8	1.05	[320]
δ-Iron	^{55}Fe, ^{59}Fe	1405–1520	177.5	42.4	0.019	[321]
		1415–1510	238.6	57.0	1.9	[322]
		1405–1515	258.3	61.7	6.8	[315]
		1428–1492	240.7	57.5	2.01	[317]
Molybdenum	^{99}Mo	1600–2200	422.0	100.8	0.38	[323]
		1850–2350	386.0	92.2	0.1	[324]
		1850–2350	405.7	96.9	0.5	[324]
		2155–2540	460.5	110.0	1.8	[325]
Nickel	^{63}Ni	680–830	291.8	69.7	5.8	[326]
		950–1020	279.7	66.8	1.9	[327]

4.1 *(cont.)*.

Material	Tracer	Temperature range °C	Activation energy E_D		Frequency factor D_0 cm²/s	Ref.
			kJ/mole	kcal/mole		
Nickel	^{63}Ni	1085–1300	291.0	69.5	2.59	[328]
		950–1250	280.5	67.0	2.4	[329]
		475–650	279.7	66.8	1.9	[330]
		675–750	279.7	66.8	1.9	[331]
		1000–1400	253.3	60.5	9.96	[332]
Niobium	^{95}Nb	1700–2100	397.7	95.0	1.3	[333]
		878–2395	401.9	96.0	1.1	[334]
		1700–2100	481.5	115.0	19.0	[335]
		1700–2100	460.5	110.0	17.0	[335]
Palladium	^{103}Pd	1060–1500	266.3	63.6	0.205	[336]
Platinum	^{193}Pt	960–1160	276.3	66.0	0.57	[337]
	^{195}Pt, ^{199}Pt	1325–1600	285.5	68.2	0.33	[338]
		1250–1725	278.4	66.5	0.22	[339]
Silicon	^{31}Si	1225–1400	460.5	110.0	1.8×10^3	[340]
		1100–1300	496.1	118.5	9×10^3	[341]
Silver	^{110}Ag	500–900	192.4	45.95	0.895	[342]
		857–951	184.6	44.1	0.46	[343]
		640–955	189.2 ± 0.8	45.2 ± 0.2	0.67	[344]
		500–950	192.6	46.0	0.9	[345]
		670–940	190.5	45.50	0.724	[346]
		640–905	170.8	40.8	0.11	[347]
		725–925	198.5	47.4	1.8	[348]
		750–925	188.4	45.0	0.53	[349]
		550–900	184.4	44.05	0.40	[350]

4.1 *(cont.)*.

Material	Tracer	Temperature range °C	Activation energy E_D		Frequency factor D_0 cm^2/s	Ref.
			kJ/mole	kcal/mole		
Silver	^{110}Ag	707–880	188.0	44.9	0.834	[351]
		630–940	184.60	44.09	0.395	[352]
		715–940	183.0	43.7	0.27	[353]
		500–900	186.7	44.6	0.86	[354]
		750–950	173.9	41.53	0.094	[355]
		650–900	187.6	44.8	1.08	[356]
		250–380	182.1	43.5	0.34	[357]
		750–900	185.9	44.4	1.06	[358]
		668–958				[359]
		720–950	181.67	43.39	0.278	[360]
β-Tin	^{113}Sn, ^{123}Sn	130–255	93.8	22.4	0.78	[361]
		160–226	105.1	25.1	10.7	[362]
		160–226	107.2	25.6	7.7	[362]
		416–490	32.2	7.7	1.73×10^{-6}	[363]
α-Titanium	^{44}Ti	690–850	122.7	29.3	6.4×10^{-8}	[364]
β-Titanium	^{44}Ti	900–1540	130.6	31.2	3.58×10^{-4}	[365]
		900–1580	152.8	36.5	1.9×10^{-3}	[366]
Tungsten	^{185}W	2000–2700	504.5	120.5	0.54	[367]
		2670–3225	641.0	153.1	42.8	[368]
		1740–2100	389.8	93.1	1.8×10^{-3}	[369]
α-Uranium	^{234}U	580–650	167.5	40.0	2×10^{-3}	[370]
β-Uranium	^{234}U	700–760	175.760	42.0	0.0135	[371]
γ-Uranium	^{234}U	800–1070	119.3	28.5	2.33×10^{-3}	[372]
Vanadium	^{48}V	880–1360	308.36	73.65	0.36	[373]

4.1 *(cont.)*.

Material	Tracer	Temperature range °C	Activation energy E_D		Frequency factor D_o cm²/s	Ref.
			kJ/mole	kcal/mole		
Vanadium	^{48}V	1360–1830	394.15	94.14	214.0	[373]
Zinc	^{65}Zn	240–418	96.3	23.0	0.18	[374]
α-Zirconium	^{95}Zr	750–850	190.5	45.5	5.6×10^{-4}	[375]
β-Zirconium	^{95}Zr	1100–1500	126.0	30.1	2.4×10^{-4}	[376]
		900–1750	82.1	19.6	3×10^{-6}	[377]

4.2 Diffusion of dopants in metals

Material	Tracer dopant	Temperature range °C	Activation energy E_D		Frequency factor D_0 cm^2/s	Ref.
			kJ/mole	kcal/mole		
Aluminium	^{110}Ag	450–632	116.8	27.9	0.08	[378]
	^{115}Cd	400–630	90.9	21.7	7.94×10^{-8}	[379]
	^{64}Cu	350–630	126.4	30.20	0.15	[380]
	^{51}Cr	360–630	83.3	19.9	1.1×10^{-6}	[381]
	^{59}Fe	360–630	58.2	13.9	4.1×10^{-9}	[381]
	^{114}In	400–630	92.9	22.2	1.42×10^{-8}	[379]
	^{54}Mn	450–650	120.6	28.8	0.22	[382]
	^{99}Mo	400–630	54.8	13.1	1.04×10^{-9}	[383]
	^{63}Ni	360–630	65.7	15.7	2.9×10^{-8}	[381]
	^{103}Pd	400–630	84.6	20.2	1.92×10^{-7}	[379]
	^{124}Sb	448–620	121.8	29.1	0.09	[378]
	^{113}Sn	400–630	84.6	20.2	3.05×10^{-7}	[379]
	^{48}V	400–630	82.1	19.6	6.05×10^{-8}	[384]
	^{57}Fe	20–642	134.8	32.2	0.12	[385]
	Si	344–631	123.9	29.6	0.346	[386]
	^{65}Zn	340–640	120.6	28.8	0.2	[387]
	^{54}Mn	500–650	229.0	54.7	1275	[388]
	^{65}Zn	700–900	121.4	29.0	0.30	[389]
	^{60}Co	400–640	173.3	41.4	250	[390]
	^{95}Zr	531–640	241.6 ± 6.49	57.7 ± 1.55	7.28×10^2	[391]
	^{95}Zr	555–640	218.1 ± 4.13	52.1 ± 2.18	3.43×10	[391]
	^{95}Zr	531–649	289.3 ± 12.72	69.1 ± 3.04	5.48×10^5	[391]
	^{55}Mn, ^{56}Mn	460–660	20.9	50.0	104	[392]
	^{28}Mg	394–655	130.42	31.15	1.24	[393]

4.2 (*cont.*).

Material	Tracer dopant	Temperature range °C	Activation energy E_D		Frequency factor D_0 cm²/s	Ref.
			kJ/mole	kcal/mole		
Aluminium	^{64}Cu	375–619	138.2	33.0	1.3	[394]
	^{198}Au	290–650	134.0	32.0	2.2	[394]
	^{114}I	442–656	122.3	29.2	1.16	[395]
Beryllium	^{110}Ag	650–900	180.9	43.2	1.76	[396]
	^{59}Fe	700–1076	216.0	51.6	0.67	[396]
	^{63}Ni	800–1250	242.8	58.0	0.2	[397]
Cadmium	^{110}Ag	180–300	106.3	25.4	2.21	[300]
	^{65}Zn	180–300	79.5	19.0	0.0016	[300]
Chromium	^{55}Fe	980–1420	332.0	79.3	0.47	[398]
			169.1	40.4	1.1×10^{-6}	
	^{14}C	600–1400	142.3	34.0	0.21	[399]
	^{55}Fe	930–1240	232.4	55.5	0.2	[400]
	^{63}Ni	1227–1416	279.3	71.0	3.35	[306]
	^{35}S	1150–1250	273.8	65.4	1.3	[401]
	^{193}Pt	1081–1208	279.3	66.7	0.65	[337]
Copper	^{198}Au	706–1010	191.3	45.7	0.15	[314]
	^{59}Fe	720–1060	213.32	50.95	1.01	[402]
	^{66}Ni	855–1055	232.8	55.6	1.93	[403]
	^{102}Pd	807–1056	228.47	54.37	1.71	[404]
	^{195}Pt	843–997	240.7	37.5	4.8×10^{-4}	[405]
	^{113}Sn	680–910	188.4	45.0	0.11	[406]
	^{204}Tl	785–996	181.3	43.3	0.71	[407]
	^{110}Ag	770–1070	195.02	46.58	0.574	[408]
	^{110}Ag	770–1070	263.8	64.3	0.63	[409]

4.2 (*cont.*).

Material	Tracer dopant	Temperature range °C	Activation energy E_D		Frequency factor D_0 cm^2/s	Ref.
			kJ/mole	kcal/mole		
Copper	^{109}Cd	770–1070	194.64	46.49	1.27	[408]
	^{109}Cd	770–1070	191.3	45.7	0.935	[410]
	^{114}In	770–1070	193.6	46.24	1.30	[408]
	^{67}Ga	880–1080	192.68	46.02	0.523	[411]
	^{68}Ge	840–1050	185.52	44.31	0.315	[411]
	^{59}Fe	460–1070	217.7	52.0	1.36	[412]
	^{51}Cr	800–1070	224.0	53.5	1.02	[412]
	^{95}Nb	807–906	251.2	60.0	2.04	[413]
	Pt	750–1075	233.2	55.7	0.67	[414]
	^{65}Zn	390–810	185.1	44.2	0.30	[415]
	Rh	750–1075	242.8	58.0	3.3	[416]
	^{59}Fe	732–1024	215.6	51.5	1.3	[417]
	^{103}Ru	948–1062	257.5	61.5	8.5	[417]
	^{7}Be	700–1075	195.94	46.80	0.66	[418]
	^{124}Sb	720–1015	182.5	43.6	0.61	[419]
	^{113}Sn	737–1047	188.20	44.95	0.842	[419]
	^{113}Sn	700–1075	187.6	44.8	0.82	[420]
	^{54}Mn	700–1075	195.5	46.7	0.74	[421]
	^{51}Cr	827–962	240.7	57.5	1.6	[442]
Germanium	^{124}Sb	720–900	210.2	50.2	0.22	[423]
	^{125}Te	770–900	274.49	65.56	2.6	[424]
	^{204}Tl	800–930	328.2	78.4	1.7×10^3	[425]
Gold	^{110}Ag	773–1039	169.1	40.4	0.08	[426]
	^{60}Co	702–948	174.2	41.6	0.068	[312]

4.2 (*cont.*).

Material	Tracer dopant	Temperature range °C	Activation energy E_D		Frequency factor D_0 cm^2/s	Ref.
			kJ/mole	kcal/mole		
Gold	^{59}Fe	701–948	174.2	41.6	0.082	[312]
	^{203}Hg	600–1027	158.26	37.38	0.116	[427]
	^{63}Ni	702–988	175.8	42.0	0.034	[312]
Indium	^{110}Ag	25–140	53.6	12.8	0.52	[428]
	^{198}Au	25–140	28.1	6.7	9×10^{-3}	[428]
Iron	^{195}Au	800–900	261.3	62.4	31	[429]
	^{14}C	616–844	122.7	29.3	2.2	[430]
	^{60}Co	800–905	270.5	64.6	64.4	[431]
	^{60}Co	808–884	257.1	61.4	6.38	[317]
	^{51}Cr	775–875	240.7	57.5	2.53	[432]
	^{64}Cu	800–1050	238.6	57.0	0.57	[433]
	^{99}Mo	750–875	305.6	73.0	7.8×10^{3}	[434]
	^{63}Ni	700–900	259.2	61.9	9.9	[429]
	^{124}Sb	800–900	278.8	66.6	1100	[435]
	^{185}W	700–900	265.9	63.5	69	[436]
	^{7}Be	1100–1350	241.2	57.6	0.1	[437]
	^{14}C	800–1400	142.4	34.0	0.15	[399]
	^{60}Co	1050–1250	435.4	104.0	1.2×10^{5}	[435]
	^{181}Hf	1075–1340	473.1	113.0	9×10^{4}	[320]
	^{95}Nb	1075–1340	344.6	82.3	530	[320]
	^{63}Ni	930–1050	280.5	67.0	0.77	[438]
	^{32}P	1280–1350	292.2	69.8	28.3	[439]
	^{35}S	1200–1350	202.6	48.4	1.35	[440]
	^{185}W	1050–1250	376.8	90.0	10^{3}	[441]

4.2 (*cont.*).

Material	Tracer dopant	Temperature range °C	Activation energy E_D kJ/mole	Activation energy E_D kcal/mole	Frequency factor D_0 cm^2/s	Ref.
Iron	^{110}Ag	748–888	288.9	69.0	1.95×10^3	[442]
	^{110}Ag	700–*Curie T*	297.2	71.0	230	[443]
	^{110}Ag	*Curie T*–900	277.2	66.2	150	[443]
	^{60}Co	800–899	234.0	55.9	1.83	[444]
	^{54}Mn	920–1280	261.7	62.5	0.16	[445]
	^{54}Mn	800–900	219.8	52.5	0.35	[445]
	^{54}Mn	700–760	233.6	55.8	1.49	[445]
	^{113}Sn	800–910	221.9	53.0	2.4	[446]
	^{35}S	700–900	231.5	55.3	34.6	[447]
	S	950–1250	221.9	53.0	1.7	[448]
	^{193}Pt	960–1260	296.0	70.7	2.7	[337]
	chem.	930–1010	295.6	70.6	1.9	[449]
Lead	^{110}Ag	163–298	60.79	14.52	4.42×10^{-2}	[450]
	^{124}Sb	315–188	92.9	22.2	0.29	[451]
	^{65}Zn	182–299	47.3	11.3	0.016	[452]
Molybdenum	^{14}C	1200–1600	171.7	41.0	2.04×10^{-2}	[453]
	^{60}Co	1850–2350	446.7	106.7	18	[454]
	^{95}Nb	1850–2350	452.6	108.1	14	[454]
	^{35}S	2220–2470	422.8	101.0	320	[455]
	^{185}W	1700–2260	460.5	110	1.7	[456]
	^{59}Fe	900–1200	291.8	69.7	3.7×10^{-3}	[457]
	^{35}S	900–1200	297.3	71.0	0.034	[457]
	^{42}K	800–1100	57.4	13.7	1.78×10^{-10}	[458]
	^{42}K	800–1100	104.7	25.0	5.5×10^{-9}	[458]

4.2 (*cont.*).

Material	Tracer dopant	Temperature range °C	Activation energy E_D		Frequency factor D_0 cm²/s	Ref.
			kJ/mole	kcal/mole		
Molybdenum	^{24}Na	800–1100	55.3	13.2	4.2×10^{-10}	[458]
	^{24}Na	800–1100	88.8	21.2	2.95×10^{-9}	[458]
	^{59}Fe	1000–1350	346.2	82.7	0.15	[459]
	Ti	1100–1400	205.2	49.0	1.64×10^{-4}	[460]
	Y	1200–1600	214.8	51.3	1.8×10^{-4}	[461]
Nickel	^{198}Au	700–1075	230.3	55.0	0.02	[642]
	^{7}Be	1020–1400	193.4	46.2	0.019	[437]
	^{14}C	600–1400	142.3	34.0	0.12	[399]
	^{60}Co	1149–1390	275.9	65.9	1.39	[306]
	^{64}Cu	1050–1360	258.3	61.7	0.57	[309]
	^{59}Fe	940–1100	255.4	61.0	0.8	[463]
	^{63}Ni	900–1200	289.7	69.2	2.22	[464]
	^{124}Sb	1020–1220	113.0	27.0	1.8×10^{-5}	[465]
	^{48}V	800–1300	274.24	66.50	0.87	[384]
	^{185}W	1100–1300	299.4	71.5	2.0	[380]
	^{110}Ag	507–782	197.6	47.2	0.025	[466]
	^{110}Ag	780–1053	292.7	69.9	104.0	[443]
	^{193}Pt	1081–1208	286.8	68.5	2.5	[337]
	^{59}Fe	1205–1395	268.8	64.2	1.0	[467]
	U	975–1075	236.1	56.4	1.0	[468]
	^{144}Ca	700–1100	254.6	60.8	0.66	[469]
	^{147}Nd	700–1100	250.45	59.82	0.44	[469]
Niobium	^{14}C	130–2340	141.9	33.9	0.01	[470]
	^{14}C	930–1800	158.7	37.9	0.033	[471]

4.2 (*cont.*).

Material	Tracer dopant	Temperature range °C	Activation energy E_D kJ/mole	Activation energy E_D kcal/mole	Frequency factor D_0 cm^2/s	Ref.
Niobium	^{60}Co	1500–2100	295.2	70.5	0.74	[472]
	^{51}Cr	947–1493	337.41	80.59	0.13	[473]
	^{55}Fe	1400–1200	325.3	77.7	1.5	[472]
	^{95}Nb	1600–2000	473.1	113	18.18	[474]
	^{35}S	1100–1500	306.1	73.1	2.6×10^{3}	[475]
	^{113}Sn	1850–2400	330.3	78.9	0.14	[476]
	^{48}V	1000–1400	355.9	85.0	2.21	[477]
	^{42}K	900–1100	56.9	13.6	3.09×10^{-10}	[378]
	^{42}K	900–1100	92.5	22.1	2.38×10^{-9}	[478]
	^{24}Na	900–1100	41.9	10.0	1.15×10^{-9}	[478]
	^{24}Na	900–1100	77.9	18.6	1.611×10^{-9}	[478]
	^{63}Ni	988–1246	336.7	80.4	9.3	[479]
	O	1060–1650	129.8	31.0	1.5×10^{-2}	[480]
	^{91}Y	1200–1600	232.8	55.6	1.5×10^{-3}	[461]
Silicon	Ag	1000–1200	293.1	70.0	4.05	[481]
	Au	1000–1200	83.7	20.0	5×10^{-5}	[481]
	Ga	1000–1200	201.0	48.0	6×10^{-3}	[481]
	As	1167–1394	331.6	79.2	6.55×10^{-2}	[482]
	Sb	1190–1405	351.7	84.0	1.14×10^{-1}	[482]
	Bi	1190–1394	370.9	88.6	1.08	[482]
	^{71}Ge	1225–1325	453.0	108.2	1.535×10^{3}	[483]
Silver	^{198}Au	718–942	202.1	48.28	0.85	[484]
	^{60}Co	600–850	125.2	29.9	3.0×10^{-7}	[485]
	^{59}Fe	600–850	123.9	29.6	9.4×10^{-7}	[485]

4.2 (*cont.*).

Material	Tracer dopant	Temperature range °C	Activation energy E_D		Frequency factor D_0 cm^2/s	Ref.
			kJ/mole	kcal/mole		
Silver	^{109}Ca, ^{60}Co	375–650	127.7	30.5	8.8×10^{-7}	[486]
	As	642–940	149.5	35.7	0.042	[487]
	^{114}In	770–940	174.97	41.79	0.55	[488]
	^{60}Fe	800–932	205.2	49.0	2.6	[417]
	^{60}Co	700–941	204.1	48.75	1.9	[417]
	^{63}Ni	749–950	229.4	54.8	21.9	[489]
	^{102}Pd	736–939	237.6	56.75	9.56	[404]
	^{35}S	600–900	167.5	40.0	1.65	[490]
	^{124}Sb	780–950	163.58	39.07	0.234	[491]
	^{65}Zn	785–895	174.6	41.7	0.532	[492]
Gold	Sn	700–1000	143.2	34.2	0.04	[493]
	In	700–1000	153.7	36.7	0.07	[493]
Tin	^{124}Sb	425–500	121.4	29.0	71.0	[494]
	^{124}Sb	425–500	123.1	29.4	73.0	[494]
	^{109}Cd	200–500	118.1	28.2	220.0	[494]
	^{109}Cd	200–500	115.6	27.6	120.0	[494]
	^{203}Hg	174–226	112.2	26.8	30.0	[495]
	^{203}Hg	174–223	105.9	25.3	7.5	[495]
	^{65}Zn	152–225	50.2 ± 1.7	12.0 ± 0.4	$1.1 \pm 0.5 \times 10$	[494]
Tantalum	^{14}C	600–2200	99.6	23.8	2.78×10^{-3}	[796]
	^{95}Nb	921–2484	413.2	98.7	0.23	[497]
Thallium	^{14}C	190–2680	116.6	38.60	0.0067	[470]
	Ti	1100–1400	201.0	48.0	8.62×10^{-5}	[460]
	^{91}Y	1200–1500	302.3	72.2	0.12	[461]

4.2 (*cont.*).

Material	Tracer dopant	Temperature range °C	Activation energy E_D		Frequency factor D_0 cm^2/s	Ref.
			kJ/mole	kcal/mole		
Titanium	^{14}C	600–800	127.7	30.5	7.9	[498]
	^{14}C	1100–1400	83.7	20.0	3.02×10^{-3}	[453]
	^{110}Ag	940–1590	180.0	43.0	3×10^{-3}	[499]
	^{51}Cr	950–1600	147.0	35.1	5×10^{-3}	[500]
			255.4	61	4.9	
	^{60}Co	900–1600	128.1	30.6	1.2×10^{-2}	[500]
	Ga	600–860	181.7 ± 19.7	43.4 ± 4.7	4.4×10^{-4}	[501]
	^{59}Fe	700–850	110.5	26.4	1.2×10^{-4}	[502]
	^{59}Fe	1000–1200	131.0	31.3	5.6×10^{-3}	[502]
	^{55}Fe	900–1600	132.3	31.6	7.8×10^{-3}	[500]
			230.3	55	2.7	
	^{99}Mo	1100–1560	214.8	51.3	0.24	[503]
	^{54}Mn	900–1600	141.1	33.7	6.1×10^{-3}	[500]
	^{95}Nb	1000–1600	164.5	39.3	5.0×10^{-3}	[500]
	^{63}Ni	295–1600	123.9	29.6	9.2×10^{-3}	[500]
	^{113}Sn	950–1600	132.3	31.6	3.8×10^{-4}	[504]
	^{235}U		122.7	29.3	5.1×10^{-4}	[505]
	^{48}V	900–1545	134.8	32.2	3.1×10^{-4}	[506]
	^{185}W	900–1250	183.8	43.9	3.6×10^{-3}	[503]
	^{95}Zr	920–1500	148.2	35.4	4.7×10^{-3}	[503]
	^{14}C	1200–1600	224.0	53.5	$8.9<10^{-3}$	[453]
Tungsten	^{99}Mo	2000–2400	506.6	121	5×10^{-2}	[507]
	^{186}Re	2100–2400	590.3	141	19.5	[507]
	^{14}C	1500–1800	158.3	37.8	$3.45\pm0.12\times10^{-3}$	[508]

4.2 (*cont.*).

Material	Tracer dopant	Temperature range °C	Activation energy E_D		Frequency factor D_0 cm^2/s	Ref.
			kJ/mole	kcal/mole		
Tungsten	^{35}S	1880–2180	293.2	69.8	2.17×10^{-5}	[509]
Uranium	^{195}Au	785–1007	127.3	30.4	4.86×10^{-3}	[510]
	^{60}Co	783–989	52.63	12.57	3.51×10^{-4}	[511]
	^{51}Cr	797–1037	102.41	24.46	5.47×10^{-3}	[511]
	^{64}Cu	787–1039	100.73	24.06	1.96×10^{-3}	[511]
	^{55}Fe	787–990	50.28	12.01	2.69×10^{-4}	[511]
	^{54}Mn	787–939	58.11	13.88	1.81×10^{-4}	[511]
	^{63}Ni	787–1039	65.57	15.66	5.36×10^{-4}	[511]
	^{95}Nb	791–1102	160.01	39.65	4.87×10^{-2}	[511]
Vanadium	^{51}Cr	960–1200	270.5	64.6	9.54×10^{-3}	[398]
	^{55}Fe	842–1171	295.2	70.5	0.60	[373]
	^{44}Ti	1100–1800				
Zinc	^{14}C	600–808	128.5	30.7	$3.51\pm2.16\times10^{-5}$	[498]
	^{110}Ag	271–413	108.9	26.0	0.32	[512]
	^{198}Au	315–415	124.47	29.73	0.97	[513]
	^{115}Cd	225–416	86.00	20.54	0.114	[513]
	^{64}Cu	338–415	125.27	29.92	2.0	[514]
	^{72}Ga	240–403	75.99	18.15	0.018	[514]
	^{114}In	271–413	79.97	19.1	0.062	[512]
α-Zirconium	^{95}Nb	894–1160	131.8	31.5	1.23×10^{-4}	[515]
	^{54}Mn	620–850	126.4	30.2	2.4×10^{-3}	[516]
	^{54}Mn	950–1150	138.2	33.0	5.6×10^{-3}	[516]
	^{235}U	950–1500	242.8	58.0	0.36	[517]
	^{51}Cr	700–850	75.4	18.0	1.19×10^{-8}	[518]

4.2 (*cont.*).

Material	Tracer dopant	Temperature range °C	Activation energy E_D		Frequency factor D_0 cm^2/s	Ref.
			kJ/mole	kcal/mole		
α-Zirkonium	^{55}Fe	750–840	201.0	48.0	2.5×10^{-2}	[519]
	^{48}V	600–850	95.9	22.9	1.12×10^{-8}	[477]
	^{14}C	1100–1600	143.2	34.2	3.57×10^{-2}	[520]
	^{51}Cr	900–1200	134.0	32.0	4.17×10^{-3}	[521]
	^{59}Fe	900–1400	113.0	27.0	9.1×10^{-3}	[521]
	^{99}Mo	900–1200	185.9	44.4	3.63×10^{-2}	[521]
	^{95}Nb	882–1758	105.1	25.1	9×10^{-6}	[522]
	^{48}V	870–1200	191.8	45.8	7.59×10^{-3}	[521]
	^{185}W	900–1250	233.6	55.8	0.41	[521]
	^{64}Cu	653–786	75.8	18.1	8.4×10^{-2}	[523]
Praseodymium	Fe	900–1330	83.7	20	1.8×10^{-2}	[524]
	Ag	905–1095	75.4	18	5.4×10^{-3}	[524]

4.3 Diffusion in alloys

Principal metal	Alloying additive % wt/wt	Tracer	Temperature range °C	Activation energy E_D kJ/mole	Activation energy E_D kcal/mole	Frequency factor D_0 cm²/s	Ref.
Aluminium	30 Ag	^{110}Ag	400–595	121.0	28.9	0.39	[425]
	48.3 Ni	^{63}Ni	1000–1350	178.0	42.5	0.00012	[526]
	49.0 Ni	^{63}Ni	1000–1350	200.5	47.9	0.00053	[526]
	50.0 Ni	^{63}Ni	1000–1350	307.3	73.4	4.461	[526]
	54.5 Ni	^{63}Ni	1000–1350	250.4	59.8	0.150	[526]
	58.5 Ni	^{63}Ni	1000–1350	253.7.	60.6	0.096	[526]
Chromium	20 Ni	^{51}Cr	1150–1450	259.6	62.0	0.28	[304]
	65 Ni	^{51}Cr	950–1360	243.3	58.6	0.2	[304]
	65 Ni	^{63}Ni	950–1360	205.1	49.0	4×10^{-3}	[304]
	31 Fe	^{51}Cr	1000–1275	316.1	75.5	24.6	[527]
	49 Fe	^{51}Cr	1000–1275	293.1	70.0	40	[527]
	9.4 Fe	^{59}Fe	1000–1250	219.0	52.3	5×10^{-4}	[528]
	19 Fe	^{59}Fe	967–1250	432.1	103.2	3.4×10^{5}	[529]
	19.0 Fe	^{59}Fe	1000–1150	284.7	68.0	1.0	[528]
	28.5 Fe	^{59}Fe	950–1250	311.9	74.5	32.0	[528]
	38.5 Fe	^{59}Fe	957–1250	369.3	88.2	10^{4}	[529]
	38.5 Fe	^{59}Fe	1050–1150	334.9	80.0	2×10^{2}	[528]
	43.5 Fe	^{59}Fe	957–1250	340.8	81.4	1.5×10^{3}	[529]
	47.2 Fe	^{59}Fe	957–1250	354.6	84.7	7.7×10^{3}	[529]
	48.3 Fe	^{59}Fe	1000–1200	288.1	68.8	27.0	[528]
Cobalt	6.2 Ni	^{60}Co	1213–1368	305.2	72.9	7.4	[306]
	8 Ni	^{60}Co		340.0	81.2	124	[500]
	10.9 Ni	^{60}Co	864–1048	280.9	67.1	0.61	[305]
	10.9 Ni	^{60}Co	1144–1297	266.3	63.6	0.21	[305]

4.3 (*cont.*).

Principal metal	Alloying additive % wt/wt	Tracer	Temperature range °C	Activation energy E_D		Frequency factor D_0 cm^2/s	Ref.
				kJ/mole	kcal/mole		
Cobalt	11.0 Ni	^{60}Co	1163–1393	290.1	69.3	2.52	[306]
	19.4 Ni	^{60}Co	1163–1393	275.9	65.9	0.99	[306]
	20.1 Ni	^{60}Co	845–988	307.7	73.5	5.96	[305]
	20.1 Ni	^{60}Co	1090–1246	297.3	71.0	2.42	[305]
	27.4 Ni	^{60}Co	1163–1393	270.5	64.6	0.70	[306]
	0.26 Si	^{60}Co	778–1048	273.8	65.4	0.5	[305]
	0.26 Si	^{60}Co	1192–1297	260.4	62.2	0.17	[305]
	6.2 Ni	^{63}Ni	1213–1368	302.3	72.2	5.4	[306]
	10.9 Ni	^{63}Ni	864–1048	274.7	65.6	0.46	[305]
	10.9 Ni	^{63}Ni	1144–1297	262.5	62.7	0.17	[305]
	11.0 Ni	^{63}Ni	1241–1410	304.0	72.6	6.42	[306]
	19.4 Ni	^{63}Ni	1163–1410	285.0	68.2	1.89	[306]
	20.1 Ni	^{63}Ni	864–988	291.8	69.7	1.66	[305]
	20.1 Ni	^{63}Ni	1060–1246	275.1	65.7	0.41	[305]
	27.4 Ni	^{63}Ni	1163–1410	281.4	67.2	1.60	[306]
	30.1 Ni	^{63}Ni	772–899	286.8	68.5	2.01	[305]
	30.1 Ni	^{63}Ni	1048–1246	271.3	64.8	0.67	[305]
	0.26 Si	^{63}Ni	772–1048	269.2	64.3	0.34	[305]
	0.26 Si	^{63}Ni	1192–1297	252.0	60.2	0.10	[305]
	10 Ni	chem.	880–1300	298.5	71.3	1.76	[530]
	20 Ni	chem.	880–1300	295.2	70.5	1.61	[530]
	30 Ni	chem.	880–1300	295.2	70.5	1.89	[530]
	40 Ni	chem.	880–1300	289.7	69.2	1.50	[530]
	50 Ni	chem.	880–1300	273.4	65.3	0.480	[530]

4.3 (*cont.*).

Principal metal	Alloying additive % wt/wt	Tracer	Temperature range °C	Activation energy E_D		Frequency factor D_0 cm^2/s	Ref.
				kJ/mole	kcal/mole		
Cobalt	60 Ni	chem.	880–1300	257.9	61.6	0.166	[530]
	70 Ni	chem.	880–1300	254.1	60.7	0.140	[530]
	80 Ni	chem.	880–1300	273.4	65.3	0.725	[530]
	90 Ni	chem.	880–1300	284.7	68.0	1.94	[530]
	10 Pd	chem.	880–1193	234.5	56.0	2.94×10^{-2}	[531]
	10 Pd	chem.	880–1193	224.0	53.5	1.24×10^{-2}	[531]
	20 Pd	chem.	880–1193	237.0	56.6	6.06×10^{-2}	[531]
	20 Pd	chem.	880–1193	235.7	56.3	5.75×10^{-2}	[531]
	30 Pd	chem.	880–1193	232.7	55.6	8.31×10^{-2}	[531]
	40 Pd	chem.	880–1193	221.9	53.0	6.81×10^{-2}	[531]
	50 Pd	chem.	880–1193	219.8	52.5	9.32×10^{-2}	[531]
	60 Pd	chem.	880–1193	228.6	54.6	2.52×10^{-1}	[531]
	70 Pd	chem.	880–1193	241.6	57.7	6.55×10^{-1}	[531]
	80 Pd	chem.	880–1193	234.9	56.1	1.92×10^{-1}	[531]
	90 Pd	chem.	880–1193	227.3	54.3	4.59×10^{-2}	[531]
Copper	1 Ni	^{64}Cu	780–890	246.6	58.9	1.86	[310]
	21.5 Ni	^{64}Cu	863–1112	231.5	55.3	1.9	[309]
	9.8 Pt	^{64}Cu	899–1046	221.1	52.8	1.1	[405]
	24.6 Pt	^{64}Cu	947–1096	229.0	54.7	0.53	[405]
	49.4 Pt	^{64}Cu	1000–1293	213.5	51.0	0,027	[405]
	18.0 Sn	^{64}Cu	600–725	122.3	29.2	0.32	[532]
	20.5 Sn	^{64}Cu	440–575	129.4	30.9	4.7	[532]
	21.5 Ni	^{63}Ni	930–1113	208.1	49.7	0.063	[309]
	9.8 Pt	^{195}Pt	906–1058	220.2	52.6	0.093	[405]

4.3 (*cont.*).

Principal metal	Alloying additive % wt/wt	Tracer	Temperature range °C	Activation energy E_D		Frequency factor D_0 cm^2/s	Ref.
				kJ/mole	kcal/mole		
Copper	24.6 Pt	^{195}Pt	946–1094	215.2	51.4	0.019	[405]
	49.4 Pt	^{195}Pt	1034–1287	249.1	59.5	0.066	[405]
	18.0 Sn	^{113}Sn	600–725	74.5	17.8	1.4×10^{-2}	[532]
	19.8 Sn	^{113}Sn	600–725	42.7	10.2	3.6×10^{-3}	[532]
	20.5 Sn	^{113}Sn	440–575	208.1	49.7	2400	[532]
	50 Pd	chem.	800–1050	224.0	53.5	0.48	[533]
	2.5 Ga	chem.	500–700	134.0	32.0	3×10^{-4}	[534]
	4.9 Ga	chem.	500–700	142.4	34.0	1.8×10^{-4}	[534]
	7.6 Ga	chem.	500–700	154.9	37.0	1.6×10^{-2}	[534]
	10.3 Ga	chem.	500–700	167.5	40.0	1.8×10^{-1}	[534]
	13.1 Ga	chem.	500–700	157.0	37.5	1.3×10^{-1}	[534]
	15.9 Ga	chem.	500–700	146.5	35.0	8×10^{-2}	[534]
	1 Al	^{110}Ag	750–950	153.2	36.6		[535]
	2 Al	^{110}Ag	750–950	87.9	21.0		[535]
	3 Al	^{110}Ag	750–950	72.0	17.2		[535]
	4 Al	^{110}Ag	750–950	218.6	52.2		[535]
	5 Al	^{110}Ag	750–950	258.7	61.8		[535]
	0 Ag	chem.	700–1000	770.8 ± 13.0	184.1 ± 3.1	0.206	[536]
	0.005 Ag	chem.	700–1000	$774.1 \pm 10,0$	184.9 ± 2.4	0.251	[536]
	0.010 Ag	chem.	700–1000	760.3 ± 9.6	181.6 ± 2.3	0.177	[536]
	0.015 Ag	chem.	700–1000	758.6 ± 10.5	181.2 ± 2.5	0.187	[536]
	0.020 Ag	chem.	700–1000	762.0 ± 10.9	182.0 ± 2.7	0.224	[536]
Gold	6 Ag	^{110}Ag	660–961	168.8	40.26	0.72	[537]
	17 Ag	^{110}Ag	650–1010	171.7	41.02	0.09	[537]

4.3 (*cont.*).

Principal metal	Alloying additive % wt/wt	Tracer	Temperature range °C	Activation energy E_D		Frequency factor D_0 cm^2/s	Ref.
				kJ/mole	kcal/mole		
Gold	34 Ag	^{110}Ag	654–971	174.7	41.73	0.11	[537]
	6 Ag	^{198}Au	718–1010	176.18	42.08	0.09	[537]
	17 Ag	^{198}Au	712–1001	180.24	43.05	0.12	[537]
	34 Ag	^{198}Au	715–1001	186.35	44.51	0.17	[537]
	50 Cd	^{198}Au	303–604	116.8	27.9	0.17	[538]
	47.5 Cd	^{198}Au	354–600	117.6	28.1	0.23	[539]
	49.0 Cd	^{198}Au	440–550	125.6	30.0	0.61	[539]
	50.0 Cd	^{198}Au	440–550	116.8	27.9	0.17	[539]
	50.5 Cd	^{198}Au	440–550	109.7	26.2	0.12	[539]
	50 Cd	^{115}Cd	330–620	117.2	28.0	0.23	[538]
	47.5 Cd	^{115}Cd	354–600	129.8	31.0	1.36	[539]
	49.0 Cd	^{115}Cd	440–550	130.6	31.2	1.50	[539]
	50.0 Cd	^{115}Cd	440–550	117.2	28.0	0.23	[539]
	50.5 Cd	^{115}Cd	440–550	113.5	27.1	0.22	[539]
Hafnium	1.5 Zr +0.15 Fe	^{14}C	1120–1760	312.3	74.6	74	[540]
	1.5 Zr + 0.15 Fe	^{14}C	1820–2130	167.5	40.0	4.2×10^{-2}	[540]
	2.7 Zr	^{181}Hf	1785–1258	183.4	43.8	4.8×10^{-3}	[540]
Iron	12.3 Cr	^{51}Cr	900–1400	230.7	55.1	1.29	[541]
	15 Cr	^{51}Cr		238.5	57.0	2.0	[542]
	17.4 Cr	^{51}Cr	900–1400	219.8	52.5	0.46	[541]
	1.7 V	^{51}Cr		238.6	57.0	2.0	[542]
	0.0001 Bi	^{55}Fe	750–890	294.5	70.34	414.0	[543]
	0.0006 Bi	^{55}Fe	750–890	277.3	66.23	80.8	[543]
	0.0010 Bi	^{55}Fe	750–890	260.7	62.27	13.97	[543]

4.3 (*cont.*).

Principal metal	Alloying additive % wt/wt	Tracer	Temperature range °C	Activation energy E_D kJ/mole	Activation energy E_D kcal/mole	Frequency factor D_0 cm^2/s	Ref.
Iron	0.0013 Bi	^{55}Fe	750–890	248.1	59.27	3.65	[543]
	9.1 Cr	^{59}Fe	575–725	230.7	55.1	9.27	[544]
	9.1 Cr	^{59}Fe	777–825	219.4	52.4	0.42	[544]
	15.2 Cr	^{59}Fe	595–677	226.5	54.1	1.25	[544]
	15.2 Cr	^{59}Fe	725–777	215.6	51.5	0.27	[544]
	17 Cr	^{59}Fe	900–1400	233.6	55.8	1.34	[544]
	19.75 Cr	^{59}Fe	575–654	217.3	51.9	0.65	[544]
	19.75 Cr	^{59}Fe	700–825	215.6	51.5	0.22	[544]
	21.0 Cr	^{59}Fe	1000–1200	218.6	52.2	0.115	[528]
	41.5 Cr	^{59}Fe	1000–1200	265.0	63.3	5.0	[528]
	46.7 Cr	^{59}Fe	1000–1200	275.9	65.9	15.8	[528]
	50.0 Cr	^{59}Fe	1000–1200	287.6	68.7	16.6	[528]
	11.8 Ni	^{59}Fe	1000–1200	278.8	66.6	1.21	[545]
	1.8 V	^{59}Fe	700–1500	236.6	56.5	1.39	[546]
	0.7 Mo	^{99}Mo	750–875	314.0	75.0	1.3×10^4	[432]
	0.006 C	^{185}W	1050–1250	358.8	85.7	1.9×10^3	[436]
	0.23 C	^{185}W	1050–1250	336.2	80.3	164	[436]
	0.56 C	^{185}W	1050–1250	324.5	77.5	38	[436]
	0.71 C	^{185}W	1050–1250	315.7	75.4	13	[436]
	1.12 C	^{185}W	1050–1250	286.4	68.4	0.41	[436]
	0.0 Cr	^{53}Fe	680–820	253.3	60.5	4.4	[547]
	0.87 Cr	^{53}Fe	680–820	252.9	60.4	4.3	[547]
	1.43 Cr	^{53}Fe	680–820	252.5	60.3	4.2	[547]
	3.09 Cr	^{53}Fe	680–820	251.6	60.1	4.0	[547]

4.3 (*cont.*).

Principal metal	Alloying additive % wt/wt	Tracer	Temperature range °C	Activation energy E_D		Frequency factor D_0 cm^2/s	Ref.
				kJ/mole	kcal/mole		
Iron	5.05 Cr	^{53}Fe	680–820	250.0	59.7	3.7	[547]
	6.68 Cr	^{53}Fe	680–820	249.1	59.5	3.6	[547]
	8.18 Cr	^{53}Fe	680–820	247.9	59.2	3.3	[547]
	11.83 Cr	^{53}Fe	680–820	244.9	58.5	2.9	[547]
	6.8 Cr	^{59}Fe	650–800	309.2	74.0	3.56×10^3	[548]
	8.9 Cr	^{59}Fe	650–800	305.6	73.0	1.11×10^3	[548]
	17 Cr, 12 Ni	^{14}C	500–996	171.7	41.0	0.96	[549]
	12 Cr	chem.	900–1100	219.4	52.4	5.6×10^{-3}	[550]
	68 Cr	^{59}Fe	650–800	309.2	74.0	3.56×10^3	[550]
	2 Cr	^{51}Cr	800–1400	244.5	58.4	3.21	[551]
	6 Cr	^{51}Cr	800–1400	237.4	56.7	1.21	[551]
	13 Cr	^{51}Cr	800–1400	231.9	55.4	0.64	[551]
	16 Cr	^{51}Cr	800–1400	218.1	52.1	0.19	[551]
	19 Cr	^{51}Cr	800–1400	216.9	51.8	0.18	[551]
	10 Ni	^{59}Fe	900–1200	217.7	52.0	2.5×10^{-3}	[552]
	30 Ni	^{59}Fe	900–1200	278.4	66.5	1	[552]
	45 Ni	^{59}Fe	900–1200	255.4	61.0	1.2	[552]
	55 Ni + 19 Cr	^{59}Fe	900–1200	255.4	61.0	4.5	[552]
	65 Ni	^{59}Fe	900–1200	272.1	65.0	1	[552]
	75 Ni	^{59}Fe	900–1200	343.3	82.0	4×10^2	[552]
	10 Ni	^{63}Ni	900–1200	301.4	72.0	1.4	[552]
	30 Ni	^{63}Ni	900–1200	301.4	72.0	1.4	[552]
	45 Ni	^{63}Ni	900–1200	401.9	96.0	2×10^4	[552]
	55 Ni ± 19 Cr	^{63}Ni	900–1200	279.3	71.0	2.8	[552]

4.3 (*cont.*).

Principal metal	Alloying additive % wt/wt	Tracer	Temperature range °C	Activation energy E_D kJ/mole	Activation energy E_D kcal/mole	Frequency factor D_0 cm^2/s	Ref.
Iron	65 Ni	^{63}Ni	900–1200	279.3	71.0	2.6	[552]
	75 Ni	^{63}Ni	900–1200	253.0	60.5	7.2×10^{-2}	[552]
	10 Ni	chem.	1000–1426	11.47	2.74	0.20	[553]
	20 Ni	chem.	1000–1426	11.43	2.73	0.15	[553]
	30 Ni	chem.	1000–1426	11.39	2.72	0.26	[553]
	40 Ni	chem.	1000–1426	11.22	2.68	0.30	[553]
	50 Ni	chem.	1000–1426	11.14	2.66	0.38	[553]
	60 Ni	chem.	1000–1426	11.05	2.64	0.41	[553]
	70 Ni	chem.	1000–1426	11.05	2.64	0.56	[553]
	80 Ni	chem.	1000–1426	11.09	2.65	0.71	[553]
	90 Ni	chem.	1000–1426	11.05	2.64	0.63	[553]
	3 Co	^{60}Co	630–750	266.3 ± 9.6	63.6 ± 2.3	9.17	[444]
	6.8 Co	^{60}Co	880–920	146.5 ± 8.4	35.0 ± 2.0	5.72×10^{-5}	[444]
	6.8 Co	^{60}Co	630–800	187.1 ± 8.0	44.7 ± 1.9	4.69×10^{-1}	[444]
	28.6 Co	^{60}Co	630–810	72.4 ± 0.6	17.3 ± 2.3	1.25×10^{-3}	[444]
	49.6 Co	^{60}Co	750–850	247.0 ± 10.9	59.0 ± 2.6	6.59×10^{-2}	[444]
	67.2 Co	^{60}Co	630–880	190.9 ± 5.4	45.6 ± 1.3	6.04×10^{-3}	[444]
	6.8 Co	^{60}Co	1010–1310	326.2	77.9	1.09×10^{-1}	[444]
	28.6 Co	^{60}Co	1060–1310	266.3	63.6	1.36×10^{-2}	[444]
	49.6 Co	^{60}Co	1060–1310	349.6	83.5	1.54×10^{-1}	[444]
	89.6 Co	^{60}Co	1060–1310	234.0	55.9	1.61×10^{-2}	[444]
	89.6 Co	^{60}Co	800–1010	251.2	60.0	6.44×10^{-2}	[444]
		^{60}Co	1192–1297	260.4	62.2	0.17	[444]
		^{60}Co	772–1048	273.8	65.4	0.50	[444]

4.3 (*cont.*).

Principal metal	Alloying additive % wt/wt	Tracer	Temperature range °C	Activation energy E_D		Frequency factor D_0 cm^2/s	Ref.
				kJ/mole	kcal/mole		
Iron	1.04 Mn	^{59}Fe	990–1240	265.4	63.4	9.0×10^{-2}	[554]
	2.03 Mn	^{59}Fe	990–1240	262.9	62.8	1.05×10^{-2}	[554]
	2.97 Mn	^{59}Fe	990–1240	255.8	61.1	5.9×10^{-2}	[554]
	4.90 Mn	^{59}Fe	990–1240	255.0	60.9	6.6×10^{-2}	[554]
	7.04 Mn	^{59}Fe	990–1240	262.1	62.6	1.10×10^{-1}	[554]
	10.41 Mn	^{59}Fe	990–1240	275.5	65.8	3.5×10^{-1}	[554]
	18.15 Mn	^{59}Fe	990–1240	282.2	67.4	6.4×10^{-1}	[554]
	25.50 Mn	^{59}Fe	990–1240	277.6	66.3	8.5×10^{-1}	[554]
	33.98 Mn	^{59}Fe	990–1240	276.1	66.1	6.0×10^{-1}	[554]
	18.15 Mn	^{54}Mn	990–1240	261.7	62.5	1.90×10^{-1}	[554]
	25.50 Mn	^{54}Mn	990–1240	251.6	60.1	1.20×10^{-1}	[554]
	33.98 Mn	^{54}Mn	990–1240	242.0	57.8	7.3×10^{-2}	[554]
	5 Mn	chem.	850–1300	250.8	59.9	7.2×10^{-2}	[554]
	10 Mn	chem.	850–1300	263.8	63.0	1.75×10^{-2}	[554]
	15 Mn	chem.	850–1300	269.6	64.4	3.0×10^{-1}	[554]
	20 Mn	chem.	850–1300	262.1	62.6	1.63×10^{-1}	[554]
	25 Mn	chem.	850–1300	248.7	59.4	7.2×10^{-2}	[554]
	30 Mn	chem.	850–1300	251.2	60.0	1.20×10^{-1}	[554]
	0.54 Mo	^{59}Fe	700–890	315.7	75.4	4600	[555]
	1.06 Mo	^{59}Fe	700–890	345.4	82.5	1.1×10^{5}	[555]
	1.5 Mo	^{59}Fe	700–890	336.2	80.3	4.1×10^{4}	[555]
	2.5 Mo	^{59}Fe	700–890	312.3	74.6	2.7×10^{3}	[555]
	7.64 Si	^{59}Fe	900–1100	228.2	54.5	1.38	[556]
	11.1 Si	^{59}Fe	900–1100	212.3	50.7	0.63	[557]

4.3 (*cont.*).

Principal metal	Alloying additive % wt/wt	Tracer	Temperature range °C	Activation energy E_D kJ/mole	Activation energy E_D kcal/mole	Frequency factor D_0 cm^2/s	Ref.
Iron	3 Si	^{59}Fe	900–1250	282.6	67.5	63.2	[551]
	2 Ti	^{59}Fe	1000–1400	216.5±6.3	51.7±1.5	0.56	[551]
	4 Ti	^{59}Fe	1000–1400	204.7±4.2	48.9±1.0	0.27	[551]
	6 Ti	^{59}Fe	1000–1400	208.9±23.4	49.9±5.6	0.40	[551]
	18.17 Ti	^{59}Fe	850–1100	268.0	64.0	12.0	[558]
	20.34 Ti	^{59}Fe	850–1100	273.0	65.2	4.0	[558]
	25.00 Ti	^{59}Fe	850–1100	260.4	62.2	5.6×10^{-1}	[558]
	27.98 Ti	^{59}Fe	850–1100	278.4	66.5	8.0×10^{-1}	[558]
	32.2 Ti	^{59}Fe	850–1100	287.2	68.6	2.8	[558]
	33.32 Ti	^{59}Fe	850–1100	314.0	75.0	11.7	[558]
	38.57 Ti	^{59}Fe	850–1100	304.8	72.8	24.0	[558]
	41.68 Ti	^{59}Fe	850–1100	287.2	68.6	2.5	[558]
	45.78 Ti	^{59}Fe	850–1100	235.7	56.3	6.3×10^{-3}	[558]
	50.00 Ti	^{59}Fe	850–1100	239.5	57.2	1.35×10^{-2}	[558]
	17 Cr 0.15 C 1.0 Si 0.80 Mn	^{63}Ni	1050–1200	280.5	67.0	1.44	[559]
	25 Cr 0.15 C 1.0 Si 0.80 Mn	^{63}Ni	1050–1200	293.9	70.2	2.17×10^{-2}	[559]
	1.25 Si ± 0.53 Cr	chem.	1000–1200	150.7	36.0	0.30	[560]
	1.17 Si + 0.86 Cr	chem.	1000–1200	162.4	38.8	0.13	[560]

4.3 (*cont.*).

Principal metal	Alloying additive % wt/wt	Tracer	Temperature range °C	Activation energy E_D		Frequency factor D_0 cm^2/s	Ref.
				kJ/mole	kcal/mole		
Iron	1.17 Si + 2.10 Cr	chem.	1000–1200	174.2	41.6	0.333	[560]
	1.22 Si + 2.86 Cr	chem.	1000–1200	179.2	42.8	0.44	[560]
	1.17 Si + 4.22 Cr	chem.	1000–1200	188.2	45.1	0.89	[560]
	1.01 Si + 0.80 Co	chem.	1000–1200	88.8	21.2	0.0018	[560]
	1.02 Si + 1.85 Co	chem.	1000–1200	81.6	19.5	0.0098	[560]
	1.06 Si + 2.82 Co	chem.	1000–1200	95.9	22.9	0.0035	[560]
	1.10 Si + 3.90 Co	chem.	1000–1200	88.8	21.2	0.0018	[560]
	1.03 Si + 4.30 Co	chem.	1000–1200	76.6	18.3	0.0006	[560]
	0.50 Co + 0.55 Cr	chem.	1000–1200	137.3	32.8	0.029	[560]
	0.51 Co + 0.83 Cr	chem.	1000–1200	143.3	34.2	0.032	[560]
	0.80 Co + 1.73 Cr	chem.	1000–1200	148.6	35.5	0.048	[560]
	1.02 Co + 2.87 Cr	chem.	1000–1200	155.3	37.1	0.076	[560]

4.3 (*cont.*).

Principal metal	Alloying additive % wt/wt	Tracer	Temperature range °C	Activation energy E_D		Frequency factor D_0 cm^2/s	Ref.
				kJ/mole	kcal/mole		
Iron	1.00 Co + 4.00 Cr	chem.	1000–1200	167.9	40.1	0.20	[560]
Lead	5 Tl	^{210}Pb	207–323	107.8	25.75	1.108	[561]
	10 Tl	^{210}Pb	207–323	106.6	25.45	0.88	[561]
	20 Tl	^{210}Pb	207–323	104.9	25.05	0.647	[561]
	34 Tl	^{210}Pb	207–323	102.7	24.53	0.367	[561]
	50 Tl	^{210}Pb	207–323	102.3	24.44	0.213	[561]
	60 Tl	^{210}Pb	207–231	105.9	25.29	0.287	[561]
	62 Tl	^{210}Pb	207–231	107.3	25.64	0.393	[561]
	74 Tl	^{210}Pb	207–231	112.3	26.82	0.691	[561]
	76 Tl	^{210}Pb	207–231	113.6	27.13	0.862	[561]
	82 Tl	^{210}Pb	207–231	118.2	28.24	2.575	[561]
	87 Tl	^{210}Pb	207–231	124.4	29.71	17	[561]
Magnesium	42.8 Ag	^{110}Ag	500–600	120.2	28.7	0.051	[474]
	47.2 Ag	^{110}Ag	500–700	153.7	36.7	0.33	[562]
Molybdenum	9.9 Cr	^{51}Cr	1200–1350	304.4	72.7	4.3	[563]
	15 W	^{185}W	1500–2200	305.6	73.0	1.4	[564]
	25 W	^{185}W	1500–2200	322.4	77.0	2.4	[565]
	35 W	^{185}W	1500–2400	385.2	92	28	[566]
	61 Pd	chem.	1000–1600	188.4	45.0	5.5×10^{-5}	[567]
	66 Pd	chem.	1000–1600	165.4	39.5	4.0×10^{-5}	[567]
	71 Pd	chem.	1000–1600	177.9	42.5	5.0×10^{-5}	[567]
	75 Pd	chem.	1000–1600	200.5	47.9	2.4×10^{-4}	[567]
	80 Pd	chem.	1000–1600	218.6	52.2	1.6×10^{-3}	[567]

4.3 (*cont.*).

Principal metal	Alloying additive % wt/wt	Tracer	Temperature range °C	Activation energy E_D		Frequency factor D_0 cm^2/s	Ref.
				kJ/mole	kcal/mole		
Molybdenum	85 Pd	chem.	1000–1600	253.3	60.5	1.6×10^{-2}	[567]
	90 Pd	chem.	1000–1600	293.1	70.0	9.0×10^{-1}	[567]
	95 Pd	chem.	1000–1600	286.2	67.5	1.4×10^{-1}	[567]
	85–95 Ni	chem.	1000–1300	277.6	66.3	0.97	[568]
	85–97 Co	chem.	800–1300	293.1	70.0	2.03	[568]
Nickel	4.3 Co	^{60}Co	1213–1368	261.3	62.4	0.49	[306]
	10.8 Co	^{60}Co	1163–1393	263.8	63.0	0.66	[306]
	21.1 Co	^{60}Co	1163–1393	255.8	61.1	0.33	[306]
	43.3 Co	^{60}Co	1163–1393	262.1	62.6	0.52	[306]
	48.6 Co	^{60}Co	701–819	257.5	61.5	0.096	[305]
	48.6 Co	^{60}Co	899–1192	252.0	60.2	0.12	[305]
	49.4 Co	^{60}Co	1110–1319	250.4	59.8	0.18	[306]
	6.35 Cr	^{51}Cr	950–1200	211.4	50.5	0.01	[329]
	11.69 Cr	^{51}Cr	950–1200	229.0	54.7	0.037	[329]
	19.8 Cr	^{51}Cr					
	0.66 Al+ 46 Fe	^{51}Cr	700–1000	242.8	58	0.1	[329]
	19.8 Cr+ 2.6 Ti	^{51}Cr	900–1300	249.1	59.5	0.21	[563]
	10 Cr+ 0.19 Mn	^{51}Cr	1100–1260	278.4	66.5	1.4	[309]
	19.9 Cr+ 0.16 Mn	^{51}Cr	1100–1270	283.4	67.7	1.9	[309]
	20 Cr+ 6 W	^{59}Fe	950–1250	334.9	80.0	160	[569]

4.3 (*cont.*).

Principal metal	Alloying additive % wt/wt	Tracer	Temperature range °C	Activation energy E_D		Frequency factor D_0 cm^2/s	Ref.
				kJ/mole	kcal/mole		
Nickel	13 Cu	^{64}Cu	1054–1260	263.8	63.0	1.5	[309]
	45.4 Cu	^{64}Cu	985–1210	252.5	60.3	2.3	[309]
	10 Au	^{63}Ni	999–1057	213.5	51.0	0.04	[538]
	20 Au	^{63}Ni	940–999	206.0	49.2	0.05	[538]
	35 Au	^{63}Ni	882–940	165.8	39.6	0.005	[538]
	50 Au	^{63}Ni	876–925	186.3	44.5	0.09	[538]
	10.8 Co	^{63}Ni	1166–1410	251.2	60.0	1.47	[306]
	21.1 Co	^{63}Ni	1166–1410	262.1	62.6	0.45	[306]
	43.3 Co	^{63}Ni	1163–1410	267.1	63.8	0.69	[306]
	49.4 Co	^{63}Ni	1104–1347	255.8	61.1	0.25	[306]
	10 Cr	^{63}Ni	1040–1275	293.9	70.2	3.3	[309]
	19.9 Cr	^{63}Ni	1040–1275	285.5	68.2	1.6	[309]
	29.7 Cr	^{63}Ni	1040–1275	295.2	70.5	2.9	[309]
	13 Cu	^{63}Ni	1054–1360	313.6	74.9	35	[309]
	45.4 Cu	^{63}Ni	985–1210	279.7	66.8	17	[309]
	20 Fe	^{63}Ni	814–1200	219.0	52.3	7.19×10^{-3}	[570]
	27 Fe	^{63}Ni	814–1200	203.9	48.7	1.95×10^{-3}	[570]
	30 Fe	^{63}Ni	814–1200	217.3	51.9	6.8×10^{-3}	[570]
	35 Fe	^{63}Ni	814–1200	229.4	54.8	1.67×10^{-2}	[570]
	1.2 Nb	^{63}Ni	1030–1230	254.6	60.8	0.12	[571]
	8.0 Nb	^{63}Ni	1030–1230	260.4	62.2	0.20	[571]
	10,0 Nb	^{63}Ni	1030–1230	280.9	67.1	1.0	[571]
	1.7 W	^{185}W	1100–1300	306.1	73.1	2.2	[309]
	5.3 W	^{185}W	1100–1300	337.0	80.5	17	[309]

4.3 (*cont.*).

Principal metal	Alloying additive % wt/wt	Tracer	Temperature range °C	Activation energy E_D		Frequency factor D_0 cm²/s	Ref.
				kJ/mole	kcal/mole		
Nickel	9.2 W	^{185}W	1100–1300	311.9	74.5	1.4	[309]
	40 Cr	^{63}Ni	1050–1250	231.1	55.2	3.43	[572]
	30 Cr	^{63}Ni	950–1300	236.1	56.4	7.21	[573]
	25 Cr	^{52}Cr	1000–1350	229.4	54.8	1.12	[574]
	20 Cr + 2 ThO_2	^{51}Cr	1040–1200	269.2	64.3	0.92	[575]
	20 Cr	^{51}Cr	1040–1200	193.0	46.1	2.95×10^{-3}	[575]
		^{63}Ni	950–1400	253.3	60.5	9.96	[576]
	8 Mo	^{63}Ni	950–1400	236.1	56.4	2.55	[576]
		^{99}Mo	950–1400	229.9	54.9	1.31	[576]
	16 Mo	^{63}Ni	950–1400	218.6	52.2	0.63	[756]
		^{99}Mo	950–1400	228.6	54.6	1.30	[576]
	18 Mo	^{63}Ni	950–1400	211.0	50.4	0.34	[576]
		^{99}Mo	950–1400	218.6	52.2	0.45	[576]
	20 Mo	^{63}Ni	950–1400	204.3	48.8	0.19	[576]
		^{99}Mo	950–1400	210.6	50.3	0.25	[576]
	23 Mo	^{63}Ni	950–1400	198.5	47.4	0.12	[576]
		^{99}Mo	950–1400	207.2	49.5	0.20	[576]
	19.9 Co	^{57}Co	580–980	259.6 ± 5.0	62.0 ± 1.2	0.32	[577]
	26.3 Co	^{57}Co	580–980	261.7 ± 7.5	62.5 ± 1.8	0.37	[577]
	30.3 Co	^{57}Co	580–980	271.7 ± 5.0	64.9 ± 1.2	0.98	[578]
	50.7 Co	^{57}Co	580–980	254.1 ± 16.3	60.7 ± 3.9	0.094	[578]
	19.9 Co	^{60}Co	580–980	250.8 ± 15.1	59.9 ± 3.6	0.22	[579]
	26.3 Co	^{60}Co	580–980	240.3 ± 15.9	57.4 ± 3.8	0.09	[579]

4.3 (*cont.*).

Principal metal	Alloying additive % wt/wt	Tracer	Temperature range °C	Activation energy E_D		Frequency factor D_0 cm^2/s	Ref.
				kJ/mole	kcal/mole		
Nickel	30.3 Co	^{60}Co	580–980	258.7±11.7	61.8±2.8	0.40	[579]
	50.7 Co	^{60}Co	580–980	252.0±8.8	60.2±2.1	0.18	[579]
Niobium	10.0 Mo	^{95}Nb	1600–2100	527.5	126.0	26.9	[580]
	20.0 Mo	^{95}Nb	1700–2100	468.9	112.0	60.3	[580]
	30.0 Mo	^{95}Nb	1700–2100	439.6	105.0	5.26	[580]
	45.0 Mo	^{95}Nb	1700–2100	393.6	94.0	0.27	[580]
	11 Ti	^{95}Nb	1720–2160	381.0	91.0	1.0	[581]
	34 Ti	^{95}Nb	1490–1960	305.6	73.0	0.15	[581]
	47 Ti	^{95}Nb	1240–1790	271.3	64.8	0.08	[581]
	5 W	^{95}Nb	1600–2000	544.2	130	2334	[474]
	10 W	^{95}Nb	1600–2000	489.8	117	164	[474]
	30 W	^{95}Nb	1600–2000	355.9	85	2.57×10^{-2}	[474]
	65 U	Nb	1050–1220	288.1	68.8	3.1×10^{-2}	[582]
	50 U	Nb	1200–1450	304.4	72.7	3.1×10^{-2}	[582]
	35 U	Nb	1200–1500	324.5	77.5	7.6×10^{-2}	[582]
	20 U	Nb	1550–1800	383.1	91.5	1.1	[582]
	10 U	Nb	1700–1900	421.2	100.6	5.2	[582]
	95 U	U	950–1050	142	34	3.2×10^{-6}	[582]
	90 U	U	950–1050	147.4	35.2	3.5×10^{-6}	[582]
	80 U	U	950–1050	165.8	39.6	1.0×10^{-5}	[582]
	65 U	U	1050–1220	222.7	53.2	1.25×10^{-4}	[582]
	50 U	U	1200–1450	239.5	57.2	2.5×10^{-4}	[582]
	35 U	U	1200–1500	274.7	65.6	4×10^{-4}	[582]
	20 U	U	1550–1800	280.5	67.0	6.3×10^{-4}	[582]

4.3 (*cont.*).

Principal metal	Alloying additive % wt/wt	Tracer	Temperature range °C	Activation energy E_D		Frequency factor D_0 cm²/s	Ref.
				kJ/mole	kcal/mole		
Niobium	10 U	U	1700–1900	305.6	73.0	2.5×10^{-3}	[582]
Platinum	25.5 Cu	^{64}Cu	1098–1385	269.6	64.4	0.67	[405]
	25.5 Cu	^{64}Pt	1140–1382	252.5	60.3	0.022	[405]
Silver	8 Au	^{110}Ag	654–945	187.53	44.79	0.52	[537]
	17 Au	^{110}Ag	634–952	184.43	44.05	0.32	[537]
	35 Au	^{110}Ag	635–956	182.29	43.54	0.23	[537]
	50 Au	^{110}Ag	634–972	180.49	43.11	0.19	[537]
	41.1 Mg	^{110}Ag	500–700	139.0	33.2	0.095	[562]
	43.6 Mg	^{110}Ag	500–700	147.8	35.3	0.15	[562]
	48.5 Mg	^{110}Ag	500–700	165.4	39.5	0.37	[562]
	1.1 Tl	^{110}Ag	640–870	182.1	43.5	0.24	[583]
	2.6 Tl	^{110}Ag	640–870	188.0	44.9	0.35	[583]
	5.5 Tl	^{110}Ag	640–870	157.4	37.6	0.10	[583]
	5 Zn	^{110}Ag	640–925	174.6	41.7	0.54	[854]
	30 Zn	^{110}Ag	773–971	150.68	35.99	0.29	[585]
	6.5 Cd	^{115}Cd	571–922	169.48	40.58	0.328	[586]
	13.6 Cd	^{115}Cd	522–867	161.65	38.61	0.218	[586]
	27.5 Cd	^{115}Cd	526–795	150.52	35.95	0.253	[586]
	4.7 In	^{114}In	573–975	168.73	40.30	0.453	[586]
	12.4 In	^{114}In	573–888	160.73	38.39	0.566	[586]
	16.6 In	^{114}In	474–729	153.28	36.61	0.537	[586]
	0.21 Pb	^{210}Pb	700–810	177.9	42.5	0.22	[587]
	0.25 Pb	^{210}Pb	700–850	158.3	37.8	0.22	[588]
	0.52 Pb	^{210}Pb	700–850	162.0	38.7	0.38	[588]

4.3 (*cont.*).

Principal metal	Alloying additive % wt/wt	Tracer	Temperature range °C	Activation energy E_D		Frequency factor D_0 cm^2/s	Ref.
				kJ/mole	kcal/mole		
Silver	0.71 Pb	^{210}Pb	700–810	187.1	44.7	0.89	[587]
	1.30 Pb	^{210}Pb	700–810	182.1	43.5	0.70	[587]
	1.32 Pb	^{210}Pb	700–850	161.5	38.5	0.46	[588]
	0.1 Cu	chem.	550–760	192.2	45.9	1.2	[589]
	2.0 Cu	chem.	550–760	200.1	47.8	2.5	[589]
	6.0 Cu	chem.	550–760	210.2	50.2	6.8	[589]
	99.0 Cu	chem.	550–760	208.0	49.7	3.1	[589]
	50 Pd	chem.	600–900	103.0	24.6	1.5×10^{-6}	[533]
	61 Zn	^{110}Ag	440–640	122.2 ± 1.3	29.2 ± 0.3	4.06	[590]
	61 Zn	^{65}Zn	440–640	108 ± 1.09	25.8 ± 0.26	1.55	[590]
Titanium	10 Cr	^{51}Cr	900–1200	168.3	40.2	0.02	[591]
	18 Cr	^{51}Cr	900–1200	186.3	44.5	0.09	[591]
	5 Fe	^{55}Fe	850–1250	165.8	39.6	0.092	[592]
	10 Fe	^{55}Fe	850–1250	203.5	48.6	2.14	[592]
	15 Fe	^{55}Fe	850–1100	243.3	58.1	52.2	[592]
	5 Nb	^{55}Fe	850–1250	138.6	33.1	7.9×10^{-3}	[592]
	10 Nb	^{55}Fe	850–1300	146.1	34.9	1.15×10^{-3}	[592]
	15 Nb	^{55}Fe	850–1250	146.5	35.0	7.9×10^{-3}	[592]
	5 Fe	^{95}Nb	1000–1250	146.1	34.9	1.82×10^{-4}	[592]
	10 Fe	^{95}Nb	900–1500	172.9	41.3	2.9×10^{-2}	[592]
	15 Fe	^{95}Nb	900–1100	234.5	56.0	9.9	[592]
	5 Nb	^{95}Nb	950–1250	125.2	29.9	1.2×10^{-4}	[592]
	10 Nb	^{95}Nb	800–1250	151.1	36.1	6.8×10^{-4}	[592]
	15 Nb	^{95}Nb	880–1260	164.5	39.3	1.5×10^{-3}	[592]

4.3 (*cont.*).

Principal metal	Alloying additive % wt/wt	Tracer	Temperature range °C	Activation energy E_D		Frequency factor D_0 cm^2/s	Ref.
				kJ/mole	kcal/mole		
Titanium	34 Nb	^{95}Nb	1060–1750	211.4	50.5	9.6×10^{-3}	[581]
	90 Zr	chem.	830–1050	164.5	39.3	1.4×10^{-2}	[593]
	75 Zr	chem.	830–1050	146.5	35.0	3.3×10^{-3}	[593]
	60 Zr	chem.	830–1050	142.8	34.1	2.7×10^{-3}	[593]
	50 Zr	chem.	830–1050	140.7	33.6	2.4×10^{-3}	[593]
	35 Zr	chem.	830–1050	136.5	32.6	1.6×10^{-3}	[593]
	20 Zr	chem.	830–1050	134.8	32.2	1.5×10^{-3}	[593]
	10 Zr	chem.	830–1050	131.9	31.5	1.3×10^{-3}	[593]
	75 Zr	chem.	650–830	65.7	15.7	5.0×10^{-7}	[593]
	60 Zr	chem.	650–830	71.6	17.1	1.2×10^{-6}	[593]
	50 Zr	chem.	650–830	74.1	17.7	1.7×10^{-6}	[593]
	35 Zr	chem.	650–830	75.4	18.0	2.0×10^{-6}	[593]
	20 Zr	chem.	650–830	75.4	18.0	2.2×10^{-6}	[593]
	2 Co	^{44}Ti	800–1450	147.4	35.2	1.4×10^{-3}	[594]
	4 Co	^{44}Ti	800–1450	147.4	35.2	3.2×10^{-3}	[594]
	6 Co	^{44}Ti	800–1450	163.3	39.0	1.8×10^{-2}	[594]
	9 Co	^{44}Ti	800–1450	173.8	41.5	7.9×10^{-2}	[594]
	11 Mn	^{44}Ti	800–1500	177.5	42.4	3.1×10^{-2}	[594]
	15 Mn	^{44}Ti	800–1500	157.0	37.5	6.1×10^{-3}	[594]
	20 Mn	^{44}Ti	800–1500	196.4	46.9	2.6×10^{-1}	[594]
	23 Mn	^{44}Ti	800–1500	196.4	46.9	2.6×10^{-1}	[594]
	1.6 Co	^{44}Ti		145.3 ± 6.7	34.7 ± 1.6	1.26×10^{-3}	[595]
	3.3 Co	^{44}Ti		140.3 ± 3.8	33.5 ± 0.9	1.58×10^{-3}	[595]
	4.9 Co	^{44}Ti		160.4 ± 5.0	38.3 ± 1.2	1.41×10^{-2}	[595]

4.3 (*cont.*).

Principal metal	Alloying additive % wt/wt	Tracer	Temperature range °C	Activation energy E_D kJ/mole	Activation energy E_D kcal/mole	Frequency factor D_0 cm^2/s	Ref.
Titanium	7.4 Co	^{44}Ti		162.9±7.5	38.9±1.8	2.50×10^{-2}	[595]
	9.7 Mn	^{44}Ti		171.2±9.6	40.9±2.6	1.90×10^{-2}	[595]
	13.3 Mn	^{44}Ti		172.1±5.4	41.1±1.3	2.06×10^{-2}	[595]
	17.9 Mn	^{44}Ti		176.3±4.2	42.1±1.0	2.60×10^{-2}	[595]
	20.6 Mn	^{44}Ti		208.1±7.5	49.7±1.8	5.47×10^{-1}	[595]
	90 V	chem.	650–1050	197.6	47.2	8.3×10^{-4}	[596]
	80 V	chem.	650–1050	199.3	47.6	1.5×10^{-3}	[596]
	70 V	chem.	650–1050	203.9	48.7	4.4×10^{-3}	[596]
	60 V	chem.	650–1050	206.8	49.4	1.3×10^{-2}	[596]
	50 V	chem.	650–1050	203.5	48.6	2.4×10^{-2}	[596]
	40 V	chem.	650–1050	153.2	36.6	1.1×10^{-2}	[596]
	30 V	chem.	650–1050	152.2	36.6	8.1×10^{-4}	[596]
	20 V	chem.	650–1050	139.8	33.4	4.1×10^{-4}	[596]
	10 V	chem.	650–1050	123.9	26.9	1.6×10^{-4}	[596]
	2.8 Fe	^{59}Fe	1000–1300	172.5	41.2	0.15	[502]
	4.8 Fe	^{59}Fe	1000–1300	201.0	48.0	1.7	[502]
	9.5 Fe	^{59}Fe	1000–1300	216.0	51.6	3.6	[502]
	10 W	chem.	1000–1200	240.7	57.5	0.05	[597]
	20 W	chem.	1000–1200	314.0	75.0	0.85	[597]
	75 Zr	chem.	900–1168	131.5	31.4	4.75	[598]
	60 Zr	chem.	900–1168	129.0	30.8	5.11	[598]
	50 Zr	chem.	900–1168	108.4	25.9	9.85	[598]
Tungsten	25 Mo	^{99}Mo	1900–2500	348	83	0.18	[565]
	27 Re	^{186}Re	2000–2400	393.6	94.0	9.2×10^{-3}	[507]

4.3 (*cont.*).

Principal metal	Alloying additive % wt/wt	Tracer	Temperature range °C	Activation energy E_D		Frequency factor D_0 cm²/s	Ref.
				kJ/mole	kcal/mole		
Tungsten	2 Be	^{185}W	900–1200	141.5	33.8	1.18×10^{-3}	[599]
	12 Be	^{185}W	900–1200	280.30	66.95	2.36	[599]
	24 Be	^{185}W	900–1200	138.67	33.12	1.1×10^{-4}	[599]
	25 Mo	^{185}W	2100–2600	486	116	26	[566]
	4 Mo	^{235}U	800–1040	138.2	33.0	2.5×10^{-3}	[600]
Uranium	4 Nb	^{235}U	800–1040	181.1	28.2	1.66×10^{-4}	[600]
	4 Zr	^{235}U	800–1040	92.1	22.0	1.26×10^{-4}	[600]
	0.57 Ag	^{65}Zn	345–420	97.13	23.12	0.22	[601]
Zinc	1.4 Ag	^{65}Zn	345–420	92.40	22.07	0.17	[601]
	0.5 Al	^{65}Zn		74.5	19.0	1.0×10^{-4}	[602]
	37.1 Al	^{65}Zn	325–405	83.7	20.0	0.012	[603]
	0.7 Sn	^{113}Sn	1000–1250	180	43	2×10^{-2}	[604]
Zirconium	3.3 Sn	^{113}Sn	1000–1250	209	50	0.2	[604]
	5.6 Sn	^{113}Sn	1000–1250	218	52	0.4	[604]
	1.33 Nb	^{95}Zr	900–1200	138	33	3×10^{-4}	[520]
	2 Nb	^{95}Zr	900–1200	130	31	1.5×10^{-4}	[520]
	5 Nb	^{95}Zr	900–1200	138	33	3×10^{-4}	[520]
	10 Nb	^{95}Zr	900–1200	147	35	4×10^{-4}	[520]
	0.7 Sn	^{95}Zr	1000–1250	151	36	2×10^{-3}	[605]
	1.3 Sn	^{95}Zr	740–827	260	62	5.0	[605]
	2.39 Sn	^{95}Zr	740–827	314	75	2100	[605]
	3.3 Sn	^{95}Zr	1000–1250	163	39	3×10^{-3}	[605]
	3.54 Sn	^{95}Zr	740–827	268.0	64.0	10	[605]
	5.6 Sn	^{95}Zr	1000–1250	209	50	0.2	[604]

4.3 (*cont.*).

Principal metal	Alloying additive % wt/wt	Tracer	Temperature range °C	Activation energy E_D		Frequency factor D_0 cm^2/s	Ref.
				kJ/mole	kcal/mole		
Zirconium	2.3 Nb	^{95}Nb	900–1160	162.4	38.8	1.63×10^{-3}	[515]
	1.5 Cr	^{51}Cr	960–1100	191.3	45.7	0.19	[515]
		^{95}Nb	900–1200	144.4	34.5	9.86×10^{-5}	[606]
	1.05 Nb	^{95}Nb	900–1200	128.1	30.6	5.07×10^{-5}	[606]
	3.25 Nb	^{95}Nb	900–1200	128.1	30.6	9.75×10^{-5}	[606]
	1.05 Nb	^{95}Zr	900–1200	110.5	26.4	3.83×10^{-5}	[606]
	3.25 Nb	^{95}Zr	900–1200	107.6	25.7	3.0×10^{-5}	[606]
Palladium	10 V	chem.	600–1300	193.8	46.3	37.2×10^{-3}	[607]
	20 V	chem.	600–1300	198.9	47.5	7.94×10^{-3}	[607]
	25 V	chem.	600–1300	203.9	48.7	9.55×10^{-3}	[607]
	30 V	chem.	600–1300	204.7	48.9	10.5×10^{-3}	[607]
	33 V	chem.	600–1300	225.2	53.8	46.8×10^{-3}	[607]
	40 V	chem.	600–1300	216.9	51.8	22.4×10^{-3}	[607]
	70 V	chem.	600–1300	145.3	34.7	1.91×10^{-3}	[607]
	75 V	chem.	600–1300	160.8	38.4	1.51×10^{-3}	[607]
	80 V	chem.	600–1300	197.6	47.2	14.1×10^{-3}	[607]
	85 V	chem.	600–1300	212.7	50.8	14.8×10^{-3}	[607]
Tungsten	90 Mo	chem.	2000–2500	490.7	117.2	4.48	[608]
	80 Mo	chem.	2000–2500	481.1	114.9	2.41	[608]
	70 Mo	chem.	2000–2500	458.9	109.6	0.64	[608]
	60 Mo	chem.	2000–2500	457.6	109.3	0.48	[608]
	50 Mo	chem.	2000–2500	450.9	107.7	0.30	[608]
	40 Mo	chem.	2000–2500	441.3	105.4	0.17	[608]
	30 Mo	chem.	2000–2500	438.4	104.7	0.14	[608]

4.3 (*cont.*).

Principal metal	Alloying additive % wt/wt	Tracer	Temperature range °C	Activation energy E_D kJ/mole	Activation energy E_D kcal/mole	Frequency factor D_0 cm^2/s	Ref.
Tungsten	20 Mo	chem.	2000–2500	430.0	102.7	0.08	[608]
	10 Mo	chem.	2000–2500	422.0	100.8	0.05	[608]
	90 Ta	dyf.	2000–2500	309.4	73.9	7.0×10^{-4}	[608]
	80 Ta	chem.	2000–2500	301.9	72.1	1.0×10^{-4}	[608]
	70 Ta	chem.	2000–2500	262.5	62.7	1.22×10^{-5}	[608]
	60 Ta	chem.	2000–2500	253.7	60.6	6.08×10^{-6}	[608]
	50 Ta	chem.	2000–2500	243.7	58.2	3.34×10^{-6}	[608]
	40 Ta	chem.	2000–2500	231.1	55.2	1.66×10^{-6}	[608]
	30 Ta	chem.	2000–2500	231.9	55.4	1.5×10^{-6}	[608]
	20 Ta	chem.	2000–2500	229.4	54.8	1.2×10^{-6}	[608]
	10 Ta	chem.	2000–2500	198.0	47.3	2.2×10^{-7}	[608]
	90 Nb	dyf.	2000–2400	439.6	105	81.45	[608]
	80 Nb	chem.	2000–2400	419	100	22.2	[608]
	70 Nb	chem.	2000–2400	376.4	89.9	1.97	[608]
	60 Nb	chem.	2000–2400	280.1	66.9	1.4×10^{-2}	[608]
	50 Nb	chem.	2000–2400	272.1	65.0	7.4×10^{-3}	[608]
	40 Nb	chem.	2000–2400	255.0	60.9	3.0×10^{-3}	[608]
	30 Nb	chem.	2000–2400	247.4	59.1	1.8×10^{-3}	[608]
	20 Nb	chem.	2000–2400	236.1	56.4	1.0×10^{-3}	[608]
	10 Nb	chem.	2000–2400	228.2	54.5	6.0×10^{-4}	[608]
	3.9 Re	^{14}C	1500–1800	154.9 ± 1.38	37.0 ± 0.33	3.00 ± 0.26 $\times10^{-3}$	[508]
	6.8 Re	^{14}C	1500–1800	151.1 ± 0.71	36.1 ± 0.17	2.24 ± 0.09 $\times10^{-3}$	[508]

4.4 Self-diffusion, heterodiffusion and chemical diffusion in metal oxides

Material	Tracer	Temperature range °C	Activation energy E_D kJ/mole	Activation energy E_D kcal/mole	Frequency factor D_0 cm^2/s	Ref.
Al_2O_3	^{26}Al	1670–1905	160.5	114.0	28	[609]
	^{241}Am	1200–1430	565.2	135.0	4.94×10^6	[610]
	^{59}Fe	900–1200	343.3	82.0	1.13	[611]
	^{59}Fe	900–1100	113.0	27.0	9.18×10^{-8}	[612]
	^{147}Pr	1350–1540	657	157	1.18×10^8	[613]
	^{239}Pu	1200–1450	595.4	142.2	1.54×10^8	[610]
	chem.	975–1075	598.7	143.0	4.6×10^7	[614]
	chem.	1300–1800	569.4	136	$\sim 10^4$	[615]
	^{22}Na	1223–1750	209.3	50.0	$2.0 \pm 0.6 \times 10^{-2}$	[616]
B_2O_3	^{18}O	350–650	137.3 ± 33.5	32.8 ± 8.0	2.49×10^{-2}	[617]
	chem.	710–1200	36.4	8.7	3×10^{-5}	[618]
BaO	^{140}Ba	1080–1250	1059	253	10^{29}	[619]
BeO	^{7}Be	1725–2000	151.4	36.15	1.1×10^{-6}	[620]
	^{7}Be	1550–1725	384.8	91.9	1.37	[620]
	^{7}Be	1730–1930	276.7	66.1	6.14×10^{-2}	[621]
	^{7}Be	1570–1730	467.2	111.6	5560	[621]
	^{7}Be	1760–2000	268	64	1.27×10^{-3}	[621]
	^{7}Be	1490–1720	151	36	1.23×10^{-6}	[621]
	^{7}Be	1500–1725	385.2	92.0	1.35	[621]
	^{7}Be	1760–2000	150.7	36.0	1.07×10^{-6}	[621]
	^{7}Be	1150–1800	261.7	62.5	2.49×10^{-3}	[622]
	^{7}Be	1500–2000	223.6	53.4	5.8×10^{-5}	[623]
	^{7}Be	1100–1800	263.8	63.0	3.2×10^{-3}	[624]
	^{14}C	1250–1800	252.9	60.4	$(4 \pm 1) \times 10^{-3}$	[625]

4.4 (*cont.*).

Material	Tracer	Temperature range °C	Activation energy E_D		Frequency factor D_0 cm²/s	Ref.
			kJ/mole	kcal/mole		
Bi_2O_3	^{210}Bi	600–700	86.7	20.7	4.29×10^{-6}	[626]
	^{210}Bi	710–780	276.3	66.0	0.45	[626]
CaO	^{45}Ca	900–1400	339.1	81.0	0.4	[627]
	^{45}Ca	1000–1400	144.9	34.6	8.75×10^{-8}	[628]
CdO	^{115}Cd	1780–2270	14.2 ± 1.3	3.4 ± 0.3	0.017	[629]
CeO	^{18}O	850–1150	49.8 ± 16.3	11.9 ± 3.9	1.5×10^{-5}	[630]
$CeO_{1,8}$	^{18}O	850–1150	15.1 ± 17.6	3.6 ± 4.2	6.2×10^{-6}	[630]
CeO_2	^{18}O	850–1150	103.8 ± 14.7	24.8 ± 3.5	1.9×10^{-4}	[630]
Cr_2O_3	^{51}Cr	1000–1350	419	100	4×10^{3}	[631]
	^{51}Cr	900–1100	92.1	22	4.29×10^{-8}	[611]
	^{51}Cr	1100–1550	255.4	61.0	0.137	[632]
	^{59}Fe	900–1100	184.2	44	4.95×10^{-6}	[611]
	chem.	1260–1400	468.9	112.0	0.5	[614]
	chem.	1260–1400	468.9	112.0	4.8×10^{4}	[614]
CoO	^{60}Co	800–1350	144.4	34.5	2.15×10^{-3}	[272]
	^{60}Co	900–1400	160.8	38.4	5.3×10^{-3}	[273]
	Co	900–1200	158.3	37.8	5.2×10^{-3}	[633]
	Co	900–1350	158.3	37.8	5.2×10^{-3}	[189]
	chem.	900–1350	100.5	24.0	4.33×10^{-3}	[270]
	chem.	900–1350	100.5	24.0	8.3×10^{-3}	[271]
	chem.	900–1350	102.2	24.4	6.6×10^{-3}	[187]
	^{35}S	1000–1090	244.5	58.4	5.0×10^{-3}	[634]
Cu_2O	^{110}Ag	800–1050	115.68	27.63	6×10^{-3}	[635]
	^{110}Ag	700–800	58.6	14.0	3.8×10^{-3}	[635]

4.4 (*cont.*).

Material	Tracer	Temperature range °C	Activation energy E_D kJ/mole	Activation energy E_D kcal/mole	Frequency factor D_0 cm²/s	Ref.
Cu_2O	^{110}Ag	800–1000	118.36	28.27	5.6×10^{-3}	[635]
	^{64}Cu	800–1050	151.1	36.1	4.36×10^{-2}	[636]
	^{114}In	800–1050	140.3	33.5	160	[637]
	^{114}In	600–780	51.9	12.4	2.4×10^{-8}	[628]
	^{114}In	780–1050	103.8	24.8	8.9×10^{-6}	[637]
	^{65}Zn	600–900	37.7	9.0	1.7×10^{-8}	[638]
	^{65}Zn	900–1050	129.4	30.9	$5{:}2\times10^{-4}$	[638]
	Cu	900–1050	100 ± 8	24 ± 2	2.15×10^{-3}	[639]
FeO	^{59}Fe	700–1000	124.3	29.7	0.118	[640]
	^{59}Fe	700–1000	126.4	30.2	0.014	[272]
	^{55}Fe	600–910	105 ± 13	25 ± 3	6×10^{-5}	[641]
$Fe_{0.94}O$	^{52}Fe, ^{59}Fe	700–1340	122.88 ± 1.3	29.35 ± 0.3	$(8.6\pm0.5)\times10^{-3}$	[642]
$Fe_{0.94}O$	chem.	900–1100	174		32.3	[643]
$Fe_{0.915}O$	chem.	900–1100	154		6.5	[643]
$Fe_{0.89}O$	chem.	900–1100	141		2.5	[643]
Fe_2O_3	^{59}Fe	760–1300	469	112	4×10^{4}	[627]
	^{59}Fe	950–1050	419.5	100.2	1.3×10^{6}	[628]
Fe_3O_4	^{59}Fe	750–1000	230.3	55.0	5.2	[626]
	^{59}Fe	1000–1220	469	112	4×10^{-5}	[626]
	^{59}Fe	770–1200	225.7	53.9	0.25	[631]
	^{59}Fe	850–1075	351.7	84.0	6×10^{5}	[632]
	^{59}Fe	850–1075	312.8	74.7	104	[632]
PbO	^{212}Pb	400–590	276	66	10^{5}	[272]
	^{210}Pb	600–680	268	64	—	[635]

4.4 (*cont.*).

Material	Tracer	Temperature range °C	Activation energy E_D		Frequency factor D_0 cm^2/s	Ref.
			kJ/mole	kcal/mole		
PbO	^{18}O	787			10^{10}	[644]
Pr_7O_{12}	^{18}O	735–865	77.04±1.05	18.40±0.25	$5.5 \pm 0.6 \times 10^{-6}$	[645]
		830–930	75.764±1.97	18.096±0.47	$1.3 \pm 0.3 \times 10^{-5}$	[645]
		730–800	76.242±3.140	18.210±0.750	$9 \pm 3 \times 10^{-8}$	[645]
MgO	^{133}Ba	1000–1725	325.52	77.75	0.07	[646]
	^{133}Ba	1900–2450	340.0	81.2	$0.028^{+0.132}_{-0.023}$	[647]
	^{133}Ba	1228–1422	319.9±0.8	76.4±0.2	$0.042^{+0.054}_{-0.023}$	[647]
	^{7}Be	1000–1700	154.1	36.8	1.41×10^{-5}	[648]
	^{45}Ca	900–1700	205.2	49.0	2.95×10^{-5}	[649]
	Co	1000–1800	206.8	49.4	5.78×10^{-5}	[650]
	Fe	1050–1720	174.2	41.6	8.83×10^{-5}	[650]
	^{28}Mg	1400–1600	330.8	79.0	0.249	[651]
		1100–1750	154.9	37.0	1.2×10^{-5}	[652]
	^{63}Ni	1000–1800	202.2	48.3	1.8×10^{-5}	[653]
	^{63}Ni	1900–2450	318±4	76±1	$0.014^{+0.048}_{-0.012}$	[647]
	^{63}Ni	1500–1700	155.3	37.1	1.3×10^{-6}	[647]
	^{63}Ni	1900–2460	203.5	48.6	1.80×10^{-5}	[654]
	^{63}Ni	1200–1300	173.8	41.5	6×10^{-6}	[655]
	^{91}Y	1400–1760	298.9+6.7	71.4±1.6	2.11±0.18	[656]
	^{18}O	1090–1420	252.0	60.2	4.5×10^{-7}	[657]
	^{18}O	1090–1420	185.9	44.4	5.2×10^{-7}	[657]

4.4 (*cont.*).

Material	Tracer	Temperature range °C	Activation energy E_D kJ/mole	Activation energy E_D kcal/mole	Frequency factor D_0 cm^2/s	Ref.
MgO	^{18}O	1020–1450	233.6	55.8	2.4×10^{-7}	[658]
	^{18}O	1020–1260	252.0	60.2	1.6×10^{-7}	[658]
	^{18}O	1260–1450	427.1	102.0	9.9×10^{-1}	[658]
	^{133}Ba	1900–2450	337.5	80.6	0.028	[659]
	^{133}Ba	1228–1422	318.2	76.0	0.042	[659]
	^{7}Be	636–2341	7.03 ± 0.08	1.68 ± 0.02	$\left(1.99 {+0.24 \atop -0.22}\right) \times 10^{-5}$	[660]
	^{85}Sr	1000–1600	12.18 ± 0.54	2.91 ± 0.13	$\left(6.0 {+10.7 \atop -3.8}\right) \times 10^{-4}$	[661]
MoO_3	^{99}Tc	480–700	121.4 ± 12.6	29.0 ± 3.0	2.0×10^{-6}	[662]
NiO	^{63}Ni	900–1000	230	55	0.041	[663]
	^{63}Ni	1140–1400	500.3	119.5	2.8×10^{-6}	[664]
	^{63}Ni	1000–1400	185.1	44.2	3.9×10^{-4}	[201]
	^{63}Ni	1000–1400	185.1	44.2	5.0×10^{-4}	[201]
	^{63}Ni	700–1400	234.5	56.0	1.7×10^{-2}	[665]
	^{63}Ni	1300–1700	224.0	53.5	8.01×10^{-4}	[666]
	^{63}Ni	1000–1400	190.9	45.6	1.83×10^{-3}	[667]
	^{63}Ni	1200–1400	202.6	48.4	4.8×10^{-4}	[668]
	^{63}Ni, ^{57}Ni	1180–1760	254.6 ± 1.3	60.8 ± 0.3	4.77 ± 1.3	[669]
	^{63}Ni, ^{57}Ni	1180–1760	262.5 ± 1.3	62.7 ± 0.3	2.06 ± 0.4	[669]
	Cr^{+3}	1330–1580	230	55	4×10^{-3}	[670]
	^{51}Cr	1100–1250	197	47	9.36×10^{-5}	[571]
	^{55}Co, ^{60}Co	1085–1649	226.770 ± 1.947	54.163 ± 0.465	$9.12 \pm 0.58 \times 10^{-3}$	[672]

4.4 (*cont.*).

Material	Tracer	Temperature range °C	Activation energy E_D		Frequency factor D_0 cm^2/s	Ref.
			kJ/mole	kcal/mole		
NiO	^{55}Co	1085–1649	226.9	54.2	9.1×10^{-3}	[673]
	Co	900–1200	226.9	54.2	9.12×10^{-3}	[674]
	^{60}Co	1085–1649	226.76	54.16	$9.12\pm0.58\times10^{-3}$	[674]
	Ni	900–1270	226.0	54.0	—	[675]
	chem.	975–1350	103.0	24.6	2.2×10^{-3}	[271]
	^{51}Cr	1192–1642	282.2	67.4	$8.6\pm1.2\times10^{-3}$	[676]
	Fe	800–1100	186.3	44.5	1.8×10^{-3}	[677]
	Fe	1000–1200	193.4	46.2	2.2×10^{-3}	[678]
NiO (PO_2 = 0.01 atm = 10.1325 hPa)	Fe	1061–1241	76.2	18.2	1.6×10^{-8}	[679]
NiO (PO_2 = 0.21 atm = 212.78 hPa)	Fe	1061–1241	108.4	25.9	5.92×10^{-7}	[679]
NiO (PO_2 = 1 atm = 1013.25 hPa)	Fe	1061–1241	108.0	25.8	6.74×10^{-7}	[679]
NiO–Al_2O_3	Ni	1533–1743	203.5	48.6	8.54×10^{-4}	[680]
	Ni	1533–1743	85.8	20.5	2.7×10^{-1}	[680]
	Ni	900–1400	210.6	50.3	1.1×10^{-3}	[200]
Nb_2O_5	^{95}Nb	500–900	118.1	28.2	0.38	[681]
SnO_2	^{113}Sn	1000–1260	467.0	118.7	10^6	[682]
SrO	^{85}Sr	1450–1600	448.8	107.2	$2.52\times10^{-2}\ {}^{+15.18}_{-2.16}$	[683]

4.4 (*cont.*).

Material	Tracer	Temperature range °C	Activation energy E_D kJ/mole	Activation energy E_D kcal/mole	Frequency factor D_0 cm^2/s	Ref.
SrO	^{85}Sr	1200–1400	268.4	64.1	$4.48\times10^{-4}\,{}^{+11.42}_{-3.22}$	[683]
TiO_2	^{59}F	770–1000	231.9	55.4	0.192	[612]
	^{59}Fe	770–1000	230.3	55.0	0.0198	[612]
	^{18}O	950–1450	251.2	60.0	2.0×10^{-3}	[684]
	^{18}O	800–1100	234.5 ± 1.3	56.0 ± 0.3	91	[685]
	^{18}O	950–1400	276.3	66.0	1.7×10^{-2}	[684]
	^{44}Ti	1197–1510	250.8	59.9	0.046	[687]
	^{44}Ti	1197–1510	203.1	48.5	0.0024	[687]
ThO_2	^{233}Pa	1800–2000	315.7	75.4	2.91×10^{-5}	[686]
	^{230}Th	1600–2100	246.2	58.8	1.25×10^{-7}	[688]
	^{237}U	1800–2000	319.9	76.4	1.10×10^{-4}	[689]
	^{228}Th	1846–2045	625.1	149.5	0.35	[690]
UO_2	^{241}Am	1200–1500	385.2	92.0	0.03	[691]
	^{237}Np	1200–1500	456.4	109.0	2.9	[691]
	Pa	1200–1500	450.5	107.6	2.5	[691]
	^{147}Pm	1120–1410	237.8	56.8	3.5×10^{-6}	[692]
	^{239}Pu	1200–1500	407.4	97.3	0.34	[691]
	Th	1200–1500	410.3	98.0	0.16	[691]
	^{233}U	1450–1785	368.4	88	4.3×10^{-4}	[693]
	^{233}U	1300–1600	437.9	104.6	0.23	[694]
	^{233}U	1200–1500	435.4	104.0	0.9	[691]
	^{233}U	1450–1700	452.2	108.0	1.2	[695]
	^{233}U	1900–2150	304.4	72.7	5.82×10^{-5}	[696]

4.4 (*cont.*).

Material	Tracer	Temperature range °C	Activation energy E_D		Frequency factor D_0 cm²/s	Ref.
			kJ/mole	kcal/mole		
UO_2	^{91}Y	1150–1450	194.3	46.4	6.8×10^{-8}	[693]
	^{95}Zr	1120–1419	247.9	59.2	1.6×10^{-6}	[693]
$UO_{2.1}$	^{18}O	500–800	90.0	21.5	2.7×10^{-4}	[697]
$UO_{2.16}$		800–910	92.1	22	6×10^{-4}	[697]
$UO_{2.002}$		500–780	273.4	65.3	120	[698]
$UO_{2.063}$		317–600	124.3	29.7	2.06×10^{-3}	[698]
Y_2O_3	^{91}Y	1400–1800	183.8	43.9	2.41×10^{-4}	[699]
ZnO	^{65}Zn	800–1330	309.8	74.0	1.3	[700]
	^{65}Zn	850–940	83.7	20.0	3×10^{-9}	[701]
	^{65}Zn	900–1025	305.6	73.0	4.8	[702]
	^{65}Zn	910–1170	309.8	74.0	30	[703]
	^{65}Zn	800–1300	372.6	89.0	0.1	[704]
	^{65}Zn	1000–1200	184.2	44.0	1.3×10^{-5}	[705]
	^{65}Zn	800–840	305.6	73.0	10	[706]
	^{65}Zn	720–780	104.7	25.0	3×10^{-7}	[707]
	Cu	1100–1250	463.5	110.7	2×10^{-7}	[708]
	^{18}O	940–1141	123.9	29.6	1.2×10^{-10}	[709]
ZrO_2	^{18}O	600–1000	189.7	45.3	2.34×10^{-2}	[710]

4.5 Self-diffusion, heterodiffusion and chemical diffusion in metal sulphides

Material	Tracer	Temperature range, °C	Activation energy E_D		Frequency factor D_0 cm²/s	Ref.
			kJ/mole	kcal/mole		
FeS	Fe	600–950	86.2	20.6	—	[711]
FeS	Fe	550–970	89.2	21.3	3.0×10^{-3}	[712]
Ag_2S	^{108}Ag	180–280	14.4	3.45	2.8×10^{-7}	[713]
Ag_2S	Ag	200–450	11.43	2.73	3.9×10^{-4}	[714]
Ag_2S	^{108}Ag	280–400	14.24	3.40	2.3×10^{-7}	[715]
Ag_2S	Cu	170–720	6.66	1.59	1.2×10^{-4}	[716]
Cu_2S	Cu	350–550	33.5	8.0	1.4×10^{-2}	[717]
Cu_2S	Ag	310–920	19.18	4.58	3.0×10^{-6}	[718]
Cu_2S	^{64}Cu	200–450	20.5	4.9	2.5×10^{-4}	[719]
$4Cu_2S$–Sb_2S_3	^{64}Cu	200–450	21.4	5.1	2.0×10^{-4}	[719]
$3Cu_2S$–Sb_2S_3	^{64}Cu	200–450	26.4	6.3	2.3×10^{-4}	[719]
Cu_2S–Sb_2S_3	^{64}Cu	200–450	48.1	11.5	1.7×10^{-5}	[719]
CuS	Cu	150–310	79.5	19.0	—	[720]
CdS	^{65}Zn	720–1000	82.9	19.8	1.3×10^{-9}	[721]
CdS	^{64}Cu	400–550	69.9	16.7	8.4×10^{-4}	[722]
CdS	^{32}P	800–1000	154.1	36.8	6.5×10^{-4}	[723]
CdS	Cu	—	118.5	28.3	9.2×10^{-3}	[724]
CdS	Ag	450–600	61.1	14.6	7.0×10^{-3}	[725]
CdS	Cu	400–750	74.5	17.8	1.6×10^{-3}	[726]
CdS	Cu	300–700	101.3	24.2	1.2×10^{-2}	[727]
CdS	Ag	300–700	142.8	34.1	13	[727]
CdS	Au	300–700	202.6	48.4	91	[727]
PbS	^{22}Na	249–313	118.5	28.3	6.3	[728]
PbS	N	—	92.1	22.0	17.8	[729]

4.5 (*cont.*).

Material	Tracer	Temperature range °C	Activation energy E_D		Frequency factor D_0 cm^2/s	Ref.
			kJ/mole	kcal/mole		
PbS	Pb	480–880	175.8	42.0	1.3	[730]
PbS	Cu	400–900	29.7	7.1	5.0×10^{-4}	[731]
ZnS	Zn	950–1030	314.0	75.0	1.5×10^{4}	[732]
PbS–0.5Bi_2S_3	Pb	500–800	102.6	24.5	4.5×10^{-6}	[733]
MnS	Fe	980–1200	268.0	64.0	29.4	[734]
NiS	Ni	450–600	89.2	21.3	3.2×10^{-4}	[735]

4.6 Self-diffusion, heterodiffusion and chemical diffusion in metal halides

Material	Tracer	Temperature range °C	Activation energy E_D		Frequency factor D_0 cm^2/s	Ref.
			kJ/mole	kcal/mole		
AgCl	Fe	232–449	122.3	29.2	15.1	[736]
AgCl	^{54}Mn	320–430	115.6	27.6	2.9	[737]
AgCl	^{54}Mn	76–320	56.1	13.4	1.8×10^{-5}	[737]
AgCl	^{114}In	300–370	154.5	36.9	1.6×10^{-4}	[738]
AgCl	^{65}Zn	352–441	97.6	23.3	4.5	[739]
AgCl	^{65}Zn	209–350	74.1	17.7	5.3×10^{-2}	[739]
AgI	Cu	180–430	9.6	2.3	1.6×10^{-4}	[740]
AgCl–$MnCl_2$	^{54}Mn	200–400	54.8	13.1	2.3×10^{-4}	[741]
BaF_2	F	350–940	154.1	36.8	3.1	[742]
CaF_2	F	350–940	193.0	46.1	50.0	[742]
CaF_2	Ca	987–1246	400.3	95.6	5.4×10^{3}	[743]
CaF_2	Ar	700–940	578.6	138.2	1×10^{10}	[744]
CaF_2	Kr	700–940	578.6	138.2	1×10^{10}	[745]
CaF_2	Xe	700–940	578.6	138.2	1×10^{10}	[745]
CaF_2	Ar	940–1320	289.3	69.1	—	[745]
CaF_2	Kr	940–1320	327.8	78.3	1×10^{4}	[745]
CaF_2	Xe	940–1320	578.6	138.2	—	[745]
CaF_2	F	600–900	62.8	15.0	—	[746]
LiF	F	730–830	289.3	69.1	5.6	[747]
LiF	F	580–730	193.0	46.1	0.8	[747]
LiF	^{6}Li	650–800	205.6	49.1	90.9	[748]
KCl	K	400–700	82.1	19.6	—	[749]
KCl	^{36}Cl	500–730	216.9	51.8	178	[750]
KCl	^{42}K	600–730	207.2	49.5	137	[750]

4.6 (*cont.*).

Material	Tracer	Temperature range °C	Activation energy E_D		Frequency factor D_0 cm^2/s	Ref.
			kJ/mole	kcal/mole		
KCl	^{203}Hg	449–559	54.8	13.1	8.2×10^{-5}	[751]
NaI	chem. 0.05 KI	420–560	129.0	30.8	0.3	[752]
NaI	chem. 0.10 KI	420–560	134.0	32.0	0.8	[752]
NaI	chem. 0.15 KI	420–560	136.9	32.7	1.3	[752]
NaI	chem. 0.20 KI	420–560	142.4	34.0	3.2	[752]
NaI	chem. 0.25 KI	420–560	144.4	34.5	4.7	[752]
NaI	chem. 0.30 KI	420–560	145.3	34.7	4.8	[752]
NaI	chem. 0.35 KI	420–560	150.3	35.9	14.7	[752]
NaI	chem. 0.40 KI	420–560	152.0	36.3	19.8	[752]
NaI	chem. 0.45 KI	420–560	154.1	36.8	30.7	[752]
NaI	chem. 0.50 KI	420–560	152.0	36.3	25.5	[752]
NaI	chem. 0.55 KI	420–560	154.1	36.8	33.7	[752]
NaI	chem. 0.60 KI	420–560	156.2	37.3	42.4	[752]
NaI	chem. 0.65 KI	420–560	157.8	37.7	49.2	[752]
NaI	chem. 0.70 KI	420–560	154.1	36.8	28.7	[752]
NaI	chem. 0.75 KI	420–560	148.2	35.4	11.3	[752]
NaI	chem. 0.80 KI	420–560	144.4	34.5	5.7	[752]
NaI	chem. 0.85 KI	420–560	138.6	33.1	2.1	[752]
NaI	chem. 0.90 KI	420–560	121.4	29.0	0.1	[752]
NaI	chem. 0.95 KI	420–560	111.8	26.7	0.03	[752]
NaCl	Cd	530–655	24.7	5.9	3.9×10^{-3}	[753]
NaCl	Au	550–700	106.3	25.4	0.2	[754]
NaCl	Au	600–850	103.0	24.6	0.2	[755]
NaCl	Zn	540–720	102.2	24.4	4.0×10^{-2}	[752]

4.6 (*cont.*).

Material	Tracer	Temperature range °C	Activation energy E_D		Frequency factor D_0 m^2/s	Ref.
			kJ/mole	kcal/mole		
NaCl	^{24}Na	> 550	173.3	41.4	3.13	[756]
NaCl	^{24}Na	< 550	74.1	17.7	1.6×10^{-5}	[756]
NaF	F	630–940	298.9	71.4	—	[756]
NaBr	^{22}Na	471–730	124.3	29.7	0.06	[757]
NaF	F	> 960	131.9	31.5	—	[758]
NaF	F	630–940	298.5	71.3	4.1×10^{4}	[759]

References

[1] J. Chojnacki, *Elementy krystalografii chemicznej i fizycznej*, ed. III, PWN, Warszawa 1973.

[2] A. F. Wells, *Models in Structural Inorganic Chemistry*, Clarendon Press, Oxford 1970.

[3] P. Kofstad, *Nonstoichiometry, Diffusion and Electrical Conductivity in Binary Metal Oxides*, J. Wiley-Interscience, New York 1972, p. 2.

[4] L. V. Azaroff, *Introduction to Solids*, McGraw-Hill, New York 1960.

[5] C. Kittel, *Introduction to Solid State Physics*, ed. III, J. Wiley-Interscience, New York 1966.

[6] N. B. Hannay, *Solid State Chemistry*, Prentice-Hall Inc., Englewood Cliffs 1967.

[7] J. Oding, *Teoria dislokacyi w metallakh i eyi primenene*, Izd. Akad. Nauk SSSR, Moscow 1959.

[8] S. Amelinck, *Direct Observation of Dislocations*, Academic Press, New York 1964.

[9] J. Friedel, *Dislocations*, Addison–Wesley, 1964.

[10] J. Weertman, J. R., Weertman *Elementary Dislocation Theory*, The Macmillan Company, New York 1964.

[11] F. Frank, *Adv. Phys.*, **1**, 91 (1952).

[12] J. Dereń, J. Haber, R. Pampuch, *Chemia ciala stalego*, PWN, Warszawa 1975, p. 265.

[13] K. Hauffe, *Reaktionen in und an festen Stoffen*, Springer, Berlin 1966.

[14] J. Gray, *The Defect Solid State*, J. Wiley-Interscience, New York 1957.

[15] F. Kröger, *The Chemisty of Imperfect Crystals*, North-Holland, Amsterdam 1974

[16] S. Mrowec, Z. Jarzębski, *Półprzewodniki tlenkowe*, WNT, Warszawa 1969.

[17] W. Van Gool, *Principles of Defect Chemistry of Crystalline Solids*, Academic Press, New York 1966.

[18] N. Grenwood, *Ionic Crystals, Lattice Defects and Nonstoichiometry*, Butterworth, London 1968.

[19] M. S. Seltzer, R. I. Jaffee, *Defects and Transport in Oxides*, Plenum Press, New York 1974.

[20] F. A. Kröger, H. J. Vink, *Solid State Phys.*, Academic Press, New York, **3**, 307 (1956).

[21] C. Wagner, *Z. phys. Chem.*, **22**, 181 (1933).

[22] H. Baumbach, C. Wagner, *Z. phys. Chem.*, **24**, 59 (1934).

[23] W. Schottky, *Z. phys. Chem.* **29**, 335 (1935).

[24] A. L. G. Rees, *Chemistry of Defect Solid State*, J. Wiley, New York 1954.

[25] F. Kröger, H. Vink, J. Boomgaard, *Z. phys. Chem.* **203**, 1 (1954).
[26] J. Besson, *J. Chim. Phys.* **62**, 937 (1965).
[27] W. Schottky, F. Stockmann, *Halbleiterprobleme, Vieweg, Braunschweig*, **1**, 80, 135 (1954).
[28] J. J. Lander, *Semiconductors*, Reinhold Publ. Co., New York 1959, chapt. 2.
[29] D. Thomas, *Semiconductors*, Reinhold Publ. Co., New York 1959, chapt. 7.
[30] W. Schottky, *Halbleiterprobleme, Vieweg, Braunschweig*, **4**, 235 (1959).
[31] W. Albers, *Physical Chemistry of 2–6 Compounds*, North-Holland, Amsterdam 1967, chapt. 4.
[32] J. W. Ong, *J. Electrochem. Soc.*, **114**, 1080 (1967).
[33] P. Manca, *J. Phys. Chem. Solids*, **20**, 268 (1961).
[34] A. K. Vijh, *J. Phys. Chem. Solids*, **30**, 1999 (1969).
[35] A. *K.* Vijh, *J. Phys. Chem. Solids*, **29**, 2233, (1968).
[36] A. K. Vijh, *J. Mater. Sci.*, **5**, 379 (1970).
[37] P. Kofstad, *Nonstoichiometry, Diffusion and Electrical Conductivity in Binary Metal Oxides*, J. Wiley-Interscience, New York 1972, p. 65.
[38] B. Fisher, J. Tannhauser, *J. Electrochem. Soc.*, **111**, 1194 (1964).
[39] B. Fisher, J. Tannhauser, *J. Chem. Phys.* **44**, 1963 (1966).
[40] L. W. Barr, D. K. Dawson, "Correlations between Melting Temperatures and Formation and Cation Motion Energies of Schottky Defects in Ionic Crystals", in *Mass Transport in Non-Metallic Solids, Proc. Brit. Ceramic Soc.*, No **19**, March 1971, p. 151.
[41] P. Kofstad, *Nonstoichiometry, Diffusion and Electrical Conductivity in Binary Metal Oxides*, J. Wiley-Interscience, New York 1972, p. 55.
[42] P. Kofstad, *J. Phys. Chem. Solids*, **28**, 1842 (1967).
[43] P. Kofstad, *J. Electrochem. Soc.*, **116**, 869 (1969).
[44] P. Kofstad, *Nonstoichiometry, Diffusion and Electrical Conductivity in Binary Metal Oxides*, J. Wiley-Interscience, New York 1972, p. 63.
[45] K. Hauffe, H. Pfeiffer, *Z. Metallkunde*, **44**, 27 (1957).
[46] P. Kofstad, *J. Phys. Chem. Solids*, **23**, 1579 (1962).
[47] K. S. Förland, *Acta Chem. Scand.*, **18**, 1267 (1959).
[48] J. S. Anderson, *Defects in Oxides*, National Bureau of Standards, Special Publication, No **364**, 1972, p. 295.
[49] J. S. Anderson, "The Thermodynamics and Theory of Nonstoichiometric Compounds" in *Problems of Nonstoichiometry*, collective work, Ed. A. Rabenau, North-Holland, Amsterdam 1970, p. 1.
[50] J. S. Anderson, *Rev. Pure and Appl. Chem.*, **21**, 67 (1971).
[51] N. Greenwood, *Ionic Crystals, Lattice Defects and Nonstoichiometry*, Butterworth, London 1968, p. 130.
[52] E. F. Bertaut, *Acta Cryst.*, **6**, 557 (1953).
[53] J. H. Santen, *Phyllips Res. Repts.*, **5**, 282 (1950).
[54] L. M. Atlas, *J. Phys. Chem. Solids*, **29**, 91, 1349 (1968).
[55] M. O'Keeffe, "Extended Defects and Their Possible Role in Mass Transport", in *Proc. Brit. Ceramic Soc.*, No **19**, 1971, p. 1.
[56] M. Brunel, F. Bergevin, *Solid State Comm.*, **4**, 165 (1966).

[57] E. F. Westrum, E. Gronvold, *J. Chem. Thermodynamics*, **1**, 543 (1969).

[58] N. Greenwood, *Ionic Crystals, Lattice Defects and Nonstoichiometry*, Butterworth, London 1968, p. 65.

[59] D. Watanabe, O. Tarasaki, J. Castles, "Electron Microscopic Study on Structure of Low Temperature Modification of Titanium Monoxide Phase", in *The Chemistry of Extended Defects in Non-Metallic Solids*, collective work, Ed. L. Eyring and M. O'Keefe, North-Holland, Amsterdam 1970, p. 238.

[60] M. D. Banus, T. B. Reed, "Structural, Electrical and Magnetic Properties of Vacancy Stabilized Cubic TiO and VO" in *The Chemistry of Extended Defects in Non-Metallic Solids*, collective work, Ed. L. Eyring and M. O'Keefe, North-Holland, Amsterdam 1970, p. 488.

[61] M. D. Banus, *Met. Res. Bull.*, **2**, 563 (1967); **3**, 723, (1968).

[62] W. Van Gool, *J. Mater. Sci.*, **1**, 261 (1966).

[63] B. T. M. Willis, *Nature*, **197**, 755 (1963).

[64] B. T. M. Willis, *Proc. Roy. Soc.*, **274**, 122 (1963).

[65] J. S. Anderson, *The Real Structure of Defect in Oxides*, collective work, Ed. M. Seltzer and R. Jaffee, Plenum Press, New York 1973, p. 25.

[66] T. C. Parks, M. R. Thornber, D. J. M. Bevan, *J. Solid State Chem.*, **1**, 536 (1970).

[67] M. R. Thornber, D. J. M. Bevan, E. Summerville, *J. Solid State Chem.*, **1**, 545 (1970).

[68] N. Greenwood, *Ionic Crystals, Lattice Defects and Nonstoichiometry*, Butterworth, London 1968, p. 139.

[69] W. L. Roth, *Acta Cryst.*, **25**, 140 (1969).

[70] F. Koch, J. B. Cohen, *Acta Cryst.*, **25**, 275 (1969).

[71] P. Vallet, P. Raccah, *Mem. Sci. Rev. Metallurg.*, **62**, 1 (1968).

[72] J. Smuts, *J. Iron Steel Inst.*, **204**, 237 (1966).

[73] A. Cheetham, B. Fender, R. Taylor, *Solid State Phys.*, **4**, 2160 (1970).

[74] B. Fender, F. Riley, *J. Phys. Chem. Solids*, **30**, 793 (1969).

[75] P. Kofstad, A. Z. Hed, *J. Electrochem. Soc.*, **115**, 102 (1968).

[76] J. Janowski, M. Jaworski, R. Benesch, *Arch. Eisenhüttenwesen*, **44**, 721 (1973).

[77] E. Jette, F. Foote, *J. Chem. Phys.*, **1**, 29 (1933).

[78] M. Settzer, A. Hed, *J. Electrochem. Soc.*, **117**, 815 (1970).

[79] D. Neuschütz, N. Towhidi, *Arch. Eisenhüttenwesen*, **41**, 303 (1970).

[80] P. Childs, L. Lanb, J. B. Wagner, "Chemical Diffusion in Nonstoichiometric Compounds" in *Proc. Brit. Ceramic Soc.*, No **19**, March 1971.

[81] P. Childs, J. B. Wagner, "Chemical Diffusion in Wustite and Chromium-Doped Manganeous Oxide" in collective work: *Heterogenous Kinetics at Elevated Temperatures*, Ed. G. Belton and W. Worrell, Plenum Press, New York 1970, p. 269.

[82] J. B. Wagner, "Chemical Diffusion Coefficients for Some Nonstoichiometric Metal Oxides", *Proceedings of a Symposium Held at Gaithersburg*, Maryland 22–25 October 1967, National Bureau of Standards, Special Publication No. **269**, 1969, p. 264.

[83] D. Johnson, *Solid State Comm.*, **7**, 1785 (1967).

[84] P. Kofstad, *Nonstoichiometry, Diffusion and Electrical Conductivity in Binary Metal Oxides*, J. Wiley-Interscience, New York 1972, p. 137 and 182.

[85] A. Magneli, *Acta Cryst.*, **2**, 501 (1948).

[86] A. Magneli, *Acta Cryst.*, **6**, 495 (1953).
[87] A. Magneli, G. Hägg, *Rev. Pure and Appl. Chem.*, **4**, 235 (1954).
[88] A. D. Wadsley, *Rev. Pure and Appl. Chem.*, **1**, 203 (1951).
[89] A. D. Wadsley, *Rev. Pure and Appl. Chem.*, **5**, 165, 190 (1955).
[90] A. D. Wadsley, *Nonstoichiometric Compounds*, collective work, Ed. L. Mandelcorn, Academic Press, New York 1964, p. 99.
[91] S. Anderson, B. Collen, U. Kuylenstierna, A. Magneli, *Acta Chem. Scand.*, **11**, 164 (1957).
[92] S. Anderson, A. Sundholru, A. Magneli, *Acta Chem. Scand.*, **13**, 989 (1959).
[93] S. Anderson, L. Jahnberg, *Arkiv. Kemi.*, **21**, 413 (1963).
[94] S. Anderson, *Acta Chem. Scand.*, **18**, 2339 (1964).
[95] A. Magneli, "Structural Order and Disorder in Oxides of Transition Metals of the Titanium, Vanadium and Chromium Groups", in collective work: *The Chemistry of Extended Defects in Non-Metallic Solids*, Ed. L. Eyring and M. O'Keefe, North-Holland, Amsterdam 1970, p. 148.
[96] G. Taylor, *Proc. Roy. Soc.*, **145**, 362 (1934).
[97] E. Orowan, *Z. Phys.*, **89**, 605, 614, 634 (1934).
[98] J. Polanyi, *Z. Phys.*, **89**, 660 (1934).
[99] A. D. Wadsley, J. S. Anderson, *Nature*, **211**, 581 (1966).
[100] J. S. Anderson, B. G. Hyde, *J. Phys. Chem. Solids*, **28**, 1393 (1967).
[101] J. S. Anderson, R. J. D. Tilley, *J. Solid State Chem.*, **2**, 472 (1970).
[102] J. G. Allpress, R. J. D. Tilley, M. J. Sienko, *J. Solid State Chem.*, **3**, 440 (1971).
[103] J. G. Allpress, J. V. Sanders, A. D. Wadsley, *Acta Cryst.*, **25**, 1156 (1969).
[104] J. G. Allpress, *J. Solid State Chem.*, **1**, 28, 66 (1970).
[105] J. G. Allpress, R. S. Roth, *J. Solid State Chem.*, **3**, 209 (1971).
[106] J. Landuyt, R. Gevers, S. Amelinck, *Phys. Stat. Solidi*, **7**, 519 (1964).
[107] J. Landuyt, S. Amelinck, *Phys. Stat. Solidi*, **31**, 589 (1969).
[108] S. Amelinck, J. Landuyt, "The Use of Electron Microscopy in the Study of Extended Defects Related to Non-Stoichiometry", in collective work: *The Chemistry of Extended Defects in Non-Metallic Solids*, Ed. L. Eyring and M. O'Keefe, North-Holland, Amsterdam 1970, p. 295.
[109] L. A. Bursil, B. G. Hyde, *Phys. Mag.*, **23**, 3 (1971).
[110] R. S. Roth, *Discussion to the paper of S. Anderson: Defects in Oxides*, National Bureau of Standards, Special Publication, No. **364**, 1972, p. 295.
[111] S. Mitoff, *Progr. Cer. Sci.*, **4**, 217 (1966).
[112] H. Baumbach, C. Wagner, *Z. phys. Chem.*, **22**, 199 (1933).
[113] P. Kofstad, *J. Phys. Chem. Solids*, **23**, 1571 (1962).
[114] R. R. Heikes, W. D. Johnston, *J. Chem. Phys.*, **26**, 582 (1957).
[115] T. Hollstein, *Ann. Phys.*, **8**, 325 (1959).
[116] G. L. Sewell, *Phys. Rev.*, **129**, 597 (1963).
[117] L. Friedman, *Phys. Rev.*, **133**, 668 (1964).
[118] L. V. Azaroff, J. J. Brophy, *Electronic Processes in Materials*, McGraw-Hill, New York 1963.
[119] D. Adler, *Solid State Phys.*, **21**, 1 (1968).
[120] J. Appel, *Solid State Phys.*, **21**, 193, (1968).
[121] E. Koch, C. Wagner, *Z. phys. Chem.*, **38**, 295 (1937).

[190] S. Mrowec, K. Przybylski, *Oxidation of Metals*, **11** (1977), p. 365, 383.
[191] N. G. Eror, J. B. Wagner, *J. Phys. Chem. Solids*, **29**, 1597 (1968).
[192] K. Przybylski, *Dissertation*, AGH, Kraków 1976.
[193] J. Dereń, Z. Jarzębski, S. Mrowec, T. Walec, *Bull. Acad. Polon. Sci., ser. chim.*, **19**, 147 (1971); **19**, 153 (1971).
[194] S. Mrowec, J. Dereń, *J. Mater. Sci.*, **8**, 545 (1973).
[195] J. E. Stroud, I. Bransky, N. M. Tallan, *J. Chem. Phys.*, **58**, 1263 (1973).
[196] S. Mitoff, *J. Chem. Phys.*, **35**, 882 (1961).
[197] M. Volpe, J. Reddy, *J. Chem. Phys.*, **53**, 1117 (1970).
[198] J. Choi, W. J. Moore, *J. Phys. Chem.*, **66**, 1308 (1963).
[199] S. Klotzman, A. Timofeev, J. Trachtenberg, *Fiz. met. i metalloved.*, **14**, 91 (1962).
[200] K. Fueki, J. B. Wagner, *J. Elektrochem. Soc.*, **112**, 384 (1965).
[201] M. T. Shim, W. J. Moore, *J. Chem. Phys.*, **26**, 802 (1957).
[202] S. Mrowec, A. Stokłosa, K. Godlewski, *Crystal Lattice Defects*, **5**, 239 (1974).
[203] R. S. Toth, R. Kilkson, D. Trivich, *Phys. Rev.*, **122**, 482 (1961).
[204] O. Böttger, *Ann. Phys.*, **10**, 232 (1952).
[205] K. Stecker, *Ann. Phys.*, **3**, 55 (1959).
[206] M. O'Keefe, W. J. Moore, *J. Chem. Phys.*, **35**, 1324 (1961).
[207] A. Kröger, *The Chemistry of Imperfect Crystal*, North-Holland, Amsterdam 1964, p. 375.
[208] R. P. Zulew, *Fiz. tverd. tela*, **1**, 111 (1959).
[209] R. P. Zulew, *Fiz. tverd. tela*, **1**, 113 (1959).
[210] J. Bloem, *Phyllips Res. Repts.*, **13**, 167 (1958).
[211] G. F. Garlick, "Photo-Conductivity", *Handbuch der Physik*, **19**, 377 (1956).
[212] M. O'Keefe, W. J. Moore, *J. Chem. Phys.*, **36**, 3009 (1962).
[213] J. Słoczyński, Z. Kowalski, T. Wójcikiewicz, *J. Mater. Sci.*, **7**, 1369 (1972).
[214] S. Mrowec, J. Nowotny, T. Walec, T. Wójcikiewicz, *Bull. Acad. Polon. Sci., Ser. sci. chim.*, **21**, 829 (1973).
[215] S. Mrowec, T. Walec, T. Wójcikiewicz, *Prace Komisji Ceramicznej Oddziału Krakowskiego PAN*, **21**, 81 (1974).
[216] I. Bransky, N. M. Tallan, *J. Chem. Phys.*, **49**, 1243 (1968).
[217] G. Sockel, B. Ilschner, *Z. phys. Chem., N.F.*, **74**, 284 (1971).
[218] P. Kofstad, A. Z. Hed, *J. Electrochem. Soc.*, **115**, 102 (1968).
[219] M. S. Seltzer, A. Z. Hed, *J. Elektrochem. Soc.*, **117**, 815 (1970).
[220] L. Couling, R. Smoluchowski, *J. Appl. Phys.*, **25**, 1538 (1954).
[221] W. Bukagov, F. Rybalko, *Zh. tekhn. fiz.* **2**, 617 (1935).
[222] E. Hart, *Acta Met.*, **5**, 597 (1957).
[223] J. Choi, P. Shewmon, *Trans. AIME*, **230**, 123 (1964).
[224] B. Boltaks, *Diffuzya v poluprovodnikakh*, Mir, Moscow 1961.
[225] J. Uminsky, B. Finkelshtein, M. Blanter, S. Kishkin, N. Fastov, S. Gorelik, *Fizicheskoe Metallovedene*, Metallurgizdat, Moscow 1955, p. 496.
[226] L. Pfeil, *J. Iron Steel Inst.*, **119**, 501 (1929); **183**, 237 (1931).
[227] K. Sachs, *Metallurgia*, **54**, 11 (1956).
[228] W. Arkharov, E. Blankova, *Fit. met. i metalloved.*, **8**, 452 (1959); **9**, 878 (1960); **10**, 63 (1960).
[229] R. Meussner, C. Birchenall, *Corrosion*, **13**, 677 (1957).

[230] S. Mrowec, *Z. phys. Chem. N.F.*, **29**, 47 (1961).
[231] S. Mrowec, T. Werber, *Acta Met.*, **8**, 819 (1959).
[232] S. Mrowec, *Arch. Hutn. PAN*, **6**, 61 (1961); **6**, 327 (1961).
[233] L. Czerski, *Arch. Hutn.*, **2**, 3 (1954).
[234] M. Cagnet, J. Moreau, *Rev. Met.*, **55**, 1091 (1958).
[235] H. Engell, F. Wever, *Acta Met.*, **5**, 695 (1957).
[236] M. Cagnet, J. Moreau, *Acta Met.*, **7**, 427 (1959).
[237] A. Brückman, S. Mrowec, T. Werber, *Fiz. met. i metalloved.*, **20**, 702 (1965).
[238] M. Davies, M. Simnad, C. Birchenall, *J. Metals*, **3**, 889 (1951); **5**, 1250 (1953).
[239] R. Carter, F. Richardson, *J. Metals*, **203**, 336 (1955).
[240] K. Löhberg, *Z. Matallkunde*, **38**, 320 (1947).
[241] P. Kofstad, P. B. Anderson, J. Krudtaa, *J. Less-Common. Metals*, **3**, 89 (1961).
[242] C. Wagner, "Diffusion and High Temperature Oxidation of Metals" in *Atom Movements*, ASM, Cleveland, Ohio 1951, p. 151.
[243] E. Gulbransen, *J. Electrochem. Soc.*, **81**, 327 (1942).
[244] E. Gulbransen, *Rev. Sci. Instr.*, **15**, 201 (1944).
[245] T. Rhodin, *J. Amer. Chem. Soc.*, **72**, 4343 (1950).
[246] K. Hauffe, *Oxidation of Metals*, Plenum Press, New York 1965.
[247] P. Kofstad, *High Temperature Oxidation of Metals*, J. Wiley, New York 1966.
[248] S. Mrowec, A. Stokłosa, *J. Thermal. Anal.*, **2**, 73 (1970).
[249] W. McKewan, W. Fassel, *Trans. AIME*, **197**, 1127 (1953).
[250] P. Kofstad, *High Temperature Oxidation of Metals*, J. Wiley, New York 1966, p. 8.
[251] O. Neunhöffera, K. Hauffe, *Z. anorg. Chem.*, **262**, 300 (1950).
[252] K. Hauffe, A. Rahmel, *Z. phys. Chem.*, **199**, 152 (1952),
[253] P. Geld, A. Krasowska, *Zh. fiz. khim.*, **34**, 1585 (1960); **34**, 1712 (1960).
[254] S. Mrowec, A. Stokłosa, M. Danielewski, *Zesz. Nauk. AGH.*, **27**, 49 (1976).
[255] S. Mrowec, A. Stokłosa, M. Danielewski, *Oxidation of Metals*, **11**, 355 (1977).
[256] M. Danielewski, S. Mrowec, A. Stokłosa, J. Przybyszewska, *Bull. Acad. Polon. Sci., Ser. sci. chim.*, **26**, 175 (1978).
[257] S. Mrowec, A. Stokłosa, K. Godlewski, *Prace Komisji Ceramicznej Krakowskiego Oddz. PAN*, **21**, 95 (1974).
[258] P. Kofstad, P. B. Anderson, *J. Phys. Chem. Solids*, **21**, 280 (1961).
[259] P. Kofstad, *High Temperature Oxidation of Metals*, J. Wiley, New York 1966, p. 68.
[260] H. Engell, K. Hauffe, B. Ilschner, *Z. Elektrochem.*, **58**, 478 (1954).
[261] A. Jenkins, *J. Inst. Met.*, **84**, 1 (1955).
[262] S. Mrowec, M. Lasoń, E. Fryt, K. Przybylski, A. Ciembroniewicz, *J. Thermal. Anal.*, **6**, 193 (1974).
[263] H. Engell, *Z. Elektrochem.*, **60**, 905 (1956).
[264] H. Engell, *Arch. Eisenhüttenwesen*, **28**, 109 (1957).
[265] G. Jander, K. Jahr, *Massanalyse*, Springer, Berlin 1940.
[266] J. Dereń, A. Kowalska, *Chem. Anal.*, **7**, 563 (1962).
[267] J. Dereń, E. Fryt, *Chem. Anal.*, **8**, 365 (1963).
[268] A. Rosenburg, *J. Electrochem. Soc.*, **107**, 795 (1960).

[269] S. Mrowec, "Oxidation Kinetic Methods in Studying the Defect Concentration and Diffusion Coefficients in Transition Metal Oxides", *Proceedings of 25th Internatinal Meeting of the Société de Chimie Physique: Reaction Kinetics in Heterogeneous Chemical Systems*, Elsevier, Amsterdam 1975, p. 381.
[270] J. Price, J. B. Wagner, *Z. phys. Chem., N.F.* **49**, 257 (1966).
[271] G. Koel, P. Gellings, *Oxidation of Metals*, **5**, 185 (1972).
[272] R. Carter, R. Richardson, *J. Metals*, **6**, 1244 (1954).
[273] W. Chen, N. Peterson, W. Reeves, *Phys. Rev.*, **186**, 897 (1969).
[274] R. Rapp, *Physicochemical Measurements in Metals Research*, vol. II, Interscience, New York, 1970.
[275] Y. Tretiakov, R. Rapp, *Trans. AIME*, **242**, 1235 (1969).
[276] B. Steel, *Mass Transport in Oxides*, Ed. J. B. Wachtman, A. D. Franklin, National Bureau of Standards, Special Publication No **296**, Washington 1968.
[277] H. Sockel, H. Schmalzried, *Ber. Buns. Phys. Chem.*, **72**, 745 (1968).
[278] E. Fryt, S. Mrowec, A. Podgórecka, *Bull. Acad. Polon. Sci., Ser. sci. chim.*, **23**, 263 (1975).
[279] J. Fisher, *J. Appl. Phys.*, **22**, 74 (1951).
[280] H. Letaw, L. Slifkin, W. Portnoy, *Rev. Sci. Instr.*, **25**, 865 (1954).
[281] H. J. de Brium, R. L. Clark, *Rev. Sci. Instr.*, **35**, 227 (1964).
[282] P. Kofstad, *Passivierende Filme und Deckschichten*, Springer, Berlin 1956, p. 22.
[283] T. S. Lundy, "Volume Diffusion in Solids", in *Physicochemical Measurements in Metals Research*, Ed. R. Rapp, Interscience, New York 1970.
[284] F. Pettit, *J. Electrochem. Soc.*, **113**, 1250 (1966).
[285] K. Fueki, J. B. Wagner, *J. Electrochem. Soc.*, **112**, 384 (1965).
[286] S. Mrowec, A. Stokłosa, *Werkstoffe u. Korrosion*, **21**, 934 (1970).
[287] S. Mrowec, A. Stokłosa, *Oxidation of Metals*, **8**, 379 (1974).
[288] S. Mrowec, A. Stokłosa, *Oxidation of Metals* **3**, 3 (1971).
[289] W. Moore, B. Selikson, *J. Chem. Phys.*, **19**, 1539 (1951); **20**, 927 (1952).
[290] S. Mrowec, *Bull. Acad. Polon. Sci., Ser. sci. chim.*, **15**, 517 (1967).
[291] S. Mrowec, *Bull. Acad. Polon. Sci., Ser. sci. chim.*, **15**, 373 (1967).
[292] S. Mrowec, T. Walec, T. Werber, *Corrosion Sci.*, **6**, 287 (1966).
[293] P. B. Childs, L. Laub, J. B. Wagner, "Chemical Diffusion in Non-Stoichiometric Compounds", *Proc. Brit. Ceramic Soc.*, No **19**, (1971).
[294] M. Beyeler, Y. Adda, *J. phys.* (*Paris*), **29**, 345 (1968).
[295] J. Burke, T. R. Ramachandran, "Self-diffusion in Al at Low Temperatures", *Met. Trans.*, **3**, No 1 (1972), p. 147.
[296] A. Hassner, W. Lange, *Phys. Stat. Solidi*, **11**, 575 (1965).
[297] J. M. Dupony, J. Mathie, Y. Adda, *Mem. Sci. Rev. Met.*, **63**, 481 (1966).
[298] A. Languille, M. P. Dariel, D. Calais, B. Cogblin, *Mem. Sci. Rev. Met.*, **70**, No 3, 241–9 (1973).
[299] E. S. Wajda, G. A. Shirn, H. B. Huntington, *Acta. Met.*, **3**, 39 (1955).
[300] W. Hirschwald, W. Schroedter, *Z. phys. Chem. N. F.*, **53**, 392 (1967).
[301] A. K. Iha, U. P. Sinna, *Scr. Met.*, **6**, 495 (1972).
[302] W. C. Hagel, *Trans. Met. Soc. AIME*, **224**, (1962).
[303] J. Askill, G. H. Tomlin, *Phil. Mag.*, **11** 467 (1965).
[304] J. Askill, *Phys. Stat. Solidi*, **A8**, 587 (1971).

[305] K. Hirano, R. P. Agarwala, B. L. Averbach, M. Cohen, *J. Appl. Phys.*, **33**, 3049 (1962).
[306] A. Hassner, W. Lange, *Phys. Stat. Solidi*, **8**, 77 (1965).
[307] D. B. Butrymowicz, J. R. Manning, M. E. Read, *J. Phys. Chem.*, **2**, 643 (1973).
[308] S. J. Rothman, N. L. Peterson, "Isotope Effect and Divacancies for Self-diffusion in Cu", *Phys. Stat. Solidi*, **35**, 305 (1969).
[309] K. Monma, H. Suto, H. Oikawa, *J. Japan Inst. Metals*, **28**, 188 (1964).
[310] E. A. Smirnov, L. I. Ivanov, E. A. Abranian, *Izv. Akad. Nauk SSSR*, **168**, (1967).
[311] H. Widner, G. R. Gunther-Mohr, *Helv. Phys. Acta*, **34**, 635 (1961).
[312] D. Duhl, K. Hirano, M. Cohen, *Acta Met.*, **11**, 1 (1963).
[313] H. M. Gilder, D. Lazarus, *J. Phys. Chem. Solids*, **26**, 2081 (1965).
[314] T. F. Archbold, W. H. King, *Trans. Met. Soc. AIME*, **233**, 839 (1965).
[315] D. Graham, D. H. Tomlin, *Phil. Mag.*, **8**, 1581 (1963).
[316] F. S. Buffington, K. Hirano, M. Cohen, *Acta Met.*, **9**, 434 (1961).
[317] D. W. James, G. M. Leak, *Phil. Mag.*, **14**, 701 (1966).
[318] A. A. Zhukhovitskii, V. A. Geodakyan, *Prim. Radioakt. Izotopov v Metallurg., Sb.*, **34**, 267 (1955); *AEC-tr-3100, Uses of Radioactive Isotopes in Metallurgy, Symposium XXXIV*, Pt. 2, p. 52.
[319] N. A. Bogdanov, *Russ. Met. Fuels* (English transl.) **2**, 61 (1962).
[320] B. Sparke, D. W. James, G. M. Leak, *J. Iron Steel Inst.* (*London*), **203**, 152 (1965).
[321] L. I. Staffansson, C. E. Birchenall, *AFOSR-733*, (1961).
[322] R. J. Borg, D. Y. F. Lai, O. Krikorian, *Acta Met.*, **11**, 867 (1963).
[323] W. Danneberg, E. Krautz, *Z. Naturforsch.*, **16a**, 854 (1961).
[324] J. Askill, D. H. Tomlin, *Phil. Mag.*, **8**, 997 (1963).
[325] L. V. Pavlinov, V. N. Bykov, *Fiz. met. i metalloved.*, **18**, 459 (1964).
[326] A. Messner, R. Benson, J. E. Dorn, *Trans. ASM*, **53**, 227 (1961).
[327] A. R. Wazzam, J. Mote, J. E. Dorn, *U. S. Report AD-257188* (1961).
[328] A. Y. Shinyaev, *Phys. Metals Metallog.* (*USSR*), (English transl.), **15**, 100 (1963).
[329] G. B. Fedorov, E. A. Smirnov, F. I. Zhomov, *Met. i matalloved. chistykh metallov*, **4**, 110 (1963).
[330] A. R. Wazzam, *J. Appl. Phys.*, **36**, 3596 (1965).
[331] A. R. Wazzam, J. E. Dorn, *J. Appl. Phys.*, **36**, 222 (1965).
[332] D. F. Kalinovich, I. I. Kovenski, M. D. Smolin, *Fiz. tverd. tela*, **10**, 569 (1968).
[333] R. F. Peart, D. Graham, D. H. Tomlin, *Acta Met.*, **10**, 519 (1962).
[334] T. S. Lundy, R. F. Winslow, R. E. Pawel, C. J. McHargue, *Trans. Met. Soc. AIME*, **233**, 1533 (1965).
[335] V. D. Lubimov, P. W. Geld, P. G. Shveykin, *Izv. Akad. Nauk SSSR*, **5**, 137 (1964).
[336] N. L. Peterson, *Phys. Rev.*, **136**, A568 (1964).
[337] B., Million J. Kucera, *Kovove Mat.*, **11**, 300 (1973).
[338] G. V. Kidson, R. Ross, *Proc. 1-st UNESCO Int. Conf. Radioisotopes in Sci. Res.*, (1958).
[339] F. Cattaneo, E. Germagnoli, F. Grasso, *Phil. Mag.*, **7**, 1373 (1962).
[340] R. F. Peart, *Phys. Stat. Solidi*, **15**, 119 (1966).
[341] B. J. Masters, J. M. Fairfield, *Appl. Phys. Letters*, **8**, 280 (1966).
[342] W. A. Johnson, *Trans. Met. Soc. AIME*, **143**, 107 (1941).
[343] V. Raimshisel, S. N. Kriukov, *Vestn. Mosk. Univ. Khim.*, **13**, 344 (1972).

[344] S. J. Rothman, N. L. Peterson, I. T. Robinson, "Isotope Effect for Self-diffusion in Single Crystals of Ag", *Phys. Stat. Solidi*, **39**, 635 (1970).
[345] D. Turnbull, *Phys. Rev.*, **76**, 471 (1949).
[346] L. Slifkin, D. Lazarus, C. T. Tomizuka, *J. Appl. Phys.*, **23**, 1032 (1952).
[347] R. D. Johnson, A. B. Martin, *Phys. Rev.*, **86**, 642 (1952); *NAA-SR-170* (1952).
[348] S. N. Kriukov, A. A. Zhukovitski, *Dokl. Akad. Nauk SSSR*, **90**, 379 (1953).
[349] A. A. Zhukovitski, V. A. Geodakyan, *Dokl. Akad. Nauk SSSR*, **102**, 301 (1955); *AEC-tr-2265* (1955).
[350] E. Sonder, *Phys. Rev.*, **100**, 1662 (1955).
[351] H. Kruegar, H. N. Hirsch, *Trans. Met. Soc. AIME*, **203**, 125 (1955).
[352] C. T. Tomizuka, E. Sonder, *Phys. Rev.*, **103**, 1182 (1956).
[353] N. H. Nachtrieb, J. Petit, J. Wahrenberg, *J. Chem. Phys.*, **26**, 106 (1957).
[354] S. D. Gertsriken, T. K. Jachenko, *Vopr. fiz. met. i metalloved., Akad. Nauk USSR*, **8**, 101 (1957).
[355] G. Airoldi, E. Germagnoli, *Energia Nucl. (Milan)*, **5**, 445 (1958).
[356] S. D. Gertsriken, D. D. Tsitsiliano, *Phys. Metals Metallog. (USSR)*, (English transl.), **6**, 80 (1958).
[357] A. V. Savitski, *Fiz.met. i metalloved.*, **16**, 886 (1963).
[358] Y. Imai, T. Miyzaki, *Sci. Rep. Res. Inst. Thoku Univ.*, **A2**, 59 (1966).
[359] W. N. Kaigordov, S. M. Klotsman, A. N. Timofeev, I. S. Trachtenberg, *Fiz. met. i metalloved.*, **25**, 910 (1968).
[360] P. Reimer, *Metall.*, **22**, 577 (1968).
[361] W. Lange, A. Hassner, *Phys. Stat. Solidi*, **1**, 50 (1961).
[362] C. Coston, N. H. Nachtrieb, *J. Phys. Chem.*, **68**, 2219 (1964).
[363] G. Pawlicki, *Nukleonika*, **12**, 1123 (1967).
[364] C. M. Libanati, S. F. Dyment, *Acta Met.*, **11**, 1263 (1963).
[365] J. F. Murdock, T. S. Lundy, E. E. Stansbury, *Acta Met.*, **12**, 1033 (1964).
[366] N. E. W. DeReca, C. M. Libanati, *Acta Met.*, **16**, 1297 (1968).
[367] W. Danneberg, *Metall.*, **15**, 977 (1961).
[368] R. L. Andelin, J. D. Knight, M. Kahn, *Trans. Met. Soc. AIME*, **233**, 19 (1965).
[369] G. M. Neumann, W. Hirschwald, *Z. Naturforsch.*, **21a**, 812 (1966).
[370] Y. Adda, A. Kirianenko, C. Mairy, *Compt. Rend.*, **253**, 445 (1961); *J. Nucl. Mater.*, **6**, 130 (1962).
[371] Y. Adda, A. Kirianenko, C. Mairy, *J. Nucl. Mater.*, **1**, 300 (1959).
[372] S. J. Rothman, L. T. Lloyd, A. L. Harkness, *Trans. Met. Soc. AIME*, **218**, 605 (1960).
[373] R. F. Peart, *J. Phys. Chem. Solids*, **26**, 1853 (1965).
[374] N. L. Peterson, S. J. Rothman, *Phys. Rev.*, **163**, 645 (1967).
[375] P. Flubacher, *EIR Bericht*, No **49**, May (1963).
[376] G. V. Kidson, J. McGurn, *Can. J. Phys.*, **39**, 1146 (1961).
[377] J. I. Federer, T. S. Lundy, *Trans. Met. Soc. AIME*, **227**, 592 (1963).
[378] S. Badrinarayanan, H. B. Mathers, *Inst. J. Appl. Radiat. Isotopes*, **19**, 353 (1968).
[379] M. S. Anand, R. P. Agarwala, *Trans. AIME*, **239**, 1848 (1967).
[380] M. S. Anand, S. P. Murarka, R. P. Agarwala, *J. Appl. Phys.*, **36**, 3860 (1965).
[381] K. Hirano, R. P. Agarwala, M. Cohen, *Acta Met.*, **10**, 857 (1962).
[382] T. S. Lundy, J. F. Murdock, *J. Appl. Phys.*, **33**, 1671 (1962).

[383] A. R. Paul, R. P. Agarwala, *J. Appl. Phys.*, **38**, 3790 (1967).
[384] S. P. Murarka, M. S. Anand, R. P. Agarwala, *Acta Met.*, **16**, 69 (1968).
[385] K. Sorensen, G. Trumpy, *Phys. Rev.*, **B7**, 1719 (1973).
[386] D. Bergner, E. Cyrener, *Neue Hütte*, **18**, 356 (1973).
[387] D. Bergner, *Neue Hütte*, **17**, 705 (1972).
[388] D. Bergner, E. Cyrener, *Neue Hütte*, **18**, 9 (1973).
[389] I. Goedeny, D. Beke, F. J. Kedves, *Phys. Stat. Solidi*, **A13**, K155 (1972).
[390] M. S. Anand, R. P. Agarwala, *Phil. Mag.* **26**, 297 (1972).
[391] T. Mariemo, S. Fijikawa, K. Hiramo, *J. Japan Inst. Light Met.*, **23**, 17 (1973).
[392] G. M. Hood, R. J. Schultz, *Phil. Mag.*, **23**, No 186, 1479 (1971).
[393] S. J. Rothman, N. L. Peterson, L. J. Nowicki, L. C. Robinson, *Phys. Stat. Solidi*, **B63**(1), (1974), K29–K33.
[394] S. Fujikawa, K. Hirano, *Keikinzoku*, **20**, No 6 (1970), 267–77.
[395] M. Hood, R. J. Schultz, "Diffusion in Al", *Phys. Rev.*, **B4**, 2339 (1971).
[396] M. C. Naik, J. M. Dupony, Y. Adda, *Mem. Sci. Rev. Met.*, **63**, 488 (1966).
[397] V. M. Ananyn, V. P. Gladkov, V. S. Zolov, D. M. Skorov, *At. Energ.*, **29**, 220 (1970).
[398] H. W. Paxton, R. A. Wolfe, *Trans. AIME*, **230**, 1426 (1964).
[399] I. I. Kovenski, *Fiz. met. i metalloved.*, **16**, 613 (1963).
[400] M. Aucouturier, M. O. Pinheiro, Rib DeCastro, P. Lacombe, *Acta Met.*, **13**, 125 (1965).
[401] M. M. Pavlichenko, I. F. Kononiuk, *Dokl. Akad. Nauk BSSR*, **8**, 157 (1964).
[402] J. G. Mullen, *Phys. Rev.*, **121**, 1649 (1961).
[403] K. J. Anusavice, J. J. Pinayan, H. Oikawa, T. DeHoff, *Trans. AIME*, **242**, 2027 (1968).
[404] N. L. Peterson, *Phys. Rev.*, **132**, 2471 (1963).
[405] R. D. Johnson, B. H. Faulkenberry, *ASD-TDR-63-625* (July 1963).
[406] P. P. Kuzmenko, L. F. Ostrovski, V. S. Kovalchuk, *Fiz. tverd. tela*, **4**, 490 (1962).
[407] S. Komura, N. Kunitomi, *J. Phys. Soc. Japan*, **18**, 208 (1963).
[408] V. A. Gorbacher, S. M. Klotsman, Ya. A. Rabovski, V. K. Talinski, A. N. Timofeev, *Fiz. met. i metalloved.*, **34**, 879 (1972).
[409] C. T. Tomizuka (cit. after D. Lazarus), *Solid State Phys.*, **10** (1960).
[410] T. Hirone, N. Kunitomi, M. Sakamoto, H. Yamaki, *J. Phys. Soc. Japan*, **13**, 838 (1958).
[411] S. M. Klotsman, Ya. A. Rabovski, V. K. Talinski, A. N. Timofeev, *Fiz. metal. i metalloved.*, **31**, No 2, 429 (1971).
[412] G. Barreau, G. Brunel, G. Cizeron, *C. R. Acad. Sci., Ser. C*, **272**, 618 (1971).
[413] M. C. Saxena, B. D. Sharma, *Trans. Indian Inst. Metals*, **23**, 16 (1970).
[414] R. L. Fogelson, Y. A. Ugai, A. V. Pokoev, *Fiz. metal. i metalloved.*, **33**, 1102 (1972).
[415] J. Kucera, B. Million, J. Peskova, *Phys. Stat. Solidi*, **A11**, 361 (1972).
[416] R. L. Fogelson, Ya. A. Ugai, A. V. Pokoev, *Fiz. met. i metalloved.*, **34**, 1104 (1972).
[417] J. Bernardini, J. Cubane, *Acta Met.*, **21**, 1561 (1973).
[418] R. L. Fogelson, Ya. A. Ugai, A. V. Pokoev, I. A. Akimov, V. D. Kretinin, *Fiz. metal. i metalloved.*, **35**, 1307 (1973).
[419] V. A. Gorbachev, S. M. Klotsman, Ya. A. Rabovski, V. K. Talinski, A. N. Timofeev, *Fiz. metal. i metalloved.*, **35**, 889 (1973).

[420] R. L. Fogelson, Y. A. Ugai, J. A. Akimova, *Fiz. met. i metalloved.*, **37**, 1107 (1974).

[421] R. L. Fogelson, Y. A. Ugai, A. V. Pokoev, *Izv. Vyssh. ucheb. zaved., chern. met.*, (**9**), 136–7 (1973).

[422] M. C. Saxena, *Trans. Indian Inst. Metals*, **24**, 56 (1971).

[423] B. I. Boltaks, V. P. Grabshak, T. D. Dzhafarov, *Fiz. tverd. tela*, **6**, 3181 (1964).

[424] V. D. Ignatkov, V. E. Kosenko, *Fiz. tverd. tela.*, **4**, 1627 (1962).

[425] V. I. Tagirov, A. A. Kulev, *Fiz. tverd. tela*, **4**, 272 (1962).

[426] S. M. Klotsman, N. K. Arkhipova, A. N. Timofeev, L. S. Trachtenberg, *Fiz. met. i metalloved.*, **20**, 390 (1965).

[427] S. M. Klotsman, A. N. Timofeev, *Fiz. met. i metalloved.*, **20**, 395 (1965).

[428] T. R. Anthony, D. Turnbull, *Phys. Rev.*, **151**, 495 (1966).

[429] R. J. Borg, D. Y. F. Lai, *Acta Met.*, **11**, 861 (1963).

[430] C. G. Homan, *Acta Met.*, **12**, 1071 (1964).

[431] V. W. Lange, A. Hassner, E. Dahn, *Neue Hütte*, **1**, 33 (1961).

[432] A. M. Huntz, M. Aucouturier, P. Lacombe, *C. R. Acad. Sci. Paris*, *C***265**, 554 (1967).

[433] M. S. Anand, R. P. Agarwala, *J. Appl. Phys.*, **37**, 4248 (1966).

[434] V. T. Borisov, W. M. Golikov, G. W. Shcherbedinski, *Fiz. met. i metalloved.*, **22**, 159 (1966).

[435] G. Bruggeman, J. Roberts, *J. Met.*, **20**, 54 (1968).

[436] J. Kieszniewski, *Prace Inst. Hutn.*, **19**, 253 (1967).

[437] G. V. Grigorev, L. V. Pavlinov, *Fiz. met. i metalloved.*, **25**, 836 (1968).

[438] K. Hirano, M. Cohen, B. L. Averbach, *Acta Met.*, **9**, 440 (1961).

[439] G. Seibel, *Compt. Rend.*, **256**, 4661 (1963).

[440] G. Seibel, *Compt. Rend.*, **255**, 3182 (1962).

[441] K. Sato, *Trans. Japan Inst. Metals*, **5**, 91 (1964).

[442] A. Bondy, V. Levy, *C. R. Acad. Sci., Ser. C*, **272**, 19 (1971).

[443] T. Egichi, Y. Iijima, K. Hirano, *Acta Cryst.*, **A28**, 161 (1972).

[444] K. Hirano, M. Cohen, *Trans. Japan Inst. Met.*, **13**, 96 (1972).

[445] K. Nohara, K. Hirano, *Proc. Int. Conf. Sci. Technol. Iron and Steel*, Tokyo, Sept. 7–11 (1970).

[446] D. Treheux, P. Marchive, J. Delagrange, P. Guivaldeng, *C. R. Acad. Sci., Ser. C*, **274**, 1260 (1972).

[447] P. L. Gruzin, V. V. Mural, A. P. Fokin, *Fiz. metal. i metalloved.*, **34**, 1326 (1972).

[448] A. Hoshino, T. Araki, *Trans. Nat. Res. Inst. Metals*, **13**, No 3 (1971) GS.

[449] S. Tsujis, K. Yamanaka, *Nippon Kinzoku Gakkaishi*, **38**, 415 (1974).

[450] A. Ascoli, L. Filoni, G. Poletti, G. Rossi, *Phys. Rev.*, B **10** (12), 5003 (1974).

[451] S. Nishikawa, K. Tsumuraya, *Phil. Mag.*, **26**, 941 (1972).

[452] R. A. Ross, H. B. Vanfleet, D. L. Decker, *Phys. Rev.*, B **9**, 4026 (1974).

[453] A. T. Nakonechnikov, L. V. Pavlinov, W. N. Bykov, *Fiz. met. i metalloved.*, **22**, 234 (1966).

[454] J. Askill, *Phys. Stat. Solidi*, **9**, K113 (1965).

[455] B. A. Vandyshev, A. S. Panov, *Fiz. met. i metalloved.*, **25**, 321 (1968).

[456] J. Askill, *Phys. Stat. Solidi*, **23**, K21 (1967).

[457] B. Lesage, A. M. Huntz, *J. Less-Common Met.*, **38**, 149 (1974).

[458] M. G. Karpman, G. N. Dubinin, G. V. Shcherbedinsky, G. P. Benediktova, *Trudy Moskov. Aviats. Inst.*, **228**, 51 (1971).
[459] K. Nohara, K. Hirano, *Nippon Kinzoku Gakkaishi*, **37**, 731 (1973).
[460] V. P. Shapovalov, A. N. Kurasov, *Izv. Akad. Nauk SSSR, Met.* No 2, **234** (1973).
[461] D. S. Gornyi, R. M. Altovski, *Fiz. met. i metalloved.*, **31**, 781 (1971).
[462] A. Chatterjee, D. J. Fabian, *J. Inst. Metals*, **96**, 186 (1968).
[463] P. Lacombe, *Colloque sur les Joints de Grains, Saclay 1960*, Presses Universitaires de France, Paris 1961.
[464] I. G. Ivanov, *Fiz. met. i metalloved.*, **22**, 725 (1966).
[465] P. P. Kuzmenko, G. P. Grinevich, *Fiz. tverd. tela*, **4**, 3266 (1962).
[466] A. R. Wazzam, P. Tung, L. B. Robinson, *J. Appl. Phys.* **42**, 5316 (1971).
[467] H. Bakker, I. Backus, F. Waals, *Phys. Stat. Solidi*, B **45**, 633 (1971).
[468] I. P. Zanghi, A. VanCraeynest, D. Calais, *J. Nucl. Mater.*, **39**, 133 (1971).
[469] A. R. Paul, R. P. Agarwala, *Met. Trans.*, **2**, 2691 (1971).
[470] F. A. Schmidt, O. N. Carlson, *J. Less-Common Metals*, **26**, 247 (1972).
[471] P. Son, S. Ihara, M. Miyake, T. Sano, *J. Japan. Inst. Met.*, **31**, 988 (1967).
[472] R. F. Peart, D. Graham, D. H. Tomlin, *Acta Met.*, **10**, 519 (1962).
[473] J. Pelleg, *J. Metals*, **20**, 54 (1968).
[474] V. D. Lubimov, P. W. Geld, G. P. Shveykin, J. A. Surima, *Izv. Akad. Nauk SSSR*, **2**, 84 (1967).
[475] B. A. Vandyshev, A. S. Panov, *Izv. Akad. Nauk SSSR*, **1**, 206 (1968).
[476] J. Askill, *Phys. Stat. Solidi*, **9**, K167 (1955).
[477] R. P. Agarwala, S. P. Murarka, M. S. Anand, *Acta Met.*, **16**, 61 (1968).
[478] J. S. Kirkaldy, P. N. Smith, R. C. Sharma, *Met. Trans.*, **4**, 624 (1973).
[479] R. P. Agarwala, K. Hirano, *Trans. Japan Inst. Met.*, **13**, 425 (1972).
[480] E. I. Shulgin, V. S. Batalov, V. P. Kocheshkov, *Metallofizika*, **37**, 60 (1971).
[481] V. P. Prutkin, A. S. Lyulovich, M. Yu. Kardashaubaev, *Krist. tonkich plenok*, (1970) 139–145.
[482] R. N. Ghostagore, *Phys. Rev.*, B **3**, 397 (1971).
[483] G. L. McVay, A. K. Ducharme, *J. Appl. Phys.*, **44**, 1409 (1973).
[484] W. C. Mallard, A. B. Gardner, R. F. Bass, L. M. Slifkin, *Phys. Rev.*, **129**, 617 (1963).
[485] K. Hirano, M. Cohen, B. L. Averbach, *Acta Met.*, **11**, 463 (1963).
[486] C. M. Teller, J. P. Stark, *Acta Met.*, **20**, 1077 (1972).
[487] T. Hehenkamp, R. Wübbenhorst, *Z. Metallkunde*, **66**, 275 (1975).
[488] V. N. Kaigorodov, J. A. Rabovski, W. K. Talinski, *Fiz. met. i metalloved.*, **24**, 117 (1967).
[489] T. Hirone, S. Miura, T. Suzuoka, *J. Phys. Soc. Japan*, **16**, 2456 (1961).
[490] N. Bartouth, J. Ouder, J. Cabane, *C. R. Acad. Sci. Paris*, **C264**, 1029 (1967).
[491] W. N. Kaigorodov, J. A. Rabovski, W. K. Talinski, *Fiz. met. i metalloved.*, **24**, 661 (1967).
[492] N. L. Peterson, S. J. Rothman, *Phys. Rev.*, **154**, 558 (1967).
[493] K. Dreyer, Chr. Merzing, Th. Heumann, "Determination of the Correlation Factor for the Diffusion of Sn and In in Al", in: *Atomic Transport in Solids and Liquids*, 1971, 237.
[494] F. H. Huang, H. B. Huntington, *Phys. Rev.*, B **9**, 1479 (1974).
[495] W. K. Warburton, *Phys. Rev.*, B**6**, 2161 (1972).

[496] D. F. Kalinovich, I. I. Kovenski, M. D. Smolin, *Fiz. met. i metalloved.*, **18**, 314 (1964).
[497] R. E. Pawel, T. S. Lundy, *J. Phys. Chem. Solids*, **26**, 937 (1965).
[498] A. F. Nakonechnikov, L. V. Pavlinov, *Izv. Akad. Nauk SSSR, Metal.*, **2**, 213 (1972).
[499] J. Askill, *Phys. Stat. Solidi*, **B43**, 1–2K (1971).
[500] G. B. Gibbs, D. Graham, D. H. Tomlin, *Phil. Mag.*, **8**, 1269 (1963).
[501] S. E. Shamblen, C. J. Rosa, *Met. Trans.*, **2**, 1925 (1971).
[502] L. G. Korneluk, L. M. Mirsky, B. S. Bokshtein, *Titanium Science and Technology*, vol. II, 1973, 905–914.
[503] L. V. Pavlinov, *Fiz. met. i metalloved.*, **24**, 272 (1967).
[504] J. Askill, G. B. Gibbs, *Phys. Stat. Solidi*, **11**, 557 (1965).
[505] L. V. Pavlinov, *Fiz. met. i metalloved.*, **30**, 800 (1970).
[506] J. F. Murdock, T. S. Lundy, E. E. Stansbury, *Acta Met.*, **12**, 1033 (1964).
[507] L. N. Larikov, V. M. Tyshkievich, L. F. Chorna, *Ukr. fiz. Zh.*, **12**, 983 (1967).
[508] A. Shepela, *J. Less-Common. Metals*, **26**, 33 (1972).
[509] V. P. Iovkov, A. S. Panov, A. V. Ryabenko, *Fiz. met. i metalloved.*, **34**, 1322 (1972).
[510] S. J. Rothman, *J. Nucl. Mater.*, **3**, 77 (1961).
[511] N. L. Peterson, S. J. Rothman, *Phys. Rev.*, **136**, A842 (1964).
[512] J. H. Rosolowski, *Phys. Rev.*, **124**, 1828 (1961).
[513] P. B. Ghate, *Phys. Rev.*, **130**, 174 (1963).
[514] A. P. Batra, H. B. Huntington, *Phys. Rev.*, **145**, 542 (1966).
[515] G. P. Tiwari, M. C. Saxena, R. V. Patil, *Trans. Indian Inst. Met.*, **26**, 55 (1973).
[516] R. Tendler, C. F. Vavotto, *J. Nucl. Mater.*, **46**, 107 (1973).
[517] G. B. Fedorov, E. A. Smirnov, F. I. Zhomov, V. N. Gusev, S. A. Paraev, *Met. metalloved. chist. metal.*, **9**, 30 (1971).
[518] R. P. Agarwala, S. P. Murarka, M. S. Anand, *Trans. Met. Soc. AIME*, **233**, 986 (1965).
[519] A. M. Blinkin, V. V. Vorobiov, *Ukr. fiz. Zh.*, **9**, 91 (1964).
[520] R. A. Andrevski. W. N. Zagriazkin, G. J. Meshchervakov, *Fiz. met. i metalloved.*, **21**, 140 (1966).
[521] L. V. Pavlinov, *Fiz. met. i metalloved.*, **24**, 272 (1967).
[522] T. S. Lundy, J. I. Federer, *Trans. AIME*, **227**, 592 (1963).
[523] M. P. Dariel, *J. Appl. Phys.*, **42**, 2251 (1971).
[524] J. E. Murphy, G. H. Adams, W. N. Cathey, *Met. Trans.*, A, **6a** (2), 343 (1975).
[525] T. Heumann, H. Bohmer, *J. Phys. Chem. Solids*, **29**, 237 (1968).
[526] G. F. Hancock, B. R. Donnell, *Phys. Stat. Solidi*, **A4**, 143 (1971).
[527] H. W. Paxton, T. Kunitake, *Trans. Met. Soc. AIME*, **218**, 1003 (1960).
[528] A. J. Shinyaev, *Fiz. met. i metalloved.*, **20**, 875 (1965).
[529] A. J. Shinyaev, *Metallurgy and Metallography* (1958); NP–Tr–448, 210 (1960).
[530] Y. Iijima, K. Hirano, *Nippon Kinzoku Gakkaishi*, **35**, (511 (1971).
[531] Y. Iijima, K. Hirano, *Trans. Japan Inst. Met.*, **13**, 419 (1972).
[532] R. Ebeling, H. Wever, *Z. Metallkunde*, **53**, 222 (1968).
[533] O. Neukam, *Galvanotechnik*, **61**, 626 (1970).
[534] M. Wilhelm, *Z. Naturforsch.*, A **29**, 733 (1974).

[535] A. Spasov, G. Ivanov, *God. Soffi Univ. Khim. Fak.*, **63**, 1968–1969, 73 (1971).
[536] H. Oikawa, H. Takei, S. Karashima, *Met. Trans.*, **4**, 653 (1973).
[537] W. C. Mallard, A. B. Gardner, R. F. Bass, L. M. Slifkin, *Phys. Rev.*, **129**, 617 (1963).
[538] H. B. Huntington, N. C. Miller, V. Nerses, *Acta Met.*, **9**, 749 (1961).
[539] D. Gupta, D. Lazarus, D. S. Lieberman, *Phys. Rev.*, **153**, 558 (1967).
[540] G. V. Meshchernikov, R. A. Andrevski, W. N. Zagriazkin, *Fiz. met. i metalloved.*, **25**, 189 (1968).
[541] R. A. Wolfe, H. W. Paxton, *Trans. Met. Soc. AIME*, **230**, 1426 (1964).
[542] L. V. Pavlinov, E.A. Isadzanov, V. P. Smirnov, *Fiz. met. i metalloved.*, **25**, 959 (1968).
[543] I. G. Ivanov, A. M. Blinkin, *Fiz. met. i metalloved.*, **19**, 274 (1965).
[545] S. Z. Bokshtayn, V. A. Kozakova, S. T. Kishkin, L. M. Mirski, *Izv. Akad. Nauk SSSR, otd. tekh. nauk*, **12**, 18 (1955).
[546] D. Y. F. Lai, R. J. Borg, *Trans. Met. Soc. AIME*, **233**, 1973, (1965).
[547] J. Kucera, B. Million, J. Ruzickova, V. Foldyna, A. Jakobova, *Acta Met.*, **22**, 135 (1974).
[548] F. Chaix, Mrs. A. M. Huntz, *Mem. Sci. Rev. Met.*, **71**, 115 (1974).
[549] R. A. Perkins, P. T. Carlson, *Met. Trans.*, **5**, 1511 (1974).
[550] M. Hartl, H. Wever, *Arch. Eisenhüttenwesen*, **44**, 381 (1973).
[551] A. W. Boven, G. M. Leak, *Met. Trans.*, **1**, 2767 (1970).
[552] P. Gniraldeng, P. Poyet, *Mem. Sci. Rev. Met.*, **70**, 715 (1973).
[553] T. Ustad, H. Sorun, *Phys. Stat. Solidi*, **A20**, 285 (1973).
[554] K. Nohara, K. Hirano, *J. Japan Inst. Met.*, **37**, 51 (1973).
[555] J. Kucera, B. Million, J. Ruzickova, A. Jakobova, V. Foldyna, *Kovove Mat.*, **11**, 456 (1973).
[556] H. V. M. Mirani, R. Harthoorn, T. J. Zuurendonk, S. J. Helmerhorst, G. De Vires, *Phys. Stat. Solidi*, A **29**, 115 (1975).
[557] S. V. Zemski, V. I. Grigorkin, L. N. Moskaleva, *Izv. Vyssh. uchebn. zaved., chern. met.*, **13**, 106 (1970).
[558] A. J. Shinyaev, *Izv. Akad. Nauk SSSR, Metal.*, **4**, 263 (1971).
[559] S. V. Zemski, L. S. Makashova, S. A. Golovanenko, *Izv. vyssh. uchebn. zaved., chern. met.*, **8**, 99 (1974).
[560] M. A. Krishtal, J. S. Sokolov, *Fiz. khim. obr. mater.*, **5**, 118 (1972).
[561] H. A. Resing, N. H. Nachtrieb, *J. Phys. Chem. Solids*, **21**, 40 (1961).
[562] H. A. Domain, H. I. Aaronson, *Trans. Met. Soc. AIME*, **230**, 44 (1964).
[563] D. F. Kalinovich, I. I. Kovenski, M. D. Smolin, *Fiz. metal. i metalloved.*, **11**, 307 (1961); *Phys. Met. Metallog. USSR* (English transl.).
[564] I. N. Francevich, D. F. Kalinovich, I. I. Kovenski, M. S. Smolin, *Ukr. fiz. zh.*, **8**, 1020 (1963).
[565] M. D. Smolin, *Fiz. met. i metalloved.*, **15**, 472 (1963).
[566] D. F. Kalinovich, I. I. Kovenski, M. D. Smolin, *Ukr. fiz. zh.*, **10**, 1365 (1965).
[567] W. Zaiss, S. Steeb, T. Krabichler, *Z. Metallkunde*, **63**, 180 (1972).
[568] C. P. Heijwegen, G. D. Rieck, *Acta Met.*, **22**, 1269 (1974).
[569] A. J. Shinyaev, *Metallurgy and Metallography*, NP–Tr–448, (1960).
[570] E. W. DeReca, C. Pampillo, *Acta Met.*, **15**, 1263 (1967).

[571] A. J. Shinyaev, *Izv. Akad. Nauk SSSR*, **1**, 203 (1968).
[572] D. F. Kalinovich, I. I. Kovenski, M. D. Smolin, *Ukr. fiz. zh.*, **19**, 1619 (1974).
[573] D. F. Kalinovich, I. I. Kovenski, M. D. Smolin, *Fiz. tverd. tela*, **14**, 3699 (1972).
[574] D. F. Kalinovich, I. I. Kovenski, M. D. Smolin, *Fiz. tverd. tela*, **13**, 2813 (1971).
[575] M. S. Seltzer, B. A. Wilcox, *Met. Trans.*, **3**, 2357 (1972).
[576] I. N. Frantsevich, D. F. Kalinovich, I. I. Kovenski, M. D. Smolin, *Atomic Transport in Solids and Liquids, Proc. Europhys. Conf.*, Tübingen, 1971.
[577] B. Million, *Z. Metallkunde*, **63**, 484 (1972).
[578] B. Million, B. Cihek, J. Krejci, K. Stranky, Czech, *J. Phys.*, **B22**, 534 (1972).
[579] B. Million, J. Kucera, *Acta Met.*, **17**, 339 (1969).
[580] Z. J. Velmozhny, V. D. Lubimov, G. P. Shveykin, *Izv. wyssh. ucheb. zaved., cvet. met.*, **2**, 135 (1966).
[581] G. B. Gibbs, D. Graham, D. H. Tomlin, *Phil. Mag.*, **8**, 1269 (1962).
[582] G. B. Fedorov, E. A. Smirnov, V. N. Gusev, *Atom. Eng.*, **32**, 11 (1972).
[583] R. E. Hoffman, *Acta Met.*, **6**, 95 (1958).
[584] M. Butsik, S. Gertsriken, *Zh. tekhn. fiz.*, **20**, 428 (1950).
[585] D. Lazarus, C. T. Tomizuka, *Phys. Rev.*, **103**, 1155 (1956).
[586] A. Schoen, *Self-Diffusion in Alpha Solid Solutions of Silver–Cadmium and Silver–Indium*, Convair, General Dynamics Corp. Div., Scientific Research Lab., Research Report No 2 (1958).
[587] R. E. Hoffman, D. Turnbull, *J. Appl. Phys.*, **23**, 1409 (1952).
[588] R. E. Hoffman, D. Turnbull, E. W. Hart, *Acta Met.*, **3**, 417 (1955).
[589] G. Barreau, G. Brunel, G. Cizeron, P. Lacombe, *Mem. Sci. Rev. Met.*, **68**, 357 (1971).
[590] W. Seith, Th. Heumann, H. Wever, *Z. Metallkunde*, **62**, 294 (1971).
[591] A. J. Mortlock, D. H. Tomlin, *Phil. Mag.*, **4**, 628 (1959).
[592] R. F. Peart, D. H. Tomlin, *Acta Met.*, **10**, 123 (1962).
[593] A. Brunsch, S. Steeb, *Z. Naturforsch.*, A **29**, 1319 (1974).
[594] F. Dyment, E. Santos, *Metal ABM*, **28**, 731 (1972).
[595] E. Santos, F. Dyment, *Phil. Mag.*, **31**, 809 (1975).
[596] A. Brunsch, S. Steeb, *High Temp.–High Pressures*, **6**, 155 (1974).
[597] S. G. Fedotov, G. N. Ronami, K. M. Constantnov, *Dokl. Akad. Nauk*, **201**, 628 (1928).
[598] V. S. Raghunathan, G. P. Tiwari, B. D. Sharma, *Met. Trans.*, **3**, 421 (1972).
[599] P. M. Arzhani, R. M. Volkova, D. A. Prokoshkin, *Izv. Akad. Nauk SSSR, otd. tekhn. nauk*, **6**, 162 (1962).
[600] A. Adda, A. Kirianenko, M. Bendazzoli, *Compt. Rend.*, **253**, 653 (1961).
[601] C. J. Santoro, *Amer. Phys. Soc. Bull.*, **12**, 1041 (1967).
[602] I. A. Naskidashvili, *Autoreferat kandidatskoi dissertasii*, Akad. Nauk Gruz. SSR, Tbilisi 1957.
[603] J. E. Hilliard, B. L. Averbach, M. Cohen, *Acta Met.*, **7**, 86 (1959).
[604] G. B. Fedorov, V. D. Guliakin, *Met. i metalloved. chistykh metallov*, **1**, 170 (1959).
[605] V. S. Lyashenko, V. N. Bykov, L. V. Pavlinov, *Fiz. met. i metalloved.*, **8**, 362 (1959); *Phys. Met. Metallog. USSR* (English transl.), **8**, 40 (1959).
[606] G. P. Tivari, B. D. Sharma, V. S. Raghunathan, R. V. Patil, *J. Nucl. Mater.*, **46**, 35 (1973).
[607] P. Lamparter, T. Krabichler, S. Steeb, *Z. Naturforsch.*, A **27**, 960 (1972).

[608] E. P. Nekhiporenko, V. M. Krivoruchko, A. S. Mitrofanov, J. T. Kondratov, *Fiz. metal. i metalloved.*, **32**, 89 (1971).
[609] A. E. Paladino, W. D. Kingery, *J. Chem. Phys.*, **37**, 957 (1962).
[610] W. Fiedler, D. Bobleter, *Atomkernenergie*, **13**, 57 (1968).
[611] D. V. Ignatov, I. N. Belokurova, I. N. Belyanin, *Metallurgy and Metal Research—All-Union Scientific and Technical Conference on the Use of Radioactive and Stable Isotopes and Radiation in the National Economy and in Science*, Izd. Akad. Nauk SSSR, Moskow 1958, p. 326; NP-Tr-448, p. 256 (1958).
[612] V. I. Izvekov, K. M. Gorbunova, *Fiz. met. i metalloved.*, **7**, 713 (1959).
[613] V. O. Obleter, W. Fiedler, F. Grasso, *Atomkernenergie*, **10**, 261 (1965).
[614] H. W. Hennicke, H. H. Strurhahn, *Z. Tonindustrie*, **95**, 127 (1971).
[615] I. I. Vishnevski, E. I. Akselrod, N. P. Talyankaya, *Fiz. tverd. tela*, **13**, 3446 (1971).
[616] G. H. Frischat, *Ber. Deut. Keram. Ses.*, **48**, 441 (1971).
[617] T. Tokuda, Ito T. Yamaguchi, *Z. Naturforsch.*, **26**, 2058 (1971).
[618] A. J. Grigorev, D. I. Polishchuk, *Fiz. arerodispersnykh sist.*, **8**, 87 (1973).
[619] R. W. Redington, *Phys. Rev.*, **87**, 1066 (1952).
[620] S. B. Austerman, D. G. Swarthout, NAA-SR-5893 (1961).
[621] S. B. Austerman, *J. Nucl. Mater.*, **14**, 248 (1964).
[622] H. J. De Bruin, G. M. Watson, *J. Nucl. Mater.*, **14**, 239 (1964); ORNL-3526 (1964).
[623] S. B. Austerman, J. W. Wagner, *J. Amer. Ceram. Soc.*, **49**, 94 (1966).
[624] H. J. De Bruin, G. M. Watson, C. M. Blood, *J. Appl. Phys.*, **37**, 4573 (1966).
[625] V. P. Gladkov, V. S. Zotov, D. M. Skorov, *At. Energ.*, **34**, 290 (1973).
[626] G. D. Palkar, D. N. Sitharamarao, A. K. Dasgupta, *Trans. Faraday Soc.*, **59**, 2634 (1963).
[627] R. Lindner, S. Austrumdal, A. Akerstrom, *Acta Chem. Scand.*, **6**, 468 (1952).
[628] Y. P. Gupta, L. J. Wevrick, *J. Phys. Chem. Solids*, **28**, 811 (1967).
[629] B. C. Harding, V. K. Bhalla, *Phil. Mag.*, **24**, 485 (1971).
[630] B. C. Steele, J. M. Flofd, *Proc. Brit. Ceram. Soc.*, **19**, 55 (1971).
[631] R. Lindner, A. Akerstrom, *Z. phys. Chem.*, **6**, 162 (1956).
[632] W. C. Hagel, A. U. Seybolt, *J. Electrochem. Soc.*, **108**, 1146 (1961).
[633] E. Fryt, S. Mrowec, T. Walec, *Oxidation of Metals*, **7**, 117 (1973).
[634] R. H. Chang, W. Steward, J. B. Wagner, *7-th International Symposium on the Reactivity of Solids, Bristol, England, 17–21 July 1972*, Chapman-Hall, London 1972, p. 231–239.
[635] A. I. Andrevski, A. V. Sandulova, M. I. Yurkevich, *Fiz. tverd. tela*, **2**, 624 (1960).
[636] W. I. Moore, B. Selikson, *J. Chem. Phys.*, **19**, 1539 (1951); **20**, 917 (1952).
[637] A. V. Sandalova, M. I. Droniuk, V. M. Rybak, K. S. Shcherbai, *Ukr. fiz. zh.*, **7**, 289 (1962).
[638] A. V. Sandalova, D. Ee-Tsin, *Fiz. tverd. tela*, **2**, 874 (1960).
[639] S. Mrowec, A. Stokłosa, *Oxidation of Metals*, **3**. 291 (1971).
[640] L. Himmel, N. F. Mehl, C. E. Birchenall, *J. Metals*, **5**, 827 (1953).
[641] P. M. Valov, *J. Solid State Chem.*, **1**, 215 (1970).
[642] W. K. Chen, N. L. Peterson, *J. Pyhs. (Paris)*, **9**, 303 (1073).
[643] H. Rickert, W. Weppner, *Z. Naturfcrsch.*, A **29**, 1849 (1974).
[644] L. Heyne, N. M. Beekmans, A. De Beer, *J. Electrochem. Soc.*, **119**, 77 (1972).
[645] G. R. Weber, L. Eyring, *Adv. Chem.*, **21**, 253 (1971).

[646] B. C. Harding, *Phil. Mag.*, **16**, 1039 (1967).

[647] B. C. Harding, *Phys. Stat. Solidi*, **B50**, 135 (1972).

[648] B. C. Harding, A. J. Mortlock, *J. Chem. Phys.*, **45**, 2699 (1966).

[649] J. Rungis, A. J. Mortlock, *Phil. Mag.*, **14**, 821 (1966).

[650] B. J. Wuensch, T. Vasilos, *J. Chem. Phys.*, **36**, 2917 (1962).

[651] R. Lindner, G. D. Parfitt, *J. Chem. Phys.*, **26**, 182 (1957).

[652] B. C. Harding, D. M. Price, A. J. Mortlock, *Phil. Mag.*, **23**, 399 (1971).

[653] B. J. Wuensch, T. Vasilos, *Reactivity of Solids*, Elsevier, New York 1961, p. 57.

[654] J. B. Waensch, T. Vasilos, *J. Chem. Phys.*, **54**, 1123 (1971).

[655] J. Mirnkes, M. Wuttig, *J. Amer. Ceram. Soc.*, **54**, 65 (1971).

[656] M. F. Barard, *J. Amer. Ceram. Soc.*, **54**, 58 (1971).

[657] S. Shirasaki, H. Yamamura, *Japan J. Appl. Phys.*, **12**, 1654 (1973).

[658] S. Shirasaki, M. Homa, *Chem. Phys. Letters*, **20**, No 4, 361–5 (1973).

[659] B. C. Harding, *Phys. Stat. Solidi*, B **50**, 135 (1972).

[660] B. C. Harding, *Phil. Mag.*, **27**, 481 (1973).

[661] A. J. Hartlock, D. M. Price, *J. Chem. Phys.*, **58**, 634 (1973).

[662] S. Tachimori, H. Nakamura, H. Amano, *J. Nucl. Sci. Technol.*, **8**, 295 (1971).

[663] W. J. Moore, *Complete Scientific Report on Contract Termination Rate Processes in Inorganic Solids at High Temperatures*, October 1, 1950, to August 1, 1951, ORO–78 (1951).

[664] R. Linder, A. Akerstrom, *Z. phys. Chem.*, **6**, 162 (1956).

[665] R. Linder, A. Akerstrom, *Disc. Faraday Soc.*, **23**, 133 (1957).

[666] W. J. Moore, J. J. Landers, S. R. Logan, M. O'Keeffe, J. S. Choi, J. Ebisuzaki, S. Brown, D. Mitchell, *Physical Chemistry of the Solid State*, TID-11020 (1960).

[667] J. S. Choi, W. J. Moore, *J. Phys. Chem.*, **66**, 1308 (1962).

[668] S. M. Klotsman, A. N. Timofeev, I. S. Trachtenberg, *Fiz. met. i metalloved.*, **14**, 428 (1962).

[669] M. L. Volpe, J. Reddy, *J. Chem. Phys.*, **53**, 1117 (1970).

[670] C. Greskovich, *J. Amer. Ceram. Soc.*, **53**, 498 (1970).

[671] M. S. Smeltzer, *J. Electrochem. Soc.*, **118**, 802 (1971).

[672] W. K. Chen, N. L. Peterson, *Atomic Transport in Solids and Liquids, Proc. Europhys. Conf.*, Tübingen, 1971, p. 255.

[673] W. K. Chen, N. L. Peterson, Eds., A. Lodding and T. Tagerwall, *Atomic Transport in Solids and Liquids, Proc. Europhys. Conf.*, Tübingen, 1971, p. 257.

[674] W. Chen, N. Peterson, *J. Phys. Chem. Solids*, **33**, 881 (1972), 617.

[675] M. J. Graham, D. Caplan, M. Cohen, *J. Electrochem. Soc.*, **119**, 1265, (1972).

[676] W. K. Chen, N. Peterson, L. C. Robinson, *J. Phys. Chem. Solids*, **34**, 705 (1973).

[677] K. J. Richards, E. F. Wagstaff, *J. Phys. Chem.*, **70**, 1553 (1966).

[678] W. Crow, *Ph. D. Thesis*, Ohio State University, 1969.

[679] W. D. Steward, J. B. Wagner, *J. Electrochem. Soc.*, **122**, 570 (1975).

[680] S. K. Hoon, *Kwahak Konghak*, **11**, 157 (1973).

[681] D. L. Douglas, *Proc. Conf. Corrosion Reactor Materials*, Salzburg 1962, IAEA, **2**, 223 (1962).

[682] R. Linder, O. Enqvist, *Arkiv. Kemi.*, **9**, 471 (1956).

[683] S. P. Murarka, R. A. Svalin, *J. Phys. Chem. Solids*, **32**, 1277 (1971).

[684] S. J. Derry, D. G. Lees, J. M. Calvert, *Proc. Brit. Ceram. Soc.*, **19**, 77 (1971).

[685] E. Eguchi, K. Yajima, *Trans. Japan. Inst. Met.*, **13**, 45 (1972).
[686] H. Furuya, S. Yajima, *J. Nucl. Mater.*, **25**, 38 (1968).
[687] T. S. Lundy, W. A. Coghlan, *J. Phys. (Paris)*, Colloq. 1973, (**9**), 299–302.
[688] R. J. Hawkins, C. B. Alcock, *J. Nucl. Mater.*, **26**, 112 (1968).
[689] H. Furuya, *J. Nucl. Mater.*, **26**, 123 (1968).
[690] A. P. King, *J. Nucl. Mater.*, **38**, 347 (1971).
[691] F. Schmitz, R. Linder, *J. Nucl. Mater.*, **17**, 259 (1965).
[692] F. Schmitz, R. Linder, *Z. Naturforsch.*, **16a**, 1996 (1961).
[693] A. B. Austern, J. Belle, *J. Nucl. Mater.*, **3**, 267 (1961).
[694] R. Linder, F. Schmitz, *Z. Naturforsch.*, **16a**, 1373 (1961).
[695] C. B. Alcock, R. J. Hawkins, A. W. D. Hills, P. McNamara, *Symposium on Thermodynamics with Emphasis on Nuclear Materials and Atomic Transport in Solids*, Vienna 1965.
[696] S. Yajima, H. Furuya, T. Hirai, *Sci. Rep. Res. Inst. Tohoku Univ.* (*Suppl.*) **18**, 238 (1966); S. Yajima, H. Furuya, T. Hirai, *J. Nucl. Mater.*, **20**, 162 (1966).
[697] P. Contamin, J. J. Becmann, J. F. Marin, *J. Nucl. Mater.*, **42**, 54 (1972).
[698] A. B. Auskern, J. Belle, *J. Nucl. Mater.*, **3**, 267 (1961).
[699] M. F. Berand, D. R. Wilder, *J. Appl. Phys.*, **34**, 2318 (1963).
[700] R. Linder, *Acta Chem. Scand.*, **6**, 457 (1952).
[701] F. Munnich, *Naturwiss.*, **42**, 340 (1955).
[702] E. A. Secco, W. J. Moore, *J. Chem. Phys.*, **23**, 1170 (1955).
[703] E. Spicar, "Untersuchung über Einstellung des Fehlstellen-Gleichgewichtes in Verbindungkristallen", *Thesis*, University of Stuttgart, Stuttgart 1956.
[704] J. P. Roberts. C. Wheeler, *Phil. Mag.*, **2**, 708 (1957).
[705] W. J. Moore, E. L. Williams, *Disc. Faraday Soc.*, **28**, 86 (1959).
[706] E. A. Secco, *Disc. Faraday Soc.*, **28**, 94 (1959).
[707] E. A. Secco, *Can. J. Chem.*, **39**, 1544 (1961).
[708] G. Muller, R. Helbig, *J. Phys. Chem. Solids*, **32**, 1971 (1971).
[709] R. Robin, A. R. Cooper, A. H. Hener, *J. Appl. Phys.*, **44**, 4770 (1973).
[710] F. J. Kenesha, D. L. Douglass, *Oxidation of Metals*, **3**, 1 (1971).
[711] R. Meussner, C. Birchenall, *Corrosion*, **13**, 677 (1957).
[712] Condid, C. Birchenall, *Oxidation of Metals*, **8**, 409 (1974).
[713] R. Allen, W. Moore, *J. Phys. Chem.*, **63**, 223 (1959).
[714] I. Bartkowicz, S. Mrowec, *Zesz. Nauk. AGH*, **424**, 183 (1974).
[715] S. Mrowec, H. Rickert, *Z. phys. Chem. N.F.*, **32**, 212 (1962).
[716] H. Braune, O. Kahn, *Z. phys. Chem.*, **112**, 270 (1924).
[717] I. Bartkowicz, E. Fryt, S. Mrowec, *Zesz. Nauk. AGH*, **247**, 19 (1969).
[718] H. Braune, O. Kahn, *Z. phys. Chem.*, **112**, 270 (1924).
[719] N. F. Zhugakov, A. S. Liyshevich, E. A. Pokrovski, I. E. Shymanovich, *Dokl. Akad. Nauk.*, B **18**, 825 (1974).
[720] M. Lambertin, J. C. Colson, I. C. Delafosse, *C.R. Acad. Sci. Paris, Ser. C.*, **272**, 267 (1971).
[721] J. Żmija, L. Sadosz, *Biul. Wojsk. Akad. Techn.*, **20**, 105 (1971).
[722] J. Żmija, *Acta Phys. Polon.*, **39**, 531 (1971).
[723] E. Nerbaner, *Phys. Stat. Solidi*, **60**, 57 (1973).
[724] K. H. Chung, H. J. Bac, *Kunsok Hakkol Chi*, **12**, 127 (1974).

[725] S. N. Barganovski, M. V. Denilenko, *Izv. vysshykh uchebnykh zaved., fiz.*, **13**, 12 (1970).
[726] R. L. Clark, *J, Appl. Phys.*, **30**, 957 (1959).
[727] J. Sullivan, *J. Phys.*, **6**, 552 (1973).
[728] C. W. Owens, D. Turnbull, *J. Appl. Phys.*, **43**, 3933 (1972).
[729] J. Bloem, F. A. Kröger, *Phyllips Rev. Rep.*, **12**, 281 (1957).
[730] J. S. Anderson, J. R. Richards, *J. Chem. Soc.*, **17**, 537 (1946).
[731] J. Bloem, F. A. Kröger, *Phyllips Rev. Rep.*, **12**, 281 (1957).
[732] E. A. Secco, *J. Chem. Phys.*, **29**, 406 (1958).
[733] G. Simkovich, J. B. Wagner, *J. Chem. Phys.*, **38**, 1368 (1963).
[734] J. S. Kirkaldy, *et al.*, *Can. Met. Quart.*, **12**, 53 (1973).
[735] K. Fueki, Y. Oguzi, T. Mukaibo, *Bull. Chem. Soc. Japan*, **41**, 569 (1968).
[736] D. L. Foster, A. L. Laskar, *Phys. Stat. Solidi*, **29**, 167 (1975).
[737] A. L. Laskov, L. M. Slifkin, *J. Nonmet.*, **1**, 83 (1972).
[738] J. Firsh, J. Hensel, G. Welt, *Ber. Buns. Phys. Chem.*, **77**, 157 (1973).
[739] A. P. Betra, L. M. Slifkin, *Phys. Stat. Solidi*, **19**, 17 (1977).
[740] C. Tiebandt, H. Reinhold, W. Jost, *Z. anorg. allg. Chem.*, **177**, 253 (1928).
[741] A. N. Murin, J. V. Sirkov, *Vestnik Leningrad. Univ.*, **2**, 122 (1972).
[742] H. Matzke, *J. Mater. Sci.*, **5**, 831 (1970).
[743] M. F. Berard, *J. Amer. Ceram. Soc.*, **54**, 144 (1971).
[744] F. W. Felix, S. Y. T. Lagerwal, *Phys. Stat. Solidi*, **4**, 73 (1971).
[745] F. W. Felix, S. Y. T. Lagerwal, *Phys. Stat. Solidi*, **4**, 78 (1971).
[746] R. Sternwinkel, *Z. Naturforsch.*, **29**, 278 (1974).
[747] H. Matzke, *J. Phys. Chem. Solids*, **32**, 437 (1971).
[748] J. N. Coles, Y. Long, *Phil. Mag.*, **29**, 457 (1974).
[749] D. C. Parasher, *J. Phys. Chem. Solids*, **31**, 2795 (1970).
[750] M. Beniere, F. Beniere, M. Chelma, *J. Chim. Phys. Physic-chim. Biol.*, **67**, 1312 (1970).
[751] C. A. Allen, W. J. Fredericks, *Phys. Stat. Solidi*, **55**, 615 (1973).
[752] J. Haget, N. B. Chank, P. Farin, L. Bonpunt, *Acta Met.*, **23**, 723 (1975).
[753] H. A. Dominguez, P. E. Munoz, *Rev. Met. Phys.*, **19**, 375 (1970).
[754] J. A. Parfianovich, S. A. Shipinin, *Fiz. chim. Acta*, **16**, 263 (1971).
[755] J. M. Bogomolova, *Fiz. chim. Acta*, **5**, 161 (1956).
[756] R. J. Maurer, P. Maprother, *Phys. Rev.*, **73**, 1260 (1948).
[757] A. N. Murin, I. V. Murin, V, Y. Portniagin, A. M. Andreev, *Fiz. tverd. tela*, **15**, 2838 (1973).
[758] N. Strenwinkel, *Z. Naturforsch.*, **29**, 278 (1974).
[759] H. J. Matzke, *J. Phys. Chem. Solids*, **32**, 437 (1971).
[760] J. Hedges, J. Mitchell, *Phil. Mag.*, **44**, 223 (1953).
[761] L. Azaroff, *Introduction to Solids*, McGraw-Hill, New York, 1960, p. 115.
[762] B. G. Hyde, *Reactivity of Solids*, Chapman and Hall, London, 1972, p. 23.
[763] R. Hoffman, *Acta Met.*, **4**, 98 (1956).
[764] V. Bugakov, F. Rybalko, *J. Techn. Phys. USSR*, **2**, 617, (1935).
[765] S. Mrowec, M. Danielewski, "Self-Diffusion and Defect Concentration in Ferrous Sulphide" (to be published in *Bull. Acad. Polon. Sci., Ser. sci. chim.*).
[766] S. Mrowec, *J. Mater. Sci.*, **9**, 1961 (1974).

Subject Index